MELZAK—Mathematical Ideas, Modeling and Applications (Volume II of Companion to Concrete Mathematics)

NAYFEH—Perturbation Methods

NAYFEH and MOOK—Nonlinear Oscillations

ODEN and REDDY—An Introduction to the Mathematical Theory of Finite Elements

PAGE—Topological Uniform Structures

PASSMAN—The Algebraic Structure of Group Rings

PRENTER—Splines and Variational Methods

RIBENBOIM—Algebraic Numbers

RICHTMYER and MORTON—Difference Methods for Initial-Value Problems, 2nd Edition

RIVLIN—The Chebyshev Polynomials

RUDIN—Fourier Analysis on Groups

SAMELSON—An Introduction to Linear Algebra

SCHUMAKER—Spline Functions: Basic Theory

SCHUSS—Theory and Applications of Stochastic Differential Equations

SIEGEL—Topics in Complex Function Theory

Volume 1—Elliptic Functions and Uniformization Theory

Volume 2—Automorphic Functions and Abelian Integrals

Volume 3—Abelian Functions and Modular Functions of Several Variables

STAKGOLD—Green's Functions and Boundary Value Problems

STOKER—Differential Geometry

STOKER—Nonlinear Vibrations in Mechanical and Electrical Systems

STOKER—Water Waves

WHITHAM—Linear and Nonlinear Waves

WOUK—A Course of Applied Functional Analysis

A First Course in Differential Geometry

About the author

CHUAN-CHIH HSIUNG is Professor of Mathematics at Lehigh University, a position he has held since 1960. After receiving his Ph.D. from Michigan State University in 1948, Dr. Hsiung lectured at both the University of Wisconsin and Northwestern University, and was a Research Fellow at Harvard University. He is the author of numerous research articles, and is the founder and Managing Editor of the *Journal of Differential Geometry*, the unique international journal in its field.

A First Course in Differential Geometry

CHUAN-CHIH HSIUNG
Lehigh University

A WILEY-INTERSCIENCE PUBLICATION

JOHN WILEY & SONS, New York • Chichester • Brisbane • Toronto

Library of Congress Cataloging in Publication Data:

Hsiung, Chuan-Chih, 1916-
A first course in differential geometry.

(Pure and applied mathematics)
"A Wiley-Interscience publication."
Bibliography: p.
Includes index.
1. Geometry, Differential. I. Title.

QA641.H74 516.3′6 80-22112
ISBN 0-471-07953-7

Printed in the United States of America

10 9 8 7 6 5 4 3 2 1

To my wife Wenchin Yu

Preface

According to a definition stated by Felix Klein in 1872, we can use geometric transformation groups to classify geometry. The study of properties of geometric figures (curves, surfaces, etc.) that are invariant under a given geometric transformation group G is called the geometry belonging to G. For instance, if G is the projective, affine, or Euclidean group, we have the corresponding projective, affine, or Euclidean geometry.

The differential geometry of a geometric figure F belonging to a group G is the study of the invariant properties of F under G in a neighborhood of an element of F. In particular, the differential geometry of a curve is concerned with the invariant properties of the curve in a neighborhood of one of its points. In analytic geometry the tangent of a curve at a point is customarily defined to be the limit of the secant through this point and a neighboring point on the curve, as the second point approaches the first along the curve. This definition illustrates the nature of differential geometry in that it requires a knowledge of the curve only in a neighborhood of the point and involves a limiting process (a property of this kind is said to be local). These features of differential geometry show why it uses the differential calculus so extensively. On the other hand, local properties of geometric figures may be contrasted with global properties, which require knowledge of entire figures.

The origins of differential geometry go back to the early days of the differential calculus, when one of the fundamental problems was the determination of the tangent to a curve. With the development of the calculus, additional geometric applications were obtained. The principal contributors in this early period were Leonhard Euler (1707–1783), Gaspard Monge (1746–1818), Joseph Louis Lagrange (1736–1813), and Augustin Cauchy (1789–1857). A decisive step forward was taken by Karl Friedrich Gauss (1777–1855) with his development of the intrinsic geometry on a surface. This idea of Gauss was generalized to $n(>3)$-dimensional space by Bernhard Riemann (1826–1866), thus giving rise to the geometry that bears his name.

This book is designed to introduce differential geometry to beginning graduate students as well as advanced undergraduate students (this introduction in the latter case is important for remedying the weakness of geometry in the usual undergraduate curriculum). In the last couple of decades differential geometry, along with other branches of mathematics, has been highly developed. In this book we will study only the traditional topics, namely, curves and surfaces in a three-dimensional Euclidean space E^3. Unlike most classical books on the subject, however, more attention is paid here to the relationships between local and global properties, as opposed to local properties only. Although we restrict our attention to curves and surfaces in E^3, most global theorems for curves and surfaces in this book can be extended to either higher dimensional spaces or more general curves and surfaces or both. Moreover, geometric interpretations are given along with analytic expressions. This will enable students to make use of geometric intuition, which is a precious tool for studying geometry and related problems; such a tool is seldom encountered in other branches of mathematics.

We use vector analysis and exterior differential calculus. Except for some tensor conventions to produce simplifications, we do not employ tensor calculus, since there is no benefit in its use for our study in space E^3. There are four chapters whose contents are, briefly, as follows.

Chapter 1 contains, for the purpose of review and for later use, a collection of fundamental material taken from point-set topology, advanced calculus, and linear algebra. In keeping with this aim, all proofs of theorems are self-contained and all theorems are expressed in a form suitable for direct later application. Probably most students are familiar with this material except for Section 6 on differential forms.

In Chapter 2 we first establish a general local theory of curves in E^3, then give global theorems separately for plane and space curves, since those for plane curves are not special cases of those for space curves. We also prove one of the fundamental theorems in the local theory, the uniqueness theorem for curves in E^3. A proof of this existence theorem is given in Appendix 1.

Chapter 3 is devoted to a local theory of surfaces in E^3. For this theory we only state the fundamental theorem (Theorem 7.3), leaving the proofs of the uniqueness and existence parts of the theorem to, respectively, Chapter 4 (Section 4) and Appendix 2.

Chapter 4 begins with a discussion of orientation of surfaces and surfaces of constant Gaussian curvature, and presents various global theorems for surfaces.

Most sections end with a carefully selected set of exercises, some of which supplement the text of the section; answers are given at the end of the book. To allow the student to work independently of the hints that accompany some of the exercises, each of these is starred and the hint together with the answer appear at the end of the book. Numbers in brackets refer to the items listed in the Bibliography at the end of the book.

Two enumeration systems are used to subdivide sections; in Chapters 1 (except Sections 4 and 7) and 2, triple numbers refer to an item (e.g., a theorem or definition), whereas in Chapters 3 and 4 such an item is referred to by a double number. However, there should be no difficulty in using the book for reference purposes, since the title of the item is always written out (e.g., Corollary 5.1.6 of Chapter 1 or Lemma 1.5 of Chapter 3).

This book can be used for a full-year course if most sections of Chapter 1 are studied thoroughly.

For a one-semester course I suggest the use of the following sections:

Chapter 1: Sections 3.1, 3.2, 3.3, 6.

Chapter 2: Section 1.1 (omit 1.1.4–1.1.6), Section 1.2 (omit 1.2.6, 1.2.7), Section 1.3 (omit 1.3.7–1.3.12), Sections 1.4 and 1.5 (omit 1.5.5); Section 2 (omit 2.3, 2.5, 2.6.4–2.6.6, 2.9–2.11, 2.14–2.23); Section 3 (omit 3.1.8–3.1.14).

Chapter 3: Section 1 (omit the proof of 1.6, 1.7, 1.8, the proof of 1.10, 1.11–1.13, 1.15–1.18); Section 2 (omit the proof of 2.4); Sections 3–9; Section 10 (omit the material after 10.7).

Chapter 4: Section 1 (omit the proofs of 1.3 and 1.4); Section 3 (omit 3.14); Sections 4 and 5.

For a course lasting one quarter I suggest omission of the following material from the one-semester outline above: Chapter 2: the second proof of 2.6, 3.2; Chapter 3: the details of 1.3 and 1.4, the proof of 5.7, Section 6, the proofs of 8.1 and 8.2; Chapter 4: Section 5.

I thank Donald M. Davis, Samuel L. Gulden, Theodore Hailperin, Samir A. Khabbaz, A. Everett Pitcher, and Albert Wilansky for many valuable discussions and suggestions in regard to various improvements of the book; Helen Gasdaska for her patience and expert skill in typing the manuscript; and the staff of John Wiley, in particular Beatrice Shube, for their cooperation and help in publishing this book.

CHUAN-CHIH HSIUNG

Bethlehem, Pennsylvania
September, 1980

Contents

GENERAL NOTATION AND DEFINITIONS

CHAPTER 1. EUCLIDEAN SPACES

1. Point Sets, 1

1.1. Neighborhoods and Topologies, 1
1.2. Open and Closed Sets, and Continuous Mappings, 4
1.3. Connectedness, 7
1.4. Infimum and Supremum, and Sequences, 9
1.5. Compactness, 11

2. Differentiation and Integration, 15

2.1. The Mean Value Theorems, 15
2.2. Taylor's Formulas, 17
2.3. Maxima and Minima, 18
2.4. Lagrange Multipliers, 20

3. Vectors, 23

3.1. Vector Spaces, 23
3.2. Inner Product, 24
3.3. Vector Product, 25
3.4. Linear Combinations and Linear Independence; Bases and Dimensions of Vector Spaces, 27
3.5. Tangent Vectors, 29
3.6. Directional Derivatives, 32

4. Mappings, 35

4.1. Linear Transformations and Dual Spaces, 35
4.2. Derivative Mappings, 40

5. Linear Groups, 46

5.1. Linear Transformations, 46
5.2. Translations and Affine Transformations, 52

5.3. Isometries or Rigid Motions, 54
5.4. Orientations, 59

6. Differential Forms, 64

6.1. 1-Forms, 64
6.2. Exterior Multiplication and Differentiation, 67
6.3. Structural Equations, 73

7. The Calculus of Variations, 75

CHAPTER 2. CURVES 78

1. General Local Theory, 78

1.1. Parametric Representations, 78
1.2. Arc Length, Vector Fields, and Knots, 82
1.3. The Frenet Formulas, 88
1.4. Local Canonical Form and Osculants, 100
1.5. Existence and Uniqueness Theorems, 105

2. Plane Curves, 109

2.1. Frenet Formulas and the Jordan Curve Theorem, 109
2.2. Winding Number and Rotation Index, 110
2.3. Envelopes of Curves, 112
2.4. Convex Curves, 113
2.5. The Isoperimetric Inequality, 118
2.6. The Four-Vertex Theorem, 123
2.7. The Measure of a Set of Lines, 126
2.8. More on Rotation Index, 130

3. Global Theorems for Space Curves, 139

3.1. Total Curvature, 139
3.2. Deformations, 147

CHAPTER 3. LOCAL THEORY OF SURFACES 151

1. Parametrizations, 151

2. Functions and Fundamental Forms, 170

3. Form of a Surface in a Neighborhood of a Point, 182

4. Principal Curvatures, Asymptotic Curves, and Conjugate Directions, 188

5. Mappings of Surfaces, 197

6. Triply Orthogonal Systems, and the Theorems of Dupin and Liouville, 203
7. Fundamental Equations, 207
8. Ruled Surfaces and Minimal Surfaces, 214
9. Levi-Civita Parallelism, 224
10. Geodesics, 229

CHAPTER 4. GLOBAL THEORY OF SURFACES 241

1. Orientation of Surfaces, 241
2. Surfaces of Constant Gaussian Curvature, 246
3. The Gauss-Bonnet Formula, 252
4. Exterior Differential Forms and a Uniqueness Theorem for Surfaces, 267
5. Rigidity of Convex Surfaces and Minkowski's Formulas, 275
6. Some Translation and Symmetry Theorems, 280
7. Uniqueness Theorems for Minkowski's and Christoffel's Problems, 285
8. Complete Surfaces, 292

Appendix 1. Proof of Existence Theorem 1.5.1, Chapter 2 307

Appendix 2. Proof of the First Part of Theorem 7.3, Chapter 3 309

Bibliography 313

Answers and Hints to Exercises 316

Index 335

General Notation and Definitions

NOTATION

Symbol	*Usage*	*Meaning*
$\in$	$x \in A$	x is an element of the set A
$\notin$	$x \notin A$	x is not an element of the set A
$\subset$	$B \subset A$	The set B is a subset of the set A
$\varnothing$	$\varnothing$	The empty set
$\cap$	$\begin{cases} A \cap B \\ \cap A_i \end{cases}$	Intersection of the sets A and B Intersection of all the sets A_i
$\cup$	$\begin{cases} A \cup B \\ \cup A_i \end{cases}$	Union of the sets A and B Union of all the sets A_i
$\{\ \}$	$\{x \mid \cdots\}$	The set of all x such that $\cdots$
$\Rightarrow$	$\cdots \Rightarrow$———	$\cdots$ implies———
$\Leftrightarrow$	$\cdots \Leftrightarrow$———	$\cdots$ if and only if———
$\to$	$A \to B$	Function on the set A to the set B
$\mapsto$	$x \mapsto x^2$	Function assigning x^2 to x
$[,]$	$[a, b]$	$\{x \mid a \leqslant x \leqslant b\}$
$(,)$	(a, b)	$\{x \mid a < x < b\}$

DEFINITIONS

A *function* f on a set A to a set B is a rule that assigns to each element x of A a unique element $f(x)$ of B. The element $f(x)$ is called the *value* of f at x, or the *image* of x under f. The set A is called the *domain* of f, the set B is often called the *range* of f, and the subset of B, denoted by $f(A)$, consisting of all elements of the form $f(x)$ is called the *image* of f.

If both f_1 and f_2 are functions on A to B, then $f_1 = f_2$ means that $f_1(x) = f_2(x)$ for all $x \in A$.

Let f: $A \to B$ and g: $B \to C$ be functions. Then the function $g(f)$: $A \to C$, whose value on each $x \in A$ is the element $g(f(x)) \in C$, is called the *composite function* of f and g, denoted by $g \circ f$.

If f: $A \to B$ is a function, C is a subset of A, and D is a subset of B, the *restriction* of f to C is the function $f|C$: $C \to B$ defined by the same rule as f, but applied only to elements of C, and the subset of A consisting of all $x \in A$ such that $f(x) \in D$ is called the *inverse image* of D and is denoted by $f^{-1}(D)$.

A function f: $A \to B$ is said to be *one-to-one* or *injective* if $x \neq y$ implies $f(x) \neq f(y)$. An injective function is called an *injection*. f is said to be *onto* or *surjective* if to each element $b \in B$ there exists at least one element $a \in A$ such that $f(a) = b$. A surjective function is called a *surjection*. A function that is both injective and surjective is said to be *bijective*. A bijective function is also called a *bijection*.

Note that under a bijective function f: $A \to B$, each element $b \in B$ is the image of one and only one element $a \in A$. We then have an inverse function f^{-1}, defined throughout B, which assigns to each element $b \in B$ the unique element $a \in A$ such that $b = f(a)$.

Let k be a nonnegative integer. A function on a Euclidean n-space E^n to the real line E^1 is said to be *of class* C^k (respectively, C^∞) or a C^k (respectively, C^∞) *function* if its partial derivatives of orders up to and including k (respectively, of all orders) exist and are continuous. A C^o function means merely a continuous function.

The words "set," "space," and "collection" are synonymous, as are the words "function" and "mapping."

A First Course in Differential Geometry

1

Euclidean Spaces

This chapter contains, for a Euclidean space of three dimensions (extension to higher dimensions is virtually automatic), the fundamental material that is necessary for later developments of this book. Although most students are probably familiar with a great part of the material, its placement in one chapter makes it convenient for purposes of review and also allows us to bring out more clearly relationships among certain notions. Depending on the backgrounds of the students, certain sections may be selected for more thorough study.

1. POINT SETS

1.1. Neighborhoods and Topologies. Let E^3 be a Euclidean three-dimensional space. In the usual sense, in E^3 we take a fixed right-handed rectangular trihedron $\mathbf{0x}_1\mathbf{x}_2\mathbf{x}_3$ (see Fig. 1.1), that is, a point $\mathbf{0}$, called the origin of E^3, and mutually orthogonal coordinate axes $\mathbf{x}_1, \mathbf{x}_2, \mathbf{x}_3$, whose positive directions form a right-handed trihedron. Then relative to $\mathbf{0x}_1\mathbf{x}_2\mathbf{x}_3$ a point $\mathbf{x}$ in E^3 has coordinates (x_1, x_2, x_3). More generally, we have the following definition.

1.1.1. Definition. A *Euclidean n*-dimensional space E^n is the set of all ordered n-tuples $\mathbf{x}=(x_1,\cdots,x_n)$ of real numbers. Such an n-tuple is a *point* in E^n.

In accordance with our stated purpose, here and throughout this book we limit our discussions to $n=1,2,3$.

Let $u_1,\cdots,u_n$ be real-valued functions on E^n such that for each point $\mathbf{x}=(x_1,\cdots,x_n)$,

$$u_1(\mathbf{x})=x_1,\cdots,u_n(\mathbf{x})=x_n.$$

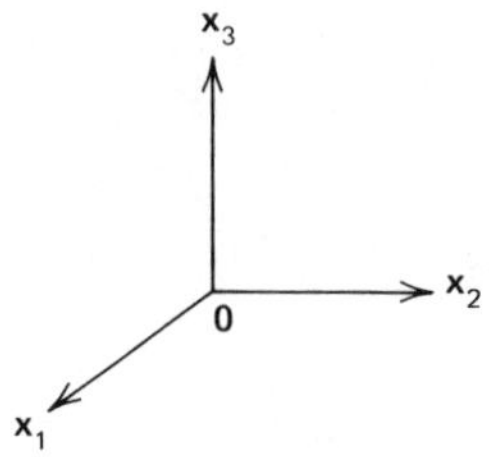

Figure 1.1

These functions $u_1, \cdots, u_n$ are called the *natural coordinate functions* of E^n.

The distance $d(\mathbf{x},\mathbf{y})$ between two points $\mathbf{x}=(x_1,\cdots,x_n)$ and $\mathbf{y}=(y_1,\cdots,y_n)$ in E^n is defined by the formula

$$d^2(\mathbf{x},\mathbf{y}) = \sum_{i=1}^{n} (x_i - y_i)^2, \qquad d(\mathbf{x},\mathbf{y}) \geqslant 0. \tag{1.1.1}$$

It is obvious that $d(\mathbf{x},\mathbf{y})=0$ if and only if $x_i = y_i$, $i=1,\cdots,n$, that is, if and only if $\mathbf{x}$ coincides with $\mathbf{y}$. Furthermore, we have $d(\mathbf{x},\mathbf{y})=d(\mathbf{y},\mathbf{x})$ and the triangle inequality for $\mathbf{z}\in E^n$

$$d(\mathbf{x},\mathbf{y}) + d(\mathbf{y},\mathbf{z}) \geqslant d(\mathbf{x},\mathbf{z}). \tag{1.1.2}$$

1.1.2. Definition. An *open spherical neighborhood* of a point $\mathbf{p}_0$ in E^n is the set of the form

$$\{\mathbf{p}\in E^n \mid d(\mathbf{p},\mathbf{p}_0) < \rho\}, \tag{1.1.3}$$

where $\rho > 0$. More generally, a *neighborhood* of $\mathbf{p}_0$ is any set that contains a spherical neighborhood of $\mathbf{p}_0$.

For $n=3$ it is convenient to use open spherical neighborhoods. However, for $n=2$ a neighborhood of $\mathbf{p}_0$ is any set that contains some open disk $\{p\in E^2 \mid \mathrm{d}(\mathbf{p},\mathbf{p}_0) < \rho\}$ about $\mathbf{p}_0$, and for $n=1$ it is an open interval containing $\mathbf{p}_0$.

We can easily obtain Lemma 1.1.3.

1.1.3. Lemma. *The neighborhoods of a point* $\mathbf{p}_0$ *in* E^n *have the following properties*:

(a) $\mathbf{p}_0$ *belongs to any neighborhood of* $\mathbf{p}_0$.

(b) *If U is a neighborhood of* $\mathbf{p}_0$, *and V a set such that $V \supset U$, then V is also a neighborhood of* $\mathbf{p}_0$.

(c) *If U and V are neighborhoods of* $\mathbf{p}_0$, *so is $U \cap V$.*

(d) *If U is a neighborhood of* $\mathbf{p}_0$, *there is a neighborhood V of* $\mathbf{p}_0$ *such that $V \subset U$ and V is a neighborhood of each of its points.*

1.1.4. Definition. In general, *a topological space* is a set S together with an assignment to each element $p_0 \in S$ of a collection of open subsets of S (see Definition 1.2.1), to be called *neighborhoods* of p_0, satisfying the four properties listed in Lemma 1.1.3, and the collection of the neighborhoods of all points of S is a *topology* for the space S.

Thus a Euclidean space E^n and the unit sphere in E^3 with center at the origin are both topological spaces.

1.1.5. Definition. Let S be a topological space, T a subset of S, and p a point of T. Then a subset U of T is a *neighborhood* of p in T if $U = T \cap V$, where V is a neighborhood of p in S. The neighborhoods U in T so defined have the four properties in Lemma 1.1.3. When T is made into a topological space by defining neighborhoods in this way, it is a *subspace* of S, and all the neighborhoods form a *relative topology* of S for the space T.

In the remainder of this section, unless stated otherwise, all spaces are supposed to be topological and all sets are to be in a general topological space, although we shall be interested only in spaces E^n for $n = 1,2,3$.

1.1.6. Definition. With respect to a subset T of a space S, each point p has one of the following three properties:

(a) p is *interior* to T if $p \in T$ and T is a neighborhood of p. The set of all the points interior to T is the *interior* of T.

(b) p is *exterior* to T if $p \notin T$ and there is a neighborhood of p that is disjoint from T, (i.e., has no points in common with T).

(c) p is a *boundary point* of T if p is neither interior nor exterior to T. The set of all boundary points of T is the *boundary* of T and is denoted by ∂T.

From the definition above it follows that an interior point of T is surrounded completely by points of T, that there are no points of T that are arbitrarily close to an exterior point of T, and that a boundary point of T may or may not belong to T.

The following is a frequently used method of obtaining new spaces from given spaces.

Let S and T be nonempty spaces. The set $S \times T$, called the *Cartesian product* of S and T, is defined to be the set of all ordered pairs (p,q) where $p \in S$ and $q \in T$. This set is made into a space as follows. If $(p,q) \in S \times T$, then a neighborhood of (p,q) is any set containing a set of the form $U \times V$, where U is a neighborhood of p in S, and V is a neighborhood of q in T. It is not hard to see that the neighborhood axioms a–d of Lemma 1.1.3 are satisfied.

1.1.7. Definition. $S \times T$, made into a space as just described, is the *topological product* of S and T.

Examples. 1. If $S = T = E^1$, then $S \times T$ is the plane with its usual topology (i.e., is E^2).

2. If $S = E^2$ and $T = E^1$, then $S \times T = E^3$. In general, $E^m \times E^n = E^{m+n}$.
3. If S is an interval on E^1, and T is a circle, then $S \times T$ is a cylinder.
4. The torus is the topological product of a circle with itself.

Exercises

1. Prove Lemma 1.1.3.

2. Let $S = \{\text{all } (x, y) \text{ with } x \text{ and } y \text{ rational numbers}\}$. (a) What is the interior of S? (b) What is the boundary of S?

1.2. Open and Closed Sets, and Continuous Mappings

1.2.1. Definition. A subset T of a space S is *open* if every point of T is interior to T; this is the same as saying that no boundary point of T belongs to T. A subset T of a space S is *closed* if every point of S that is not in T is in fact exterior to T; this is the same as saying that every boundary point of T is in T. The empty set, denoted by $\varnothing$, is an open set that contains no elements and is therefore a subset of every set.

The behavior of open and closed sets under the operations of union and intersection is of fundamental importance and is described by the following theorem.

1.2.2. Theorem. (a) *The union of any collection of open sets in a space S is open.*

(b) *The intersection of a finite collection of open sets in S is open.*
(c) *The intersection of any collection of closed sets in S is closed.*
(d) *The union of a finite collection of closed sets in S is closed.*

Proof. (a) Let $\{U_i\}$ be a collection of open sets in S, where i ranges over some set of indices. Let $U = \cup U_i$ and take p in U. Then $p \in U_i$ for some i, and by Definition 1.2.1, U_i is a neighborhood of p. Since $U \supset U_i$, U is a neighborhood of p by Lemma 1.1.3(b). Thus U is a neighborhood of each of its points and is therefore open by Definition 1.2.1.

(b) Let U_1 and U_2 be open sets in S and take $p \in U_1 \cap U_2$. Then by Definition 1.2.1, U_1 and U_2 are neighborhoods of p. From Lemma 1.1.3(c) it thus follows that $U_1 \cap U_2$ is a neighborhood of p. Since p is arbitrary, $U_1 \cap U_2$ is open by Definition 1.2.1. Using mathematical induction, we can extend this to a finite collection of open sets in S.

Parts (c) and (d) of Theorem 1.2.2 are obtained from parts (a) and (b) by considering the complementary sets.

Remark. The statement of Theorem 1.2.2b may not be true if "finite" is replaced by "infinite." For example, if S is the real line E^1, and U_n is the open interval $(-1/n, 1/n)$, then each U_n is an open set, but the intersection of all the U_n is the point 0, which is not open.

1.2.3. Definition. A point p is a *limit* (or *cluster* or *accumulation*) *point* of a set T if every neighborhood of p contains a point of T distinct from p. The *closure* of a subset T of S, denoted by $\bar{T}$, is the union of T and the set of its limit points.

Example. Consider the set

$$S = \{p \in E^2 \mid 0 < d(\mathbf{p}, \mathbf{0}) < 1\} \cup \{\text{the point } (0,2)\},$$

where $\mathbf{0}$ is the point (0,0). The boundary of S consists of the circumference, where $d(\mathbf{p},\mathbf{0}) = 1$, and the two points $\mathbf{0}$ and (0,2). The interior of S is the set of points $\mathbf{p}$ with $0 < d(\mathbf{p},\mathbf{0}) < 1$, and the closure of S is the set consisting of the point (0,2), together with the unit disk, the set of all points $\mathbf{p}$ such that $d(\mathbf{p},\mathbf{0}) \leqslant 1$.

1.2.4. Definition. Let T be a subset of a space S. The set of all points of S that are not in T is the *complement* of T in S, and is denoted by $S - T$.

The following lemma is an immediate consequence of Definitions 1.2.1 and 1.2.4.

1.2.5. Lemma. *A subset T of a space S is open (respectively, closed) if and only if $S - T$ is closed (respectively, open).*

1.2.6. Theorem. *Let T be any subset of a space S. Then T is closed if and only if $T = \bar{T}$.*

Proof. First suppose that T is closed. Then $S - T$ is open. If $S - T$ contains a limit point p of T, then every neighborhood $N(p)$ of p contains a point of T, and $N(p) \not\subset S - T$. This contradicts the fact that $S - T$ is open. Hence $T = \bar{T}$.

Conversely, suppose that $T = \bar{T}$. Let $p \in S - T$. Then $p \notin \bar{T}$, and p is not a limit point of T. Thus p has a neighborhood that contains no point of T

and belongs to $S-T$. This shows that $S-T$ is open, and therefore that T is closed.

1.2.7. Definition. Let S and T be two spaces. A mapping $f\colon S \to T$ is *continuous at a point* $p \in S$ if for any neighborhood V of $f(p)$ in T there is a neighborhood U of p in S such that $f(U) \subset V$. f is continuous if it is *continuous* at each point p in S.

1.2.8. Definition. Let S and T be two spaces, and $f\colon S \to T$ be a bijection. If both f and its inverse f^{-1} are continuous, then f is a *homeomorphism*, and S and T are *homeomorphic*.

Exercises

1. By constructing an example, show that the union of an infinite collection of closed sets may not be closed.

*2. Show that (a) the interior of a set S is the largest (in the sense of inclusion) open set contained in S, and (b) the closure of a set S is the smallest closed set that contains S.

*3. Let C be a closed set, and U an open set. Prove: (a) $C-U=\{\text{all } p \in C \text{ with } p \notin U\}$ is closed, and (b) $U-C$ is open.

4. Show that if p is a limit point of a set S in E^n, then every neighborhood of p contains infinitely many points of S.

5. If S and T both are the real line, the usual definition of continuity is that $f\colon S \to T$ is continuous at $x \in S$ if for any $\varepsilon > 0$ there is a $\delta > 0$ such that $|f(x') - f(x)| < \varepsilon$ whenever $x' \in S$ and $|x' - x| < \delta$. Prove that this is equivalent to Definition 1.2.7.

6. Let S and T be two spaces and $f\colon S \to T$ a mapping. Prove that f is continuous if and only if the inverse image of any open set in T is open in S. Use this to show that the composition of continuous mappings is continuous.

7. Let S be the union of two closed sets U and V, and $f\colon S \to T$ a mapping. Suppose that the restriction of f to U and to V are continuous mappings of U and V into T, respectively. Show that f is continuous. Given an example to show that this does not hold if U and V are not closed.

*Asterisks indicate exercises for which hints are provided at the end of the book.

1.3. Connectedness

1.3.1. Definition. A space S is *connected* if it is not the union of two nonempty disjoint open sets. A subset of a space is *connected* if it is a connected space as a subspace with the relative topology. A space is *disconnected* if it is not connected.

Examples. 1. The empty set is connected.

2. Let $S=(0,1)\cup(2,3)$ on E^1. Then $(0,1)$ is open and also closed, since $(0,1)=[0,1]\cap S$. Hence S is disconnected.

We can easily prove the following two equivalent definitions of connectedness.

1.3.2. Lemma. *A space is connected if and only if it is not the union of two nonempty disjoint closed sets.*

1.3.3. Lemma. *A space S is connected if and only if the only sets in S that are both open and closed are S and the empty set.*

Since connectedness is defined in terms of open sets, it is a topological invariant, an invariant preserved by homeomorphisms. However it is also preserved by continuous mappings, as indicated in Theorem 1.3.4.

1.3.4. Theorem. *Every continuous image of a connected set is connected.*

Proof. Suppose that S is any connected set, that $f: S\to T$ is continuous and surjective, and that T is disconnected. If $T=U\cup V$, where U and V are nonempty disjoint open sets, then $f^{-1}(U)$ and $f^{-1}(V)$ are clearly disjoint and nonempty, and are open by Exercise 6 of Section 1.2. Thus $S=f^{-1}(U)\cup f^{-1}(V)$ is disconnected, contradicting the assumption. Hence T is connected.

Example. Consider a continuous mapping of the unit interval of real numbers x such that $[0,1]$ is mapped onto the circumference of a circle, that is, x is mapped onto the point $(\cos 2\pi x, \sin 2\pi x)$ in E^2. By Theorem 1.3.4, the circumference of a circle is connected.

In Theorem 1.3.5 we have one of the most important connected spaces.

1.3.5. Theorem. *E^1 is connected.*

Proof. Let F be a closed proper subset of E^1 so that $E^1\neq F\neq\varnothing$. Then we can assume that $x\in E^1, x\notin F, y\in F$. Furthermore, without loss of generality we can assume that $y<x$, since using inf instead of sup (see

Definition 1.4.1) we can have the case by a similar argument of $y>x$. Let $z=\sup\{t\in F|t<x\}$. Then $y\leqslant z\leqslant x$. Since F is closed, every neighborhood of z meets F so that $z\in F$. Thus $z<x$. Since the open interval (z,x) does not intersect F, it follows that z is not an interior point of F and so F is not open. Hence by Lemma 1.3.3, E^1 is connected.

1.3.6. Theorem. *A subset of E^1 is connected if and only if it is an interval (open, closed, or half-open).*

Proof. If a connected subset S of E^1 is not an interval, then there exists $a\in E^1$ such that $a\notin S$ and that $A=\{x|x>a\}\cap S$ is a nonempty subset of S, so that A is open in S. Since also $A=\{x|x\geqslant a\}\cap S$, A is closed in S. By Lemma 1.3.3, S is disconnected, contradicting the assumption. Hence S is an interval.

The proof that each interval S on E^1 is connected is identical with the proof of Theorem 1.3.5 except that now $F\subset S$, and $x\in S$.

1.3.7. Theorem. *If S and T are connected spaces, then $S\times T$ is connected.*

Proof. Suppose that $S\times T$ is not connected. Then by Definition 1.3.1, $S\times T=U\cup V$, where U and V are nonempty disjoint open sets. Take (x,y) in U. The set $S\times\{y\}$ is homeomorphic to S and so is connected. It follows that $S\times\{y\}$ is contained in U, for otherwise its intersections with U and V would form a decomposition into nonempty disjoint open sets. But then a similar reasoning would show that for each $x'\in S$ the slice $\{x'\}\times T$ would be in U; thus all $S\times T$ would be in U, and V would be empty. This contradicts the assumption that V must be nonempty. Hence $S\times T$ is connected. Q.E.D.

From Theorems 1.3.5, 1.3.6, and 1.3.7 we immediately have two examples.

1.3.8. Examples. 1. The closed interval $I=[a,b]$ on E^1, the closed rectangle $I^2=I\times I$, and, therefore by mathematical induction, the closed n-dimensional cube I^n are connected. Similarly, E^1 and E^n are connected.

2. A closed disk that is homeomorphic to I^2 is connected. The surface S^2 of a 2-sphere can be expressed as the union of two closed disks with nonempty intersection. So by Exercise 5, below, this surface S^2 is connected. Similarly, the n-sphere is connected for any n.

1.3.9. Definition. If T is a connected subset of a set S and is not contained in any other connected subset of S, then T is a *component* of S.

In other words, components of S are maximal connected subsets of S.

Exercises

1. Prove that the set $(1,1,3)$ is an open set, and discuss all the properties described in Sections 1.1–1.3 for this set of points.
2. Discuss all the properties as in Exercise 1 for the set of points (x, y) in E^2 such that $x^2 - y^2 < 0$.
3. Prove Lemmas 1.3.2 and 1.3.3.
4. Let A and B be connected sets in E^2. Is $A \cap B$ or $A \cup B$ necessarily connected?
5. Let A and B be connected sets in a space S, and suppose that $A \cap \bar{B}$ is not empty. Show that $A \cup B$ is connected.

1.4. Infimum and Supremum, and Sequences

1.4.1. Definition. Let S be a set of real numbers. If M is a real number such that $x \leqslant M$ for each $x \in S$, M is an *upper bound* of S. If A is an upper bound of S, and there is no upper bound of S smaller than A, then A is the *least upper bound* or the *supremum* of S, abbreviated lub S or sup S. On the other hand, if m is a real number such that $m \leqslant x$ for each $x \in S$, then m is a *lower bound* of S. If B is a lower bound of S and there is no lower bound of S greater than B, then B is the *greatest lower bound* or the *infimum* of S, abbreviated glb S or inf S.

For the existence of a supremum and an infimum we have the following axiom and theorem.

1.4.2. Axiom of Completeness. *Any nonempty set of real numbers that has an upper bound has a supremum.*

1.4.3. Theorem. *Any nonempty set S of real numbers that has a lower bound has an infimum.*

Proof of Theorem 1.4.3. The set T, obtained by changing every element of S to its negative, has as an upper bound the negative of any lower bound of S. By the axiom of completeness, T then has a supremum, whose negative must be the infimum of S.

1.4.4. Definition. Let S be a subset of E^n, and $d(\mathbf{p},\mathbf{q})$ the distance between two points $\mathbf{p}$ and $\mathbf{q}$ in S. Then the *diameter* of S is

$$\sup_{\mathbf{p},\mathbf{q} \in S} d(\mathbf{p},\mathbf{q})$$

S is *bounded* if its diameter is finite.

From Definition 1.4.4 it follows that if a subset S of E^n is bounded, it lies entirely within some sphere of radius $r>0$ with center at the origin $\mathbf{0}=(0,\cdots,0)$ in E^n.

1.4.5. Definition. Let $\{\mathbf{p}_m\}$ be a sequence of points in E^n. Then $\{\mathbf{p}_m\}$ is *convergent* to a point $\mathbf{p}$ in E^n, written

$$\lim_{m\to\infty} \mathbf{p}_m = \mathbf{p}$$

if and only if for every real number $\epsilon>0$ there is an integer M such that whenever $m>M$, then $d(\mathbf{p}_m,\mathbf{p})<\epsilon$. A sequence of points in E^n is *convergent* if it is convergent to some point in E^n. A sequence is *divergent* if it is not convergent.

Since this definition refers to the limit of a sequence, we cannot use it to prove that a sequence is convergent unless we already know what the limit is. Theorem 1.4.6 provides a way out of this inconvenience.

1.4.6. Theorem. *A sequence $\{\mathbf{p}_m\}$ of points in E^n is convergent if and only if*

$$\lim_{m,m'\to\infty} d(\mathbf{p}_m,\mathbf{p}_{m'})=0,$$

that is, if and only if for every real number $\epsilon>0$ there exists an integer M such that $d(\mathbf{p}_m,\mathbf{p}_{m'})<\epsilon$ whenever $m\geqslant M$ and $m'\geqslant M$.

1.4.7. Definition. The condition in Theorem 1.4.6 for the sequence $\{\mathbf{p}_m\}$ to be convergent is the *Cauchy condition*, and sequences that satisfy the Cauchy condition are *Cauchy sequences.*

From Theorem 1.4.6 it follows that every Cauchy sequence is bounded; therefore an unbounded sequence necessarily diverges.

1.4.8. Theorem. *Let $\{C_m\}$ be a sequence of nonempty closed subsets of E^n such that $C_{m+1}\subset C_m$ for all m, and $\lim_{m\to\infty} d_m=0$, where d_m denotes the diameter of C_m. Then $\cap C_m\neq\varnothing$.*

Proof. Since $\lim_{m\to\infty} d_m=0$, given $\epsilon>0$, there exists an integer M such that $d_M<\epsilon$. Let $m>M$, $m'>M$, and for each m choose a point $\mathbf{p}_m\in C_m$. Then $\mathbf{p}_m\in C_M$, $\mathbf{p}_{m'}\in C_M$ so that $d(\mathbf{p}_m,\mathbf{p}_{m'})\leqslant d_M<\epsilon$. Thus $\{\mathbf{p}_m\}$ is a Cauchy sequence, and therefore by Theorem 1.4.6 it is convergent, say $\mathbf{p}_m\to\mathbf{p}$. Let m be arbitrary. For $m'>m$, $\mathbf{p}_{m'}\in C_m$. Since C_m is closed and $\mathbf{p}_{m'}\to\mathbf{p}$, $\mathbf{p}\in C_m$ by Theorem 1.2.6. Hence $\mathbf{p}\in C_m$, and the theorem is proved.

Exercises

1. Show that the sup and inf of a set are uniquely determined whenever they exist.
2. Let a sequence $\{x_n\}$ on E^1 be defined as follows: $x_1=1, x_2=2, x_{n+1}=\frac{1}{2}(x_{n-1}+x_n), n=2,3,\cdots$. Show that $d(x_m, x_n)\leqslant 1/2^{N-1}$ if $m\geqslant N, n\geqslant N$, so that the Cauchy condition is fulfilled.

1.5. Compactness

1.5.1. Definition. A space is a *Hausdorff space* if for any two distinct points p and q there are neighborhoods U of p and V of q such that $U\cap V=\varnothing$. Thus distinct points are separated by disjoint neighborhoods.

Example. Any Euclidean space is a Hausdorff space.

1.5.2. Definition. A *covering* of a space S is a collection of sets in S whose union is S. It is an *open covering* if all the sets of the collection are open.

1.5.3. Definition. Given a covering of a space, a *subcovering* is a covering whose sets all belong to the given covering.

1.5.4. Definition. A *compact space* is a Hausdorff space with the property that any open covering contains a finite covering, that is, a subcovering consisting of finitely many sets. A subset of a space is *compact* if it is a compact subspace.

Example. The real line is not compact. For example, take the collection of open intervals $(n-1, n+1)$ for all integers n. This is an open covering of the real line; but obviously no finite collection of these intervals can cover the whole line. A similar argument shows that E^n for each n is noncompact, and in fact so is any unbounded subset of E^n.

For a subset of E^n to be compact, we have the following necessary and sufficient condition.

1.5.5. Theorem. *A subset S of E^n is compact if and only if it is closed and bounded.*

Proof. The "if" part is the well-known Heine-Borel theorem. To prove it, we assume that S is closed and bounded and that there is an open covering $\{U_i\}$ of S. Then the new sets, V_i, constructed as follows, are open by Theorem 1.2.2(a): $V_1 = U_1$, $V_2 = U_1 \cup U_2$, $V_3 = U_1 \cup U_2 \cup U_3 = V_2 \cup U_3$, and, in general, $V_{i+1} = V_i \cup U_{i+1}$. These sets V_i form an increasing sequence $V_1 \subset V_2 \subset \cdots \subset V_i \subset \cdots$ and an open covering of S. From these open sets V_i, by Exercise 3(a) of Section 1.2 we have closed sets $S_i = S - V_i$. Since S is bounded, S_i is also bounded. Thus we have a decreasing sequence of closed bounded sets:

$$S \supset S_1 \supset S_2 \supset \cdots \supset S_i \supset \cdots.$$

Let d_i be the diameter of S_i. Then we have a monotonic decreasing sequence $\{d_i\}$ of positive numbers with 0 as its infimum, so that $\lim_{i\to\infty} d_i = 0$. Therefore by applying Theorem 1.4.8 to the sequence $\{S_i\}$ we have at least a point $\mathbf{p} \in \cap S_i$. This $\mathbf{p}$ must belong to S but not to any of the sets V_i. However, this cannot happen, since $\{V_i\}$ is a covering of S. Thus one of the sets S_i is empty, say S_j, so that V_j contains S, and the sets $U_1, U_2, \cdots, U_j$ form a finite covering of S. Hence S is compact.

To prove the "only if" part, we assume that S is compact and first show that S is bounded. Let $N(\mathbf{x}_0; r)$ denote the open sphere in E^n with center at $\mathbf{x}_0$ and radius r. The collection of spheres $N(\mathbf{0}; k)$, where $k = 1, 2, 3, \cdots$, and $\mathbf{0}$ denotes the origin $(0, \cdots, 0)$ in E^n, is an open covering of S. By compactness, a finite subcollection also covers S, hence S is bounded.

Now we prove that S is closed. Suppose the contrary. Then by Theorem 1.2.6 there is a limit point $\mathbf{y}$ of S such that $\mathbf{y} \notin S$. If $\mathbf{x} \in S$, let $r_x = \frac{1}{2} d(\mathbf{x}, \mathbf{y})$. Each r_x is positive, since $\mathbf{y} \notin S$ and the collection $\{N(\mathbf{x}; r_x) | \mathbf{x} \in S\}$ is an open covering of S. By compactness, a finite collection of these neighborhoods covers S, say $S \subset \cup_{k=1}^{p} N(\mathbf{x}_k; r_k)$, where $r_i = r_{x_k}$. Let r denote the smallest of the radii $r_1, r_2, \cdots, r_p$. Then it is easy to show that the neighborhood $N(\mathbf{y}; r)$ has no point in common with any of the neighborhoods $N(\mathbf{x}_k; r_k)$. In fact, if $\mathbf{x} \in N(\mathbf{y}; r)$, then $d(\mathbf{x}, \mathbf{y}) < r \leqslant r_k$ and

$$\begin{aligned} d(\mathbf{x}, \mathbf{x}_k) &= d(\mathbf{y} - \mathbf{x}_k, \mathbf{y} - \mathbf{x}) \geqslant d(\mathbf{y}, \mathbf{x}_k) - d(\mathbf{x}, \mathbf{y}) \\ &= 2r_k - d(\mathbf{x}, \mathbf{y}) > r_k, \end{aligned}$$

so that $\mathbf{x} \notin N(\mathbf{x}_k; r_k)$. Thus $N(\mathbf{y}; r) \cap S$ is empty, contradiction the fact that $\mathbf{y}$ is a limit point of S. Hence S is closed. Q.E.D.

From the definition it is obvious that compactness is a topological invariant. In fact it is preserved by any continuous mapping, as stated in the following theorem.

1.5.6. Theorem. *Let f: $S \to T$ be a continuous mapping of a compact space S onto a Hausdorff space T. Then T is compact.*

Proof. Let $\{U_i\}$ be an open covering of T. Then the sets $f^{-1}(U_i)$ form a covering of S, which is open by Exercise 6 of Section 1.2. Since S is compact, a finite subcollection of the sets $f^{-1}(U_i)$, say $f^{-1}(U_1), f^{-1}(U_2), \cdots, f^{-1}(U_n)$, will cover S. Then $U_1, U_2, \cdots, U_n$ form a finite subcovering of the given covering $\{U_i\}$ of T. Since T is assumed to be a Hausdorff space, it is compact.

1.5.7. Definition. Let f: $S \to T$ be a function on a set S into a set T. Then the *graph* of f is the set of all ordered pairs (x, y) with $x \in S$ and $y = f(x) \in T$, and is thus a special subset of $S \times T$, the Cartesian product of S and T.

In particular, if T is E^1 and S is a subset of E^1 or E^2, then $S \times T$ is a subset of E^2 or E^3, and the graph of a function f: $S \to T$ can be geometrically visualized as a curve in E^2 or E^3.

1.5.8. Definition. A function f: $S \to E^1$ on a subset S of E^n attains an *absolute maximum* (respectively, *minimum*) at a point $\mathbf{p}_0 \in S$, and $\mathbf{p}_0$ is an *absolute-maximum* (respectively, *-minimum*) *point* of f, if $f(\mathbf{p}_0) \geqslant f(\mathbf{p})$ (respectively, $f(\mathbf{p}_0) \leqslant f(\mathbf{p})$) for all $p \in S$.

By means of Theorems 1.5.5 and 1.5.6 we can easily deduce Theorem 1.5.9.

1.5.9. Theorem. *If f: $S \to E^1$ is a continuous function on a compact subset S of E^n, then f attains an absolute maximum and an absolute minimum somewhere on S.*

Proof. By Theorems 1.5.5 and 1.5.6, the continuous function f is bounded on S, that is, the image $f(S)$ is a bounded set. Then by Axiom 1.4.2 and Theorem 1.4.3, there exist both the supremum M and the infimum m for the values of f on S so that $m \leqslant f(\mathbf{p}) \leqslant M$ for all $\mathbf{p} \in S$, and M cannot be decreased, nor m increased. If there is a point $\mathbf{p}_0$ in S such that $f(\mathbf{p}_0) = M$, then M is the maximum for f on S, and it is attained there.

If there is no such point $\mathbf{p}_0$ with $f(\mathbf{p}_0) = M$, then $f(\mathbf{p}) < M$ for all $\mathbf{p} \in S$. In this case, set

$$g(\mathbf{p}) = \frac{1}{M - f(\mathbf{p})}. \tag{1.5.1}$$

Since the denominator of (1.5.1) is never zero, g is continuous on S and must therefore be bounded there. If we suppose that $g(\mathbf{p}) \leqslant A$ for all $\mathbf{p} \in S$, then from (1.5.1) we would have

$$f(\mathbf{p}) \leqslant M = \frac{1}{A}, \qquad \text{for all} \quad \mathbf{p} \in S,$$

which however contradicts the fact that M is the supremum for the values of f on S. Thus we conclude that this does not occur and there is a point $\mathbf{p}_0 \in S$ such that $f(\mathbf{p}_0)=M$. Similarly, we can show that there is a point in S where f has the value m, and a minimum is attained. Q.E.D.

When S is the closed interval $[a,b]$ on E^1, we immediately obtain Corollary 1.5.10 from Theorem 1.5.9.

1.5.10. Corollary. *A continuous function f: $[a,b]\to E^1$ attains an absolute maximum and an absolute minimum at some points of the closed interval $[a,b]$.*

The three examples that follow show that if the interval is not closed, Corollary 1.5.10 might not be true.

Examples. 1. $f(x)=x$, for $0<x\leqslant 10$.
There is an absolute maximum at $x=10$, but there is no absolute minimum.
2. $f(x)=x^2$, for $-1\leqslant x<2$.
There is an absolute minimum at $x=0$, but there is no absolute maximum.
3. $f(x)=\tan x$, for $-\frac{\pi}{2}<x<\frac{\pi}{2}$.
There is neither an absolute maximum nor an absolute minimum.

1.5.11. Theorem. *Let S be a connected set, and let f: $S\to E^1$ be a continuous function. Let $p,q\in E^1$ be any two image points of f, and suppose that $p<r<q$. Then there is a point $s\in S$ such that $f(s)=r$.*

Proof. Suppose that $f(s)$ is never r for any point $s\in S$. Then we always have either $f(s)>r$ or $f(s)<r$. Let U be the subset $\{s\in S \mid f(s)>r\}$ of S, and V the subset $\{s\in S \mid f(s)<r\}$. These two sets U and V must cover S. Since f is continuous, by Exercise 6 of Section 1.2 the inverse image $f^{-1}(I)$ of the open unbounded interval I consisting of the numbers x with $x>r$ (or with $x<r$) must be open relative to S. Thus both U and V are open relative to S. Moreover, since there is a point $t\in S$ with $f(t)=p$, the set V contains t and is therefore not empty; likewise, U is not empty. The existence of two such sets contradicts the assumption that S was connected. Hence somewhere in S, f must take the value r. Q.E.D.

In particular, when S is a closed interval $[a,b]$, $p=\inf f(S)$ and $q=\sup f(S)$, Theorem 1.5.11 becomes the following intermediate value theorem of Bolzano.

1.5.12. Theorem. *If f: $[a,b]\to E^1$ is a continuous function, then f assumes every value between its maximum and minimum.*

Exercises

*1. Show that every infinite subset of a compact space has a limit point.

2. If U is open in E^n and $C\subset U$ is compact, show that there is a compact set D such that $C\subset$ interior D and $D\subset U$.

3. Let X be a compact subset of E^n and f: $X\to E^1$ be a continuous function satisfying $f(x)>0$ for all $x\in X$. Show that there exists $\epsilon>0$ such that $f(x)\geqslant\epsilon$ for all $x\in X$.

4. Investigate the possibility of absolute extrema of $f(x)=x^4+256/x^2$ for $0<x\leqslant 4$, and find any which exist.

5. Consider the function $27/\sin x+64/\cos x$, $0<x<\pi/2$. Why must there be an absolute minimum in this open interval? Find where it occurs and the corresponding value of the function.

6. Without drawing the graph, find the absolute maximum and minimum of the function $2\sin x-1+2\cos^2 x$.

2. DIFFERENTIATION AND INTEGRATION

In this section we list some fundamental formulas and theorems in elementary calculus that we shall use later. We give here only proofs that cannot be found in ordinary calculus books.

2.1. The Mean Value Theorems. This section contains four mean value theorems of differential and integral calculus. The mean value theorem of differential calculus is one of the most important theoretical results in the subject. It expresses a geometric property of differentiable functions and has many applications. The mean value theorems of integral calculus give estimations of definite integrals.

2.1.1. Theorem (The mean value theorem of differential calculus). *Suppose that f: $[a,b]\to E^1$ is a continuous function and is differentiable in the interval (a,b). If we put $h=b-a$, there is a point $a+\theta h\in(a,b)$, $0<\theta<1$, such that*

$$f(a+h)=f(a)+hf'(a+\theta h), \tag{2.1.1}$$

where the prime denotes the derivative with respect to the argument.

2.1.2. Theorem (The first mean value theorem of integral calculus). *If f: $[a,b]\to E^1$ is a continuous function, there is a point $c\in(a,b)$ such that*

$$\int_a^b f(x)\,dx=(b-a)f(c). \tag{2.1.2}$$

Proof of Theorem 2.1.2. Putting $F(x)=\int_a^x f(t)\,dt$ and applying Theorem 2.1.1 to $F(x)$ immediately give (2.1.2).

2.1.3. Theorem (The generalized first mean value theorem of integral calculus). *Suppose that $f, g\colon [a,b]\to E^1$ are continuous functions and that g is everywhere nonnegative (or nonpositive). Then there is a point $c\in(a,b)$ such that*

$$\int_a^b f(x)g(x)\,dx=f(c)\int_a^b g(x)\,dx. \tag{2.1.3}$$

It is obvious that when $g(x)=1$ for all $x\in[a,b]$ Theorem 2.1.3 becomes Theorem 2.1.2.

Proof. We consider the case of g everywhere nonnegative and from this case, by replacing g by $-g$, we can deduce the case of g everywhere nonpositive. Denote the minimum and maximum values of f in $[a,b]$ by m and M, respectively. Then for every $x\in[a,b]$

$$mg(x)\leqslant f(x)g(x)\leqslant Mg(x). \tag{2.1.4}$$

By the definition of a definite integral we know that if $F\colon [a,b]\to E^1$ is a continuous function and $F\geqslant 0$ throughout $[a,b]$, then

$$\int_a^b F(x)\,dx\geqslant 0, \tag{2.1.5}$$

from which it follows that if $G, H\colon [a,b]\to E^1$ are continuous functions and $G\geqslant H$ throughout $[a,b]$, then

$$\int_a^b G(x)\,dx\geqslant\int_a^b H(x)\,dx. \tag{2.1.6}$$

By integrating (2.1.4) between a and b, and using (2.1.6), we obtain

$$m\int_a^b g(x)\,dx\leqslant\int_a^b f(x)g(x)\,dx\leqslant M\int_a^b g(x)\,dx. \tag{2.1.7}$$

For simplicity, denote $\int_a^b g(x)\,dx$ by I. Since $g(x)\geqslant 0$ for all $x\in[a,b]$, by (2.1.5) we have $I\geqslant 0$. If $I=0$, then (2.1.7) implies that both sides of (2.1.3) are zero, so that (2.1.3) is valid for any c. Now suppose that $I>0$. Then divide (2.1.7) by I, so that

$$m\leqslant\frac{1}{I}\int_a^b f(x)g(x)\,dx\leqslant M.$$

Hence Theorem 1.5.12 gives our formula (2.1.3) for some $c\in(a,b)$.

2.1.4. Theorem (The second mean value theorem of integral calculus). *Suppose that $f, g\colon [a,b]\to E^1$ are functions of classes C^0 and C^1, respectively, and* *that g is monotonic. Then there exists a point* $c \in [a,b]$ *such that*

$$\int_a^b f(x)g(x)\,dx = g(a)\int_a^c f(x)\,dx + g(b)\int_c^b f(x)\,dx. \qquad (2.1.8)$$

Proof. We first notice that we can assume that $g(b)=0$, since replacing $g(x)$ by $g(x)-g(b)$ changes both sides of (2.1.8) by the same amount, and gives us a function that vanishes at $x=b$. Moreover, we can assume $g(a)>0$; for if $g(a)<0$ we need only replace $g(x)$ by $-g(x)$, which changes the sign of both sides of (2.1.8). [The case $g(a)=0$ is trivial: if both $g(a)$ and $g(b)$ vanish, $g(x)$ must be identically zero, and both sides of (2.1.8) are zero also.] Thus we need only prove that if $g(x)$ is continuous and monotonic decreasing and $g(b)=0$, then

$$\int_a^b f(x)g(x)\,dx = g(a)\int_a^c f(x)\,dx. \qquad (2.1.9)$$

Now by putting $F(x)=\int_a^x f(t)\,dt$ and applying the formula for integration by parts to the left-hand side of (2.1.9), we obtain

$$\int_a^b f(x)g(x)\,dx = F(x)g(x)\Big|_a^b + \int_a^b F(x)[-g'(x)]\,dx. \qquad (2.1.10)$$

The first term on the right-hand side of (2.1.10) vanishes, since $F(a)$ and $g(b)$ are zero. The expression $-g'(x)$ is everywhere nonnegative, so that by applying Theorem 2.1.3 we find that the integral on the right-hand side of (2.1.10) has the value

$$F(c)\int_a^b [-g'(x)]\,dx \qquad \text{for some} \quad c\in[a,b].$$

Since $F(c)=\int_a^c f(x)\,dx$ and $\int_a^b [-g'(x)]\,dx = g(a)-g(b)=g(a)$, (2.1.10) is just (2.1.9), and our theorem is established.

Exercises

1. Use Theorem 2.1.1 to establish the following inequalities:

(a) $\sin x < x$, for $x>0$.

(b) $\dfrac{h}{1+h} < \ln(1+h) < h$, for $h>-1,\ h\neq 0$.

***2.** If $f'(x)=0$ for all x, $a<x<b$, show that f is constant there.

2.2. Taylor's Formulas. In this section we give a generalization of Theorem 2.1.1 (the mean value theorem of differential calculus), together with a further generalization to functions of two variables.

2.2.1. Theorem. *Let f: $[a, b] \to E^1$ be a function of class C^n, and let the $(n+1)$st derivative $f^{(n+1)}(x)$ of f exist for $x \in (a, b)$. If we put $h = b - a$, there is a point $a + \theta h \in (a, b)$, $0 < \theta < 1$, such that*

$$\begin{aligned} f(a+h) &= f(a) + f'(a)h + \cdots \\ &\quad + \frac{f^{(n)}(a)}{n!}h^n + \frac{f^{(n+1)}(a+\theta h)}{(n+1)!}h^{n+1}. \end{aligned} \tag{2.2.1}$$

When $a = 0$, (2.2.1) is called *Maclaurin's formula.*

2.2.2. Theorem. *Let D be a neighborhood of a point (a, b) in E^2 and let f: $D \to E^1$ be a function of class C^{n+1}. Then there is a point $(a+\theta h, b+\theta k) \in D$, $0 < \theta < 1$, such that*

$$\begin{aligned} &f(a+h, b+k) \\ &= f(a,b) + \sum_{i=1}^{n} \frac{1}{i!}\left[\left(h\frac{\partial}{\partial x} + k\frac{\partial}{\partial y}\right)^i f(x,y)\right]_{\substack{x=a\\y=b}} + R_{n+1}, \end{aligned} \tag{2.2.2}$$

where

$$R_{n+1} = \frac{1}{(n+1)!}\left[\left(h\frac{\partial}{\partial x} + k\frac{\partial}{\partial y}\right)^{n+1} f(x,y)\right]_{\substack{x=a+\theta h\\y=b+\theta k}}. \tag{2.2.3}$$

Exercises

1. Find the Taylor expansion for the function

$$f(x, y) = x^4 + 3x^2y^2 - xy + 2$$

at the point $(1, -2)$ through terms of degree 3.

2.3. Maxima and Minima

2.3.1. *Definition.* A function f: $E^n \to E^1$ has a *relative* (or *local*) *maximum* (respectively, *minimum*) at $\mathbf{x}_0 \in E^n$, and $\mathbf{x}_0$ is a *relative* (or *local*) *-maximum* (respectively, *-minimum*) *point* of f, if there is a neighborhood U about $\mathbf{x}_0$ in E^n such that $f(\mathbf{x}) \leqslant f(\mathbf{x}_0)$ [respectively, $f(\mathbf{x} \geqslant f(\mathbf{x}_0)$] for all $\mathbf{x} \in U$. The term *extremum* is used to refer to either maximum or minimum.

Necessary conditions for relative extremum are given in Theorem 2.3.2.

2.3.2. Theorem. *At an interior point $\mathbf{p} \in S \subset E^n$, if a function $f\colon S \to E^1$ has a relative extremum and is differentiable, that is, has first partial derivatives, these derivatives vanish at* $\mathbf{p}$.

The converse of Theorem 2.3.2 is generally not true.

2.3.3. Definition. If a function $f\colon S \subset E^n \to E^1$ is differentiable at a point $\mathbf{p} \in S$ and all the first partial derivatives of f are zero at $\mathbf{p}$, then $\mathbf{p}$ is a *critical point* of f.

In terms of the critical points we have the following criterion for the relative extrema of a function $f\colon S \subset E^2 \to E^1$.

2.3.4. Theorem. *Suppose that a function $f\colon S \subset E^2 \to E^1$ is differentiable, that is, has first partial derivatives, and that an interior point $\mathbf{p}$ of S is a critical point of f. Let x, y be the natural coordinate functions of S, and f_x, f_y and f_{xx}, f_{xy}, f_{yy} denote, respectively, the first and second partial derivatives of $f(x, y)$. Suppose further that f_x and f_y are differentiable at $\mathbf{p}$, and write*

$$A = f_{xx}(\mathbf{p}), \qquad B = f_{xy}(\mathbf{p}), \qquad C = f_{yy}(\mathbf{p}).$$

Then we have the following cases:

(a) *If $B^2 - AC < 0$ and $A > 0$, then f has a relative minimum at* $\mathbf{p}$.

(b) *If $B^2 - AC < 0$ and $A < 0$, then f has a relative maximum at* $\mathbf{p}$.

(c) *If $B^2 - AC > 0$, then f has neither a relative maximum nor a relative minimum at $\mathbf{p}$, and $\mathbf{p}$ is called a saddle point.*

(d) *If $B^2 - AC = 0$, no conclusion may be drawn, and any of the behaviors of f described in parts (a)–(c) may occur.*

A critical point $\mathbf{p}$ is said to be nondegenerate or degenerate according as $B^2 - AC \neq 0$ or $= 0$.

According to Corollary 1.5.10, a continuous function $f\colon [a, b] \to E^1$ attains an absolute maximum and an absolute minimum at some points of the interval $[a, b]$. If an absolute extremum occurs at an interior point of the interval, then it is also a relative extremum, and therefore we have $f'(x) = 0$ at the extremum point, provided f is differentiable there.

A function $g\colon E^1 \to E^1$ is periodic with period p if there is a constant integer $c > 0$ such that $g(t) = g(t + c)$ for all t, and p is the least such number c. Now let a C^1 function $f\colon E^1 \to E^1$ be periodic with period $b - a$. Then by Corollary 1.5.10, in the interval $[a, b]$ the periodic function f has an absolute maximum and an absolute minimum, which now are also a relative maximum and a relative minimum, respectively. In particular, if the interval $[a, b]$ is of length $l > 0$ with the two end points a and b identified, between any two consecutive relative maxima (respectively, minima) of f there is a relative minimum (respectively, maximum); therefore the relative maxima and minima of f occur in pairs in the interval $[a, b]$. To sum up, we hence obtain Theorem 2.3.5.

2.3.5. Theorem. *Let f: $E^1 \to E^1$ be a periodic continuous C^1 function with period $b-a$. Then in the interval $[a, b]$, every absolute extremum of f is also a relative extremum, and f has at least a relative maximum and a relative minimum. In particular, if the interval $[a, b]$ is of length $l>0$ with the two end points a and b identified, the relative maxima and minima of f occur in pairs in the interval $[a, b]$.*

2.3.6. Theorem (The chain rule). *Let f: $S \subset E^3 \to E^1$ be a function of class C^1. If x_1, x_2, x_3, and t are the natural coordinate functions of S and E^1, respectively, and g: $E^1 \to S$ defined by $g(t)=(x_1(t), x_2(t), x_3(t))$ is a differentiable function, then $fg(t)=f(x_1(t), x_2(t), x_3(t))$ is differentiable, and*

$$f'(x_1(t)x_2(t), x_3(t)) = \sum_{i=1}^{3} f_{x_i}(x_1, x_2, x_3)x_i'(t),$$

where the prime denotes the derivative with respect to t, and f_{x_i} the partial derivative with respect to x_i.

Exercises

1. Find the critical points of the function $f(x, y)=2x^3-y^3+12x^2+27y$ and determine the nature of each of them.
2. Show that all the critical points of the function $f(x, y)=\cos(xe^y)$ are degenerate.
3. Find the critical points of the function $f(x, y, z)=(1-x^2)y+(1-y^2)z$.

2.4. Lagrange Multipliers. The following theorem gives a very elegant and useful method, known as the method of Lagrange multipliers, for studying extremal problems for functions of several variables with constraints. This method avoids either the impossibility or some complication arising from direct elimination of variables, and permits the retention of symmetry when the variables enter symmetrically at the outset of the problem.

2.4.1. Theorem. *Suppose that a real-valued function $f(x, y)$ is differentiable in a neighborhood $N_{(a,b)}$ of a point (a, b) in E^2, and that it has a relative extremum at (a, b) subject to a constraint*

$$\phi(x, y)=0, \tag{2.4.1}$$

where ϕ is differentiable in $N_{(a,b)}$, and (2.4.1) defines a differentiable function

$$y=g(x) \tag{2.4.2}$$

in a neighborhood of a such that $g(a)=b$. Then the point (a, b) is a critical point of $F(x,y)=f(x,y)-\lambda\phi(x,y)$ subject to no constraint, where λ is a constant, called a Lagrange multiplier.

Proof. Since $f(x,y)$ with (2.4.2) has a relative extremum at (a,b), by Theorems 2.3.2 and 2.3.6 we obtain

$$0=f'(a,g(a))=f_x(a,b)+f_y(a,b)g'(a). \tag{2.4.3}$$

From (2.4.1), (2.4.2) and Theorem 2.3.6 we also have

$$\phi_x(a,b)+\phi_y(a,b)g'(a)=0. \tag{2.4.4}$$

Elimination of $g'(a)$ from (2.4.3) and (2.4.4) gives

$$f_x(a,b)\phi_y(a,b)-f_y(a,b)\phi_x(a,b)=0. \tag{2.4.5}$$

On the other hand, by Definition 2.3.3 a critical point (x,y) of $F(x,y)$ satisfies

$$f_x(x,y)-\lambda\phi_x(x,y)=0, \qquad f_y(x,y)-\lambda\phi_y(x,y)=0,$$

or, by elimination of λ,

$$f_x(x,y)\phi_y(x,y)-f_y(x,y)\phi_x(x,y)=0. \tag{2.4.6}$$

From (2.4.5) and (2.4.6) it follows immediately that (a,b) is a critical point of $F(x,y)$; hence our theorem is proved.

2.4.2. Definition. A *quadratic form* $f(x,y)$ in two variables x and y is a homogeneous quadratic polynomial, that is, a function of two variables x and y, which has the form

$$f(x,y)=ax^2+2bxy+cy^2, \tag{2.4.7}$$

where a, b, and c are constants. The quadratic form $f(x,y)$ is *positive* (respectively, *negative*) *definite* if it satisfies

$$f(x,y)>0\,[\text{respectively}, f(x,y)<0], \quad \text{for all} \quad (x,y)\neq(0,0), \tag{2.4.8}$$

and it is *positive* (respectively, *negative*) *semidefinite* if it satisfies

$$f(x,y)\geqslant 0\,[\text{respectively}, f(x,y)\leqslant 0], \quad \text{for all} \quad (x,y), \tag{2.4.9}$$
$$f(x_0,y_0)=0, \qquad \text{for some} \quad (x_0,y_0)\neq(0,0).$$

By completing the square we obtain

$$f(x,y)=a\left[\left(x+\frac{b}{a}y\right)^2+(ac-b^2)\frac{y^2}{a^2}\right],$$

and therefore we arrive at Lemma 2.4.3.

2.4.3. Lemma. *The function* $f(x, y)$ *of* (2.4.7) *is positive* (*respectively, negative*) *definite if and only if it satisfies*

$$ac-b^2>0, a>0 \qquad (\text{respectively}, ac-b^2<0, a<0), \qquad (2.4.10)$$

$f(x, y)$ *is positive* (*respectively, negative*) *semidefinite if and only if it satisfies*

$$ac-b^2=0, \qquad a\geqslant 0 \qquad (\text{respectively}, a\leqslant 0). \qquad (2.4.11)$$

The following example illustrates that the method of Lagrange multipliers in Theorem 2.4.1 is particularly useful when both $f(x, y)$ and $\phi(x, y)$ are quadratic forms.

2.4.4. Example. Consider the quadratic form $f(x, y)$ of (2.4.7) and

$$\phi(x, y)=x^2+y^2-1. \qquad (2.4.12)$$

From Theorem 2.3.5 it follows that every absolute extremum of $f(x, y)$ subject to $\phi(x, y)=0$ is also a relative extremum, and such $f(x, y)$ has an absolute maximum and an absolute minimum. To find those extrema we put, in accordance with Theorem 2.4.1,

$$F(x, y)=ax^2+2bxy+cy^2-\lambda(x^2+y^2-1),$$

from which we have, for a relative extremum (x_0, y_0) of $f(x, y)$ subject to $\phi(x, y)=0$,

$$\frac{1}{2}F_x(x_0, y_0)=ax_0+by_0-\lambda x_0=0, \qquad (2.4.13)$$

$$\frac{1}{2}F_y(x_0, y_0)=bx_0+cy_0-\lambda y_0=0, \qquad (2.4.14)$$

together with

$$x_0^2+y_0^2=1. \qquad (2.4.15)$$

Multiplying (2.4.13) by x_0, and (2.4.14) by y_0, and adding them together, we obtain

$$ax_0^2+2bx_0y_0+cy_0^2=\lambda(x_0^2+y_0^2),$$

or, by virtue of (2.4.7) and (2.4.15),

$$f(x_0, y_0)=\lambda. \qquad (2.4.16)$$

Thus *the value of the Lagrange multiplier* λ *is in this case precisely the extremum of the function* $f(x, y)$, *which we are seeking.*

We may now easily determine the value of λ from (2.4.13) and (2.4.14). To do so we rewrite these equations as

$$(a-\lambda)x_0+by_0=0, \qquad (2.4.17)$$

$$bx_0+(c-\lambda)y_0=0. \qquad (2.4.18)$$

These are two linear homogeneous equations in the unknowns x_0 and y_0, which by (2.4.15) have a nontrivial solution $(x_0, y_0) \neq (0,0)$. This can happen only if the determinant of the system is zero, that is, if

$$\lambda^2 - (a+c)\lambda + ac - b^2 = 0. \tag{2.4.19}$$

Thus the required extrema of $f(x, y)$ are the roots of (2.4.19). To find the corresponding extremum points we first eliminate λ from (2.4.13) and (2.4.14) to obtain

$$bx_0^2 + (c-a)x_0y_0 - by_0^2 = 0, \tag{2.4.20}$$

and then solve (2.4.15) and (2.4.20) simultaneously for x_0 and y_0, which are here considered to be variables.

Exercises

1. Use the Lagrange multipliers to find the extrema and their respective points of
   (a) $x^2 + 4xy + y^2$ on $x^2 + y^2 = 1$,
   (b) $x^2 + 12xy + 2y^2$ on $4x^2 + y^2 = 25$.

3. VECTORS

3.1. Vector Spaces. Let E^3 be a three-dimensional Euclidean space and $\mathbf{0x}_1\mathbf{x}_2\mathbf{x}_3$ a fixed right-handed rectangular trihedron in E^3 as introduced in Section 1.1. Then relative to $\mathbf{0x}_1\mathbf{x}_2\mathbf{x}_3$, a point $\mathbf{x}$ in E^3 has coordinates (x_1, x_2, x_3).

3.1.1. Definition. Two ordered pairs of points $\mathbf{x} = (x_1, x_2, x_3)$, $\mathbf{y} = (y_1, y_2, y_3)$ and $\mathbf{x}' = (x_1', x_2', x_3')$, $\mathbf{y}' = (y_1', y_2', y_3')$ are *equivalent* if the line segments $\mathbf{xy}$ and $\mathbf{x}'\mathbf{y}'$ are of the same length and are parallel in the same sense. This relation, being reflexive, symmetric, and transitive, is clearly an equivalence relation. Such an equivalence class of ordered pairs of points is a *vector*, which can also be denoted by $\mathbf{xy}$ with respect to the representative pair $\mathbf{x}, \mathbf{y}$. The components of the vector $\mathbf{xy}$ are $y_i - x_i$, $i = 1,2,3$, and two vectors are *equal* if and only if they have the same components.

For any two vectors $\mathbf{v}, \mathbf{w}$ in E^3 with the respective components v_i, w_i, $i = 1, 2, 3$, and for any scalar (real number) c, we define the addition of $\mathbf{v}$, $\mathbf{w}$, and the scalar multiplication of $\mathbf{v}$ by c, as follows:

$$(v_1, v_2, v_3) + (w_1, w_2, w_3) = (v_1 + w_1, v_2 + w_2, v_3 + w_3), \tag{3.1.1}$$

$$c(v_1, v_2, v_3) = (cv_1, cv_2, cv_3). \tag{3.1.2}$$

3.1.2. Definition. The set of all vectors in E^3 with the addition and scalar multiplication is a *vector space* (often called a *linear space*) over the field R of real numbers, which is a set satisfying the following conditions:

$$\text{The set is an Abelian group under addition with } (0,0,0) \text{ as the identity,} \tag{3.1.3}$$

$$a(\mathbf{v}+\mathbf{w})=a\mathbf{v}+a\mathbf{w}, \qquad (a+b)\mathbf{v}=a\mathbf{v}+b\mathbf{v}, \qquad a,b\in R \quad \text{(distributive laws)}, \tag{3.1.4}$$

$$a(b\mathbf{v})=(ab)\mathbf{v}, \qquad 1\mathbf{v}=\mathbf{v}. \tag{3.1.5}$$

Using the origin $\mathbf{0}$ of our coordinate system $\mathbf{0}\mathbf{x}_1\mathbf{x}_2\mathbf{x}_3$, we can set up a 1:1 correspondence between every point $\mathbf{x}$ in E^3 and the vector $\mathbf{0x}$ that will be called the *position vector* of $\mathbf{x}$. Notice that this correspondence is defined only with reference to the point $\mathbf{0}$. By this correspondence we shall be able to use the symbol for the position vector of a point also to denote the point, and we thus see that E^3 itself is a vector space over R.

3.2. Inner Product

3.2.1. Definition. The *inner* or *scalar product* $\mathbf{v}\cdot\mathbf{w}$ of any two vectors $\mathbf{v}$, $\mathbf{w}$ with the respective components $v_i, w_i, i=1,2,3$, is a scalar defined by

$$\mathbf{v}\cdot\mathbf{w}=\sum_{i=1}^{3} v_i w_i. \tag{3.2.1}$$

From (3.2.1) it follows that the inner product has the following properties:

$$\begin{aligned}
&\mathbf{v}\cdot\mathbf{w}=\mathbf{w}\cdot\mathbf{v},\\
&\mathbf{v}\cdot\mathbf{v}\geqslant 0, \quad \text{and} \quad \mathbf{v}\cdot\mathbf{v}=0 \quad \text{if and only if} \quad \mathbf{v}=0,\\
&(\mathbf{v}_1+\mathbf{v}_2)\cdot\mathbf{w}=\mathbf{v}_1\cdot\mathbf{w}+\mathbf{v}_2\cdot\mathbf{w},\\
&(c\mathbf{v})\cdot\mathbf{w}=\mathbf{v}\cdot(c\mathbf{w})=c(\mathbf{v}\cdot\mathbf{w}), \qquad c=\text{a scalar}.
\end{aligned} \tag{3.2.2}$$

Write

$$\|\mathbf{v}\|=+\sqrt{\mathbf{v}^2}=+\sqrt{\mathbf{v}\cdot\mathbf{v}}, \tag{3.2.3}$$

and call it the *norm* (or *length*) of the vector $\mathbf{v}$. A vector with unit norm is called a *unit vector*. Thus (Fig. 1.2)

$$\|\mathbf{x}-\mathbf{y}\|=\text{the distance between two points with position vectors } \mathbf{x} \text{ and } \mathbf{y}. \tag{3.2.4}$$

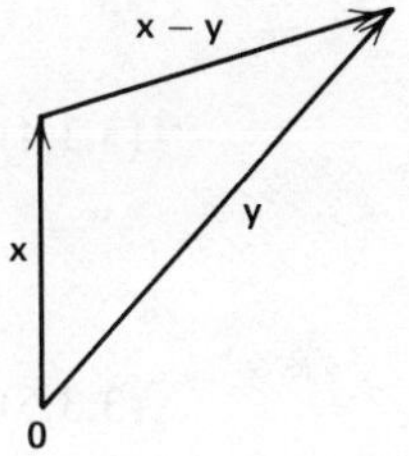

Figure 1.2

Let θ be the angle between any two vectors $\mathbf{v}$ and $\mathbf{w}$. Then by means of the trigonometric law of cosines applied to the triangle with sides $\mathbf{v}, \mathbf{w}, \mathbf{u}= \mathbf{w}-\mathbf{v}$, we obtain

$$\mathbf{v}\cdot\mathbf{w}=\|\mathbf{v}\|\,\|\mathbf{w}\|\cos\theta. \tag{3.2.5}$$

Thus for nonzero vectors $\mathbf{v}$ and $\mathbf{w}$, $\mathbf{v}\cdot\mathbf{w}=0$ implies that $\mathbf{v}$ and $\mathbf{w}$ are orthogonal.

The determinant of three vectors $\mathbf{u}, \mathbf{v}, \mathbf{w}$ with the respective components u_i, v_i, w_i, $i=1,2,3$, is defined to be

$$|\mathbf{u},\mathbf{v},\mathbf{w}|=\begin{vmatrix} u_1 & u_2 & u_3 \\ v_1 & v_2 & v_3 \\ w_1 & w_2 & w_3 \end{vmatrix}. \tag{3.2.6}$$

Then we readily have

$$\begin{gathered} |\mathbf{u}+\mathbf{u}',\mathbf{v},\mathbf{w}|=|\mathbf{u},\mathbf{v},\mathbf{w}|+|\mathbf{u}',\mathbf{v},\mathbf{w}|, \\ |c\mathbf{u},\mathbf{v},\mathbf{w}|=c\,|\mathbf{u},\mathbf{v},\mathbf{w}|, \qquad c=\text{a scalar}, \\ |\mathbf{u},\mathbf{v},\mathbf{w}|=-|\mathbf{v},\mathbf{u},\mathbf{w}|=\text{etc.} \end{gathered} \tag{3.2.7}$$

3.3. Vector Product

3.3.1. Definition. The *vector product* of two vectors $\mathbf{v}, \mathbf{w}$ is a vector $\mathbf{z}$, written as

$$\mathbf{z}=\mathbf{v}\times\mathbf{w}, \tag{3.3.1}$$

such that the relation

$$|\mathbf{v},\mathbf{w},\mathbf{x}|=\mathbf{z}\cdot\mathbf{x} \tag{3.3.2}$$

holds for all vectors $\mathbf{x}$.

It follows that $\mathbf{z}$ has the components

$$z_i=(-1)^{i-1}\det(\mathbf{vw})_i, \qquad i=1,2,3, \tag{3.3.3}$$

where $\det(\mathbf{vw})_i$ denotes the determinant of the square matrix $(\mathbf{vw})_i$ obtained by deleting the ith column from the matrix

$$(\mathbf{vw}) = \begin{pmatrix} v_1 & v_2 & v_3 \\ w_1 & w_2 & w_3 \end{pmatrix}. \tag{3.3.4}$$

Thus the vector product has the following properties:

$$\begin{aligned} &\mathbf{v}\times\mathbf{w}+\mathbf{w}\times\mathbf{v}=0,\\ &(\mathbf{v}_1+\mathbf{v}_2)\times\mathbf{w}=\mathbf{v}_1\times\mathbf{w}+\mathbf{v}_2\times\mathbf{w},\\ &(c\mathbf{v})\times\mathbf{w}=c(\mathbf{v}\times\mathbf{w}), \qquad c=\text{a scalar}. \end{aligned} \tag{3.3.5}$$

Geometrically, $\mathbf{z}$ is orthogonal to $\mathbf{v}$ and $\mathbf{w}$. Moreover, if $\mathbf{u}$ is the unit vector orthogonal to $\mathbf{v}$ and $\mathbf{w}$ such that

$$\delta = |\mathbf{v},\mathbf{w},\mathbf{u}| > 0, \tag{3.3.6}$$

it follows from (3.3.2) that $\mathbf{z}=\delta\mathbf{u}$; therefore for nonzero $\mathbf{v}$ and $\mathbf{w}$, $\mathbf{v}\times\mathbf{w}=0$ implies that $\mathbf{v}=c\mathbf{w}$, where c is a scalar, which means that $\mathbf{v}$ and $\mathbf{w}$ are parallel.

From (3.2.6), (3.3.2), and (3.3.3) we easily obtain

$$\mathbf{w}\cdot\mathbf{u}\times\mathbf{v}=\mathbf{u}\cdot\mathbf{v}\times\mathbf{w}=\mathbf{v}\cdot\mathbf{w}\times\mathbf{u}=|\mathbf{u},\mathbf{v},\mathbf{w}|. \tag{3.3.7}$$

Moreover, for a combination of the inner product and the vector product we have the *Lagrange identity*:

$$(\mathbf{v}_1\times\mathbf{v}_2)\cdot(\mathbf{w}_1\times\mathbf{w}_2)=(\mathbf{v}_1\cdot\mathbf{w}_1)(\mathbf{v}_2\cdot\mathbf{w}_2)-(\mathbf{v}_1\cdot\mathbf{w}_2)(\mathbf{v}_2\cdot\mathbf{w}_1). \tag{3.3.8}$$

In our later applications we shall consider vectors that are functions of one or more variables, that is, whose components are functions of one or more variables. In case the vectors are functions of one variable t, we have that the following formulas hold, provided the derivatives in these formulas exist:

$$\begin{aligned} &\frac{d}{dt}(\mathbf{v}\pm\mathbf{w})=\frac{d\mathbf{v}}{dt}\pm\frac{d\mathbf{w}}{dt},\\ &\frac{d}{dt}(c\mathbf{v})=\frac{dc}{dt}\mathbf{v}+c\frac{d\mathbf{v}}{dt}, \qquad c=\text{a scalar}\\ &\frac{d}{dt}(\mathbf{v}\cdot\mathbf{w})=\frac{d\mathbf{v}}{dt}\cdot\mathbf{w}+\mathbf{v}\cdot\frac{d\mathbf{w}}{dt},\\ &\frac{d}{dt}(\mathbf{v}\times\mathbf{w})=\frac{d\mathbf{v}}{dt}\times\mathbf{w}+\mathbf{v}\times\frac{d\mathbf{w}}{dt},\\ &\frac{d}{dt}|\mathbf{u},\mathbf{v},\mathbf{w}|=\left|\frac{d\mathbf{u}}{dt},\mathbf{v},\mathbf{w}\right|+\left|\mathbf{u},\frac{d\mathbf{v}}{dt},\mathbf{w}\right|+\left|\mathbf{u},\mathbf{v},\frac{d\mathbf{w}}{dt}\right|, \end{aligned} \tag{3.3.9}$$

where, for example, $d\mathbf{v}/dt$ is also a vector defined as usual by

$$\frac{d\mathbf{v}}{dt}=\lim_{\Delta t\to 0}\frac{\Delta\mathbf{v}}{\Delta t}=\lim_{\Delta t\to 0}\frac{\mathbf{v}(t+\Delta t)-\mathbf{v}(t)}{\Delta t} \tag{3.3.10}$$

Similarly, we can discuss the partial derivatives of vectors that are functions of two or three variables.

Exercises

1. Prove (3.2.4).

2. Show that for vectors $\mathbf{u}_1, \mathbf{u}_2, \mathbf{v}_1, \mathbf{v}_2$,

$$(\mathbf{u}_1 \times \mathbf{u}_2) \cdot (\mathbf{v}_1 \times \mathbf{v}_2) = |\mathbf{u}_1 \times \mathbf{u}_2, \mathbf{v}_1, \mathbf{v}_2|.$$

3. Prove (3.3.8).

4. Show that for vectors $\mathbf{u}, \mathbf{v}, \mathbf{w}$,

$$(\mathbf{u} \times \mathbf{v}) \times \mathbf{w} = (\mathbf{u} \cdot \mathbf{w})\mathbf{v} - (\mathbf{v} \cdot \mathbf{w})\mathbf{u}.$$

5. Show that for vectors $\mathbf{u}_1, \mathbf{u}_2, \mathbf{v}_1, \mathbf{v}_2$,

$$(\mathbf{u}_1 \times \mathbf{u}_2) \times (\mathbf{v}_1 \times \mathbf{v}_2) = |\mathbf{u}_1, \mathbf{v}_1, \mathbf{v}_2| \mathbf{u}_2 - |\mathbf{u}_2, \mathbf{v}_1, \mathbf{v}_2| \mathbf{u}_1.$$

6. Show that for vectors $\mathbf{u}_1, \mathbf{u}_2, \mathbf{v}_1, \mathbf{v}_2, \mathbf{w}$,

$$\begin{aligned} |\mathbf{u}_1 \times \mathbf{u}_2, \mathbf{v}_1 \times \mathbf{v}_2, \mathbf{w}| &= |\mathbf{u}_1, \mathbf{v}_1, \mathbf{v}_2|(\mathbf{u}_2 \cdot \mathbf{w}) - |\mathbf{u}_2, \mathbf{v}_1, \mathbf{v}_2|(\mathbf{u}_1 \cdot \mathbf{w}) \\ &= |\mathbf{u}_1, \mathbf{u}_2, \mathbf{v}_2|(\mathbf{v}_1 \cdot \mathbf{w}) - |\mathbf{u}_1, \mathbf{u}_2, \mathbf{v}_1|(\mathbf{v}_2 \cdot \mathbf{w}). \end{aligned}$$

***7.** Show that the volume of the parallelepiped with sides $\mathbf{u}, \mathbf{v}, \mathbf{w}$ is $\pm \mathbf{u} \cdot \mathbf{v} \times \mathbf{w}$.

3.4. Linear Combinations and Linear Independence; Bases and Dimensions of Vector Spaces. Let $\mathbf{v}_1, \cdots, \mathbf{v}_i$ be vectors of a Euclidean space E^3, and $c_1, \cdots, c_i$ scalars. Throughout this section the subscript i denotes a positive integer.

3.4.1. Definition. An indicated sum of the form $c_1\mathbf{v}_1 + \cdots + c_i\mathbf{v}_i$ is called a *linear combination* of the vectors $\mathbf{v}_1, \cdots, \mathbf{v}_i$.

It is easily seen that the set of all vectors represented by linear combinations of given vectors $\mathbf{v}_1, \cdots, \mathbf{v}_i$ of E^3 is a subspace of E^3, which is denoted by $[\mathbf{v}_1, \cdots, \mathbf{v}_i]$, and called the *subspace generated* (or *spanned*) by the vectors $\mathbf{v}_1, \cdots, \mathbf{v}_i$. Geometrically, $[\mathbf{v}_1]$ is the line containing the vector $\mathbf{v}_1$, and $[\mathbf{v}_1, \mathbf{v}_2]$ is the plane containing the two vectors $\mathbf{v}_1$ and $\mathbf{v}_2$.

3.4.2. Definition. In E^3 vectors $\mathbf{v}_1, \cdots, \mathbf{v}_i$ are *linearly independent* if, for all scalars $c_1, \cdots, c_i$,

$$c_1\mathbf{v}_1 + \cdots + c_i\mathbf{v}_i = 0 \qquad \text{implies} \qquad c_1 = \cdots = c_i = 0. \tag{3.4.1}$$

Vectors that are not linearly independent are *linearly dependent*. If the vectors $\mathbf{v}_1,\cdots,\mathbf{v}_i$ are linearly independent (respectively, dependent), we may also say that the set $\{\mathbf{v}_1,\cdots,\mathbf{v}_i\}$ is *linearly independent* (respectively, *dependent*).

From Definition 3.4.2 we can easily deduce the following results.

If one of the vectors $\mathbf{v}_1,\cdots,\mathbf{v}_i$ is the zero vector, the vectors $\mathbf{v}_1,\cdots,\mathbf{v}_i$ are linearly dependent. Any subset of a linearly independent set is linearly independent.

3.4.3. Theorem. *Let the components of the vector $\mathbf{v}_i$ be (v_{i1}, v_{i2}, v_{i3}). Then the vectors $\mathbf{v}_1,\cdots,\mathbf{v}_i$ for $1 \leqslant i \leqslant 3$ are linearly independent (respectively, dependent) if and only if the rank of the matrix*

$$\begin{pmatrix} v_{11} & v_{12} & v_{13} \\ \cdots & \cdots & \cdots \\ v_{i1} & v_{i2} & v_{i3} \end{pmatrix} \tag{3.4.2}$$

is equal to (respectively, less than) i.

From Theorem 3.4.3 it follows that two vectors $\mathbf{v}_1, \mathbf{v}_2$ are linearly dependent if and only if $\mathbf{v}_1 = c\mathbf{v}_2$ with c a nonzero scalar, and three vectors are linearly dependent if and only if one vector is a linear combination of the other two. Thus we obtain the following geometric interpretation of linear dependence.

3.4.4. Corollary. *Two vectors are linearly dependent if and only if they are parallel to each other, and three vectors are linearly dependent if and only if they are parallel to a plane.*

3.4.5. Definition. The set $\{\mathbf{v}_1,\cdots,\mathbf{v}_i\}$ of vectors of a subspace S^i of E^3 is a *basis* of S^i if it is a linearly independent set that generates S^i, and the number i is the *dimension* of S^i. We write $i = \dim S^i$. If $\mathbf{v}_1,\cdots,\mathbf{v}_i$ are mutually orthogonal unit vectors, the basis $\{\mathbf{v}_1,\cdots,\mathbf{v}_i\}$ of S^i is an *orthonormal* basis.

We can easily prove the following two theorems.

3.4.6. Theorem. *Let $\mathbf{u}_1, \mathbf{u}_2, \mathbf{u}_3$ be vectors of E^3 defined by*

$$\mathbf{u}_1 = (1,0,0), \qquad \mathbf{u}_2 = (0,1,0), \qquad \mathbf{u}_3 = (0,0,1). \tag{3.4.3}$$

Then the set $\{\mathbf{u}_1, \mathbf{u}_2, \mathbf{u}_3\}$ is a basis of E^3, which is called the natural basis of E^3.

3.4.7. Theorem. *If the set $\{\mathbf{v}_1,\cdots,\mathbf{v}_i\}$ of vectors is a basis of a subspace S^i of E^3, then every vector of S^i is uniquely expressible as a linear combination of the vectors $\mathbf{v}_1,\cdots,\mathbf{v}_i$.*

Exercises

1. Prove Theorem 3.4.3.
2. Determine whether each of the following sets of vectors of E^3 is linearly independent or dependent:
   (a) $\{(-1,2,1), (3,1,-2)\}$,
   (b) $\{(1,3,2), (2,1,0), (0,5,4)\}$,
   (c) $\{(1,0,-1), (2,1,3), (-1,0,0), (1,0,1)\}$.
3. Let $\mathbf{v}_1$ and $\mathbf{v}_2$ be linearly independent vectors of E^3, and a, b, c, d be scalars. Prove that the vectors $a\mathbf{v}_1 + b\mathbf{v}_2$ and $c\mathbf{v}_1 + d\mathbf{v}_2$ are linearly independent if and only if $ad - bc \neq 0$.
4. Let $\mathbf{v}_1, \mathbf{v}_2, \mathbf{v}_3$ be linearly independent vectors of E^3, and $a_{ij}(i, j = 1,2,3)$ be scalars. Prove that the vectors $\Sigma_{j=1}^{3} a_{ij}\mathbf{v}_j$, $i = 1,2,3$, are linearly independent if and only if $\det(a_{ij}) \neq 0$.
5. Find a basis for E^3, that contains the vectors $(1,-1,0)$ and $(2,1,3)$.
6. Find the dimension and a basis of the subspace $[(1,2,0), (2,1,-1), (3,0,-2)]$ of E^3.
7. Prove Theorem 3.4.6.
8. Prove Theorem 3.4.7.

3.5. Tangent Vectors

3.5.1. Definition. A *tangent vector* of E^3 at a point $\mathbf{p}$, denoted by $\mathbf{v}_p$, is a vector $\mathbf{v}$ with the initial point $\mathbf{p}$, or $\mathbf{v}_p$ is an ordered pair of points $\mathbf{p}$ and $\mathbf{p} + \mathbf{v}$ where $\mathbf{p} + \mathbf{v}$ is considered as the position vector of a point (Fig. 1.3); see Definition 3.1.1. p is the *point of application* of $\mathbf{v}_p$. For simplicity, we may omit "tangent" for a tangent vector of E^3 at a point.

For example, if $\mathbf{p} = (1,2,1)$ and $\mathbf{v} = (2,3,3)$, then $\mathbf{v}_p$ is a pair of points $(1,2,1)$ and $(3,5,4)$.

It should be remarked that two tangent vectors $\mathbf{v}_p$ and $\mathbf{w}_q$ are equal if and only if $\mathbf{v} = \mathbf{w}$ and $\mathbf{p} = \mathbf{q}$. Thus if $\mathbf{p} \neq \mathbf{q}$, then $\mathbf{v}_p \neq \mathbf{v}_q$, and $\mathbf{v}_p, \mathbf{v}_q$ are said to be

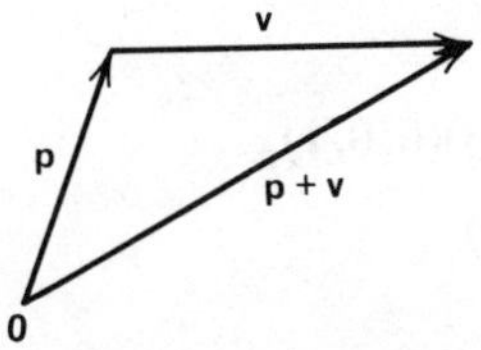

Figure 1.3

parallel. By defining

$$\mathbf{v}_p+\mathbf{w}_p=(\mathbf{v}+\mathbf{w})_p,\; c(\mathbf{v}_p)=(c\mathbf{v})_p, \tag{3.5.1}$$

where c is a scalar, we see that the set $T_p(E^3)$, consisting of all tangent vectors of E^3 at a point $\mathbf{p}$, is a vector space that is called the *tangent space* of E^3 at $\mathbf{p}$. Moreover, at each $\mathbf{p}$ this tangent space $T_p(E^3)$ is isomorphic to E^3, since the mapping f: $E^3 \to T_p(E^3)$ defined by $f(\mathbf{v})=\mathbf{v}_p$, where $\mathbf{v}$ is considered as the position vector of a point in E^3 with reference to an origin $\mathbf{0}$, is indeed a linear isomorphism, a bijective mapping preserving the addition and scalar multiplication (cf. Definition 5.1.1).

3.5.2. Definition. A *vector field* $\mathbf{v}$ on E^3 is a function that assigns to each point $\mathbf{p}$ of E^3 a tangent vector $\mathbf{v}_p$ [or written as $\mathbf{v}(\mathbf{p})$] of E^3 at $\mathbf{p}$.

Let $\mathbf{v}$ and $\mathbf{w}$ be vector fields on E^3. Then by (3.5.1) we can define $\mathbf{v}+\mathbf{w}$ to be the vector field on E^3 such that

$$(\mathbf{v}+\mathbf{w})_p=\mathbf{v}_p+\mathbf{w}_p \qquad \text{for all } \mathbf{p}. \tag{3.5.2}$$

Similarly, if g is a real-valued function on E^3, then by (3.5.1) we can define $g\mathbf{v}$ to be the vector field on E^3 such that

$$(g\mathbf{v})_p=g(\mathbf{p})\mathbf{v}_p \qquad \text{for all } \mathbf{p}. \tag{3.5.3}$$

3.5.3. Definition. Let $\mathbf{u}_1,\mathbf{u}_2,\mathbf{u}_3$ be the vector fields on E^3 such that with respect to a fixed right-handed rectangular trihedron $\mathbf{0x}_1\mathbf{x}_2\mathbf{x}_3$ in E^3

$$\mathbf{u}_1(\mathbf{p})=(1,0,0)_p, \qquad \mathbf{u}_2(\mathbf{p})=(0,1,0)_p, \qquad \mathbf{u}_3(\mathbf{p})=(0,0,1)_p \tag{3.5.4}$$

for each $\mathbf{p}$ of E^3. Then $\mathbf{u}_1\mathbf{u}_2\mathbf{u}_3$ is the *natural frame field* on E^3.

Thus $\mathbf{u}_i (i=1,2,3)$ is the unit vector field in the positive $\mathbf{x}_i$ direction.

3.5.4. Lemma. *If $\mathbf{v}$ is a vector field on E^3, there are three uniquely determined real-valued functions v_1, v_2, v_3 on E^3 such that*

$$\mathbf{v}=v_1\mathbf{u}_1+v_2\mathbf{u}_2+v_3\mathbf{u}_3. \tag{3.5.5}$$

The functions v_1, v_2, v_3 are called the Euclidean coordinate functions of $\mathbf{v}$ or the components of $\mathbf{v}$ with respect to the frame field $\mathbf{u}_1\mathbf{u}_2\mathbf{u}_3$.

Proof. By definition, $\mathbf{v}$ assigns to each point $\mathbf{p}$ a tangent vector $\mathbf{v}_p$ at $\mathbf{p}$, whose components are functions of $\mathbf{p}$, and therefore can be written as $v_1(\mathbf{p}), v_2(\mathbf{p}), v_3(\mathbf{p})$. Thus

$$\begin{aligned}
\mathbf{v}_p&=(v_1(p),v_2(p),v_3(p))\\
&=v_1(\mathbf{p})(1,0,0)_p+v_1(\mathbf{p})(0,1,0)_p+v_3(\mathbf{p})(0,0,1)_p\\
&=v_1(\mathbf{p})\mathbf{u}_1(\mathbf{p})+v_2(\mathbf{p})\mathbf{u}_2(\mathbf{p})+v_3(\mathbf{p})\mathbf{u}_3(\mathbf{p})
\end{aligned}$$

for each $\mathbf{p}$. This shows that the vector fields $\mathbf{v}$ and $v_1\mathbf{u}_1+v_2\mathbf{u}_2+v_3\mathbf{u}_3$ have the same (tangent vector) value at each point. Hence (3.5.5) is proved.

The tangent-vector identity $(a_1,a_2,a_3)_p=\Sigma_{i=1}^3 a_i\mathbf{u}_i(\mathbf{p})$ appearing in this proof will be used often.

A vector field $\mathbf{v}$ is differentiable if its Euclidean coordinate functions v_1, v_2, v_3 are differentiable. From now on we shall assume all vector fields to be differentiable.

From the paragraph preceding Definition 3.5.2, for each point $\mathbf{p}$ of E^3 there is a canonical isomorphism $\mathbf{v}\to\mathbf{v}_p$ of E^3 onto $T_p(E^3)$ at $\mathbf{p}$. Using this isomorphism, the inner product on E^3 itself may be transferred to each of its tangent spaces.

3.5.5. Definition. The *inner product of tangent vectors* $\mathbf{v}_p$ and $\mathbf{w}_p$ at the same point $\mathbf{p}$ of E^3 is the number $\mathbf{v}_p\cdot\mathbf{w}_p=\mathbf{v}\cdot\mathbf{w}$.

Evidently, this definition provides an inner product on each tangent space $T_p(E^3)$ with the same properties as the original inner product on E^3. In particular, each tangent vector v_p of E^3 at $\mathbf{p}$ has norm (or length) $\|\mathbf{v}_p\|=\|\mathbf{v}\|$, and two tangent vectors $\mathbf{v}_p$ and $\mathbf{w}_p$ of E^3 at $\mathbf{p}$ are orthogonal if and only if $\mathbf{v}_p\cdot\mathbf{w}_p=0$.

3.5.6. Definition. A set $\mathbf{e}_1(\mathbf{p}),\mathbf{e}_2(\mathbf{p}),\mathbf{e}_3(\mathbf{p})$ of three mutually orthogonal unit vectors, tangent to E^3 at a point $\mathbf{p}$, is a *frame* at $\mathbf{p}$. Vector fields $\mathbf{e}_1,\mathbf{e}_2,\mathbf{e}_3$ constitute a *frame field* on E^3 if they do so at each point $\mathbf{p}$ of E^3.

Thus $\mathbf{e}_1\mathbf{e}_2\mathbf{e}_3$ is a frame if and only if

$$\mathbf{e}_i\cdot\mathbf{e}_j=\delta_{ij},\qquad \text{for } 1\leqslant i,j\leqslant 3, \tag{3.5.6}$$

where δ_{ij} is the Kronecker delta (0 if $i\neq j$, 1 if $i=j$). For example, at each point $\mathbf{p}$ of E^3 the vectors $\mathbf{u}_1(\mathbf{p}),\mathbf{u}_2(\mathbf{p}),\mathbf{u}_3(\mathbf{p})$ given by (3.5.4) constitute a frame at $\mathbf{p}$.

Let $\mathbf{e}_i=(e_{i1},e_{i2},e_{i3})$. Similar to (3.2.6), define $\mathcal{E}=(\mathbf{e}_1,\mathbf{e}_2,\mathbf{e}_3)$ to be the matrix of the vectors $\mathbf{e}_1,\mathbf{e}_2,\mathbf{e}_3$ by

$$\mathcal{E}=\begin{bmatrix} e_{11} & e_{12} & e_{13}\\ e_{21} & e_{22} & e_{23}\\ e_{31} & e_{32} & e_{33}\end{bmatrix}. \tag{3.5.7}$$

From (3.3.7) it follows that

$$(\mathbf{e}_1\times\mathbf{e}_2)\cdot\mathbf{e}_3=(\mathbf{e}_2\times\mathbf{e}_3)\cdot\mathbf{e}_1=(\mathbf{e}_3\times\mathbf{e}_1)\cdot\mathbf{e}_2=\det\mathcal{E}, \tag{3.5.8}$$

where $\det\mathcal{E}$ denotes the determinant of the matrix $\mathcal{E}$.

By Theorem 3.4.7 and Definitions 3.5.5 and 3.2.1 we obtain the following useful result.

3.5.7. Lemma. *Let $\mathbf{e}_1\mathbf{e}_2\mathbf{e}_3$ be a frame field on E^3.*

(a) *If $\mathbf{v}$ is a vector field on E^3, then $\mathbf{v}=\sum_{i=1}^3 f_i\mathbf{e}_i$, where the functions $f_i=\mathbf{v}\cdot\mathbf{e}_i$ are called the coordinate functions of $\mathbf{v}$ with respect to $\mathbf{e}_1\mathbf{e}_2\mathbf{e}_3$.*

(b) *If $\mathbf{v}=\sum_{i=1}^3 f_i\mathbf{e}_i$ and $\mathbf{w}=\sum_{i=1}^3 g_i\mathbf{e}_i$, then $\mathbf{v}\cdot\mathbf{w}=\sum_{i=1}^3 f_ig_i$. In particular,*

$$\|\mathbf{v}\|=\left(\sum_{i=1}^{3} f_i^2\right)^{1/2}.$$

Thus a given vector field $\mathbf{v}$ has a different set of coordinate functions with respect to each choice of a frame field $\mathbf{e}_1\mathbf{e}_2\mathbf{e}_3$, and for a particular problem one such choice may be more suitable than another.

Exercises

1. Let $\mathbf{v}=(-1,-2,1)$ and $\mathbf{w}=(1,3,-2)$.
(a) At an arbitrary point $\mathbf{p}$, express the tangent vector $2\mathbf{v}_p + 3\mathbf{w}_p$ as a linear combination of $\mathbf{u}_1(\mathbf{p}),\mathbf{u}_2(\mathbf{p}),\mathbf{u}_3(\mathbf{p})$.
(b) If $\mathbf{p}=(1,-1,2)$, express each of the tangent vectors $\mathbf{v}_p$, $-2\mathbf{w}_p$ and $\mathbf{v}_p+\mathbf{w}_p$ as a pair of points.

2. Let $\mathbf{v}_1=\mathbf{u}_1-x_1^2\mathbf{u}_2+\mathbf{u}_3$, $\mathbf{v}_2=2\mathbf{u}_3$, and $\mathbf{v}_3=\mathbf{u}_1+\mathbf{u}_2$.
(a) Prove that the vectors $\mathbf{v}_1(\mathbf{p}),\mathbf{v}_2(\mathbf{p}),\mathbf{v}_3(\mathbf{p})$ are linearly independent at each point $\mathbf{p}$ of E^3.
(b) Express the vector field $x_1\mathbf{u}_1+x_2\mathbf{u}_2+x_3\mathbf{u}_3$ as a linear combination of $\mathbf{v}_1,\mathbf{v}_2,\mathbf{v}_3$.

3. Prove that the tangent vectors

$$\mathbf{e}_1=\frac{1}{\sqrt{6}}(1,2,1),\qquad \mathbf{e}_2=\frac{1}{\sqrt{8}}(-2,0,2),\qquad \mathbf{e}_3=\frac{1}{\sqrt{3}}(1,-1,1)$$

constitute a frame. Express $\mathbf{v}=(6,1,-1)$ as a linear combination of these vectors.

***4.** If $\mathbf{v}$ and $\mathbf{w}$ are vector fields on E^3 that are linearly independent at each point, show that

$$\mathbf{e}_1=\frac{\mathbf{v}}{\|\mathbf{v}\|},\qquad \mathbf{e}_2=\frac{\tilde{\mathbf{w}}}{\|\mathbf{w}\|},\qquad \mathbf{e}_3=\mathbf{e}_1\times\mathbf{e}_2$$

is a frame field, where $\tilde{\mathbf{w}}=\mathbf{w}-\mathbf{w}\cdot\mathbf{e}_2\mathbf{e}_2$.

3.6. Directional Derivatives. Let f be a differentiable function on E^3. Then $t\to f(\mathbf{p}+t\mathbf{v})$ is an ordinary differentiable function on the real line $\mathbf{p}+t\mathbf{v}$, which passes through a tangent vector $\mathbf{v}_p$ of E^3 at a point $\mathbf{p}$, and the derivative of this function f with respect to t at $t=0$ is the initial rate of change of f as $\mathbf{p}$ moves in the $\mathbf{v}$ direction. Thus we have Definition 3.6.1.

3.6.1. Definition. Let f be a differentiable real-valued function on E^3, and $\mathbf{v}_p$ a tangent vector of E^3 at a point $\mathbf{p}$. Then

$$\mathbf{v}_p[f]=\frac{d}{dt}(f(\mathbf{p}+t\mathbf{v}))|_{t=0} \tag{3.6.1}$$

is the derivative of f with respect to $\mathbf{v}_p$.

It should be noted that this definition of derivative is the same as that of the directional derivative given in elementary calculus but restricted to a unit vector $\mathbf{v}_p$. However we shall still refer to $\mathbf{v}_p[f]$ as a directional derivative.

For example, suppose that

$$f=x_1x_2x_3, \qquad \mathbf{p}=(1,-4,2), \qquad \mathbf{v}=(1,1,0), \tag{3.6.2}$$

where x_1, x_2, x_3 are the coordinate functions of E^3. Then $\mathbf{p}+t\mathbf{v}=(1+t,-4+t,2)$ and $f(\mathbf{p}+t\mathbf{v})=2(1+t)(-4+t)$. Thus by (3.6.1) we have

$$\mathbf{v}_p[f]=\frac{d}{dt}2(1+t)(-4+t)|_{t=0}=2(2t-3)|_{t=0}=-6.$$

A computation of a general $\mathbf{v}_p[f]$ in terms of the partial derivatives of f at the point $\mathbf{p}$ is given in Lemma 3.6.2.

3.6.2. Lemma. *If $\mathbf{v}_p=(v_1,v_2,v_3)_p$ is a tangent vector of E^3 at a point $\mathbf{p}$, then*

$$\mathbf{v}_p[f]=\sum_{i=1}^{3} v_i\frac{\partial f}{\partial x_i}(\mathbf{p}). \tag{3.6.3}$$

Proof. Let $\mathbf{p}=(p_1,p_2,p_3)$. Then

$$f(\mathbf{p}+t\mathbf{v})=f(p_1+tv_1,p_2+tv_2,p_3+tv_3).$$

Since $(d/dt)(p_i+tv_i)=v_i$, by putting $x_i=p_i+tv_i$ and using the chain rule (Theorem 2.3.6) and (3.6.1) we can immediately obtain (3.6.3). Q.E.D.

Using (3.6.3) we can also find $\mathbf{v}_p[f]$ for the example above. From (3.6.2) it follows that

$$\frac{\partial f}{\partial x_1}(\mathbf{p})=x_2x_3(\mathbf{p})=-8,$$

$$\frac{\partial f}{\partial x_2}(\mathbf{p})=x_1x_3(\mathbf{p})=2,$$

$$\frac{\partial f}{\partial x_3}(\mathbf{p})=x_1x_2(\mathbf{p})=-4.$$

Thus by (3.6.3) we obtain $\mathbf{v}_p[f]=1(-8)+1(2)+0(-4)=-6$.

The main properties of this notion of derivative are given in the following theorem.

3.6.3. Theorem. *Let f and g be differentiable functions on E^3, $\mathbf{v}_p$ and $\mathbf{w}_p$ tangent vectors at a point* $\mathbf{p}$, *and a and b real numbers. Then*

(a) $(a\mathbf{v}_p+b\mathbf{w}_p)[f]=a\mathbf{v}_p[f]+b\mathbf{w}_p[f]$,
(b) $\mathbf{v}_p[af+bg]=a\mathbf{v}_p[f]+b\mathbf{v}_p[g]$,
(c) $\mathbf{v}_p[fg]=\mathbf{v}_p[f]\cdot g(\mathbf{p})+f(\mathbf{p})\cdot\mathbf{v}_p[g]$.

Proof. Parts (a) and (b) follow readily from Lemma 3.6.2 and the sum formula for partial derivatives. Similarly, we can also prove (c) by noticing that

$$\sum_{i=1}^{3} v_i\frac{\partial(fg)}{\partial x_i}(\mathbf{p})=\sum_{i=1}^{3} v_i\left(\frac{\partial f}{\partial x_i}(\mathbf{p})\cdot g(\mathbf{p})+f(\mathbf{p})\cdot\frac{\partial g}{\partial x_i}(\mathbf{p})\right)$$

$$=\left(\sum_{i=1}^{3} v_i\frac{\partial f}{\partial x_i}(\mathbf{p})\right)g(\mathbf{p})+f(\mathbf{p})\left(\sum_{i=1}^{3} v_i\frac{\partial g}{\partial x_i}(\mathbf{p})\right). \qquad \text{Q.E.D.}$$

Parts (a) and (b) mean that $\mathbf{v}_p[f]$ is *linear* in both $\mathbf{v}_p$ and f, and (c) is the product formula for directional derivatives.

By the same method used in the last section, we can define the operation $\mathbf{v}[f]$ of a vector field $\mathbf{v}$ on a function f to be the real-valued function whose value at each point $\mathbf{p}$ is $\mathbf{v}(\mathbf{p})[f]$, the derivative of f with respect to the tangent vector $\mathbf{v}(\mathbf{p})$ at $\mathbf{p}$. In particular, if $\mathbf{u}_1\mathbf{u}_2\mathbf{u}_3$ is the natural frame field on E^3,

$$\mathbf{u}_i[f]=\frac{\partial f}{\partial x_i}, \qquad i=1,2,3. \tag{3.6.4}$$

In fact, since, for instance, $\mathbf{u}_1(\mathbf{p})=(1,0,0)_p$ by (3.5.4), from (3.6.3) it follows immediately that $\mathbf{u}_1(\mathbf{p})[f]=(\partial f/\partial x_1)(\mathbf{p})$ for each point $\mathbf{p}$, hence that $\mathbf{u}_1[f]=\partial f/\partial x_1$.

3.6.4. Corollary. *If* $\mathbf{v}$ *and* $\mathbf{w}$ *are vector fields on E^3, and f, g, h are real-valued differentiable functions, then*

(a) $(f\mathbf{v}+g\mathbf{w})[h]=f\mathbf{v}[h]+g\mathbf{w}[h]$,
(b) $\mathbf{v}[af+bg]=a\mathbf{v}[f]+b\mathbf{v}[g]$, *for all real numbers a and b*,
(c) $\mathbf{v}[fg]=\mathbf{v}[f]\cdot g=f\mathbf{v}[g]$.

Proof. By Theorem 3.6.3 we see that (a) and (b) are true at each point $\mathbf{p}$ of E^3. Similarly, we can prove (c). In fact, the value of $\mathbf{v}[fg]$ at $\mathbf{p}$ is $\mathbf{v}_p[fg]$, which, by Theorem 3.6.3, is equal to

$$\mathbf{v}_p[f]\cdot g(\mathbf{p})+f(\mathbf{p})\mathbf{v}_p[g]=\mathbf{v}[f](\mathbf{p})\cdot g(\mathbf{p})+f(\mathbf{p})\mathbf{v}[g](\mathbf{p})$$

$$=(\mathbf{v}[f]\cdot g+f\cdot\mathbf{v}[g])(\mathbf{p}). \qquad \text{Q.E.D.}$$

By using (3.6.4) we can easily compute $\mathbf{v}[f]$. For example, if $\mathbf{v}=x_1\mathbf{u}_1-x_2\mathbf{u}_2+x_3\mathbf{u}_3$ and $f=x_1x_2^2+x_2x_3^3$,

$$\mathbf{v}[f]=x_1\mathbf{u}_1[f]-x_2\mathbf{u}_2[f]+x_3\mathbf{u}_3[f]=-x_1x_2^2+2x_2x_3^3.$$

Remark. For simplicity, if the point of application is not crucial, we shall frequently omit the point of application $\mathbf{p}$ from the notation $\mathbf{v}_p$.

Exercises

1. Let $\mathbf{v}_p$ be the tangent vector of E^3 for which $\mathbf{v}=(1,2,2)$ and $\mathbf{p}=(2,-1,1)$. Use both Definition 3.6.1 and Lemma 3.6.2 to compute the directional derivative $\mathbf{v}_p[f]$, where $f=x_1x_2^2+x_2x_3^3$.
2. Let $\mathbf{v}=x_3\mathbf{u}_1-x_1\mathbf{u}_2+x_2\mathbf{u}_3$, $f=x_1x_3$, $g=x_2^3$. Compute the following functions: (a) $\mathbf{v}[f]$; (b) $\mathbf{v}[fg]$; (c) $\mathbf{v}[\mathbf{v}[f]]$.

*3. Prove the identity $\mathbf{v}=\Sigma_{i=1}^3\mathbf{v}[x_i]\mathbf{u}_i$, where x_1, x_2, x_3 are the natural coordinate functions.

4. If $\mathbf{v}[f]=\mathbf{w}[f]$ for every function f on E^3, show that $\mathbf{v}=\mathbf{w}$.

4. MAPPINGS

In this section we discuss mappings of E^n into E^m for general n and $m, 1\leqslant n, m\leqslant 3$. If $n=3$ and $m=1$, such a mapping is just a real-valued function on E^3. If $n=1$ and $m=3$, such a mapping is a curve in E^3. If $n=2$ and $m=3$, such a mapping is a surface in E^3. The last two particular cases are discussed in detail in later chapters.

4.1. Linear Transformations and Dual Spaces

4.1.1. Definition. Given a mapping F: $E^n\rightarrow E^m$, let $f_1, bf_2, \cdots, f_m$ denote the real-valued functions on E^n such that

$$F(\mathbf{p})=(f_1(\mathbf{p}), f_2(\mathbf{p}),\cdots, f_m(\mathbf{p}))$$

for every point $\mathbf{p}\in E^n$. Then these functions $f_1, f_2, \cdots, f_m$ are the *Euclidean coordinate functions* of F, and we write

$$F=(f_1, f_2, \cdots, f_m). \tag{4.1.1}$$

The mapping F is of class $C^k (k\geqslant 1)$ if $f_1, f_2, \cdots, f_m$ are of class C^k in the usual sense.

Note that the coordinate functions of F are the composite functions $f_i=x_i(F)$, where $x_1, x_2, \cdots, x_m$ are the coordinate functions of E^m.

Mappings may be described in many different ways. For example, suppose that $F: E^3 \to E^3$ is the mapping $F=(xy, yz, z^2)$. Then

$$F(\mathbf{p})=\left(x(\mathbf{p})y(\mathbf{p}), y(\mathbf{p})z(\mathbf{p}), z(\mathbf{p})^2\right) \qquad \text{for all } \mathbf{p}\in E^3.$$

If $\mathbf{p}=(p_1, p_2, p_3)$, then by the definition of the coordinate functions,

$$x(\mathbf{p})=p_1, \quad y(\mathbf{p})=p_2, \quad z(\mathbf{p})=p_3,$$

and therefore we obtain the following *pointwise* formula for F:

$$F(p_1, p_2, p_3)=\left(p_1p_2, p_2p_3, p_3^2\right) \qquad \text{for all} \quad p_1, p_2, p_3.$$

4.1.2. Definition. A mapping $F: E^n \to E^m$ is a *linear transformation* of E^n into E^m if, for all vectors $\mathbf{x}$ and $\mathbf{y}$ of E^n and all scalars c and d,

$$F(c\mathbf{x}+d\mathbf{y})=c(F(\mathbf{x}))+d(F(\mathbf{y})). \tag{4.1.2}$$

The *kernel* of a linear transformation $F: E^n \to E^m$, denoted by ker F, is the set

$$\ker F=\{\mathbf{x}|\mathbf{x}\in E^n, F(\mathbf{x})=0\}. \tag{4.1.3}$$

4.1.3. Examples of Mappings. 1. The mapping $F: E^3 \to E^3$ such that

$$F(x, y, z)=(2x, y-z, y+z),$$

where x, y, z are the coordinate functions of E^3. This is a linear transformation of $E^3 \to E^3$. Thus by a well-known theorem of linear algebra, F is completely determined by its value on three (linearly independent) points, say the unit points

$$\mathbf{u}_1=(1,0,0), \quad \mathbf{u}_2(0,1,0), \quad \mathbf{u}_3(0,0,1).$$

2. The mapping $F: E^2 \to E^2$ such that

$$F(u, v)=(x, y), x=u^2+v^2, y=u+v,$$

where u, v and x, y are the coordinate functions of the uv-plane and the xy-plane, respectively.

It often helps to compute the image of a number of selected curves and regions. For instance, the image curve of the horizontal line $v=c$ in the xy-plane under F is given by the equation

$$x=(y-c)^2+c^2.$$

Since x and y are symmetric in u and v, points (a, b) and (b, a) in the uv-plane both have the same image. Thus the line $v=u$ divides the uv-plane into two half-planes, which are mapped by F onto the same set in the xy-plane. To determine this set, we first find the image of the line $v=u$

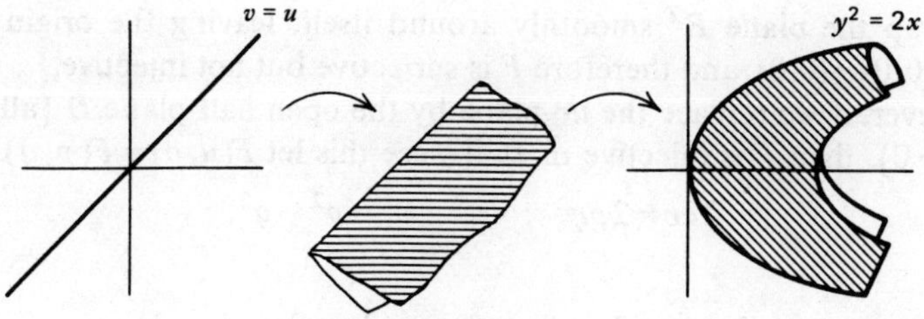

Figure 1.4

to be the parabola $y^2=2x$. The image of any point (x, y) lies within this parabola, for $2x-y^2=(u-v)^2 \geqslant 0$. Conversely, it is easy to see that every point (x, y) on or within this parabola is in turn the image of a point (u, v); see Fig. 1.4. Accordingly, one may picture the effect of F approximately as follows: first, fold the uv-plane along the line $v=u$; then fit the folded edge along the parabola $y^2=2x$ and flatten out the rest to cover the inside of the parabola smoothly. (To permit the necessary distortion, think of the uv-plane as a sheet of rubber.)

3. The mapping $F: E^2 \to E^2$ such that

$$F(u, v)=(x, y), \; x=u^2-v^2, \; y=2uv, \tag{4.1.4}$$

where u, v and x, y are the coordinate functions of the uv-plane and the xy-plane respectively.

Let us first use polar coordinates r, t in the uv-plane so that

$$u=r\cos t, \qquad v=r\sin t, \qquad 0 \leqslant t \leqslant 2\pi.$$

By using some trigonometric identities, we can express (4.1.4) as

$$F(r\cos t, r\sin t)=(r^2\cos 2t, r^2\sin 2t), \qquad 0 \leqslant t \leqslant 2\pi.$$

From this it follows that under F the image curve of the circle of radius r and center at the origin counterclockwise once is the circle of radius r^2 and center at the origin counterclockwise twice (Fig. 1.5). Thus the effect of F

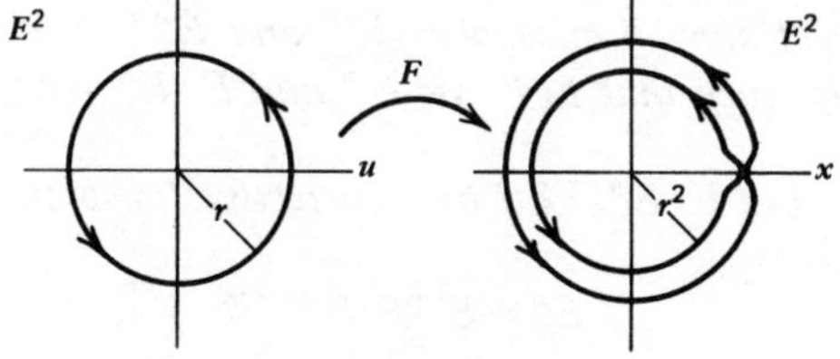

Figure 1.5

is to wrap the plane E^2 smoothly around itself, leaving the origin fixed, since $F(0,0)=(0,0)$, and therefore F is surjective but not injective.

However, if we replace the uv-plane by the open half-plane D {all (u, v) with $u>0$}, then F is injective in D. To see this let $F(u, v)=F(p, q)$. Then

$$2uv=2pq, \qquad u^2-v^2=p^2-q^2,$$

so that

$$0=u^2(u^2-v^2-p^2+q^2)=(u^2+q^2)(u^2-p^2).$$

Thus $u^2=p^2$. Since $u>0$ and $p>0$, we have $u=p$ and $v=q$. Hence F maps D onto a subset D' in the xy-plane in an injective fashion. This mapping has an inverse F^{-1}, which maps D' onto D. Solving (4.1.4) we obtain

$$u=\frac{1}{\sqrt{2}}\left(x+\sqrt{x^2+y^2}\,\right)^{1/2}, v=\frac{y}{\sqrt{2}}\left(x+\sqrt{x^2+y^2}\,\right)^{-1/2}.$$

The set D' is the set of points (x, y) for which $x+\sqrt{x^2+y^2}>0$, that is, all points (x, y) except those of the form $(c,0)$ with $c\leqslant 0$.

4.1.4. Definition. Let A and B be two subspaces of E^n. Then

$$E^n=A\oplus B \tag{4.1.5}$$

means that $E^n=A+B=\{\mathbf{x}+\mathbf{y}|\mathbf{x}\in A, \mathbf{y}\in B\}$ and $A\cap B=0$, (4.1.5) is a *decomposition* of E^n into a direct sum of the subspaces A and B, and each of the sets A and B is the *complement* of the other in E^n.

From the definitions we can easily prove the following lemmas.

4.1.5. Lemma. *Let F: $E^n\to E^m$ be a linear transformation. Then* ker *F is a subspace of E^n, $F(E^n)$ is a subspace of E^m, and F is a bijection if* ker $F=0$. *The dimension of $F(E^n)$ is called the rank of F, and the dimension of* ker F *is called the nullity of F.*

4.1.6. Lemma. *Let F: $E^n\to E^m$ be an injective but not surjective linear transformation. Then there is a decomposition*

$$E^m=E'^n\oplus E''^{m-n} \tag{4.1.6}$$

of E^m into a direct sum of subspaces E'^n and E''^{m-n} of dimensions n and $m-n$, respectively, such that $F(E^n)=E'^n$ and F: $E^n\to E'^n$ is a bijection.

4.1.7. Lemma. *Let F: $E^n\to E^m$ be a surjective linear transformation. Then there is a decomposition*

$$E^n=E'^m\oplus E''^{n-m} \tag{4.1.7}$$

of E^n into a direct sum of subspaces E'^m and E''^{n-m} of dimensions m and

$n-m$; respectively, such that $E''^{n-m}=\ker F$, and $F|E'^m$ is a bijection onto E^m.

Lemma 4.1.8 follows immediately from Definition 4.1.2.

4.1.8. Lemma. *Let V be a vector space over the field R of real numbers, and let F, G be linear transformations of V into R. Define the sum $F+G$ and the product cF of F by a scalar c as follows:*

$$(F+G)(\mathbf{x})=F(\mathbf{x})+G(\mathbf{x}), \qquad \text{for all} \quad \mathbf{x}\in V, \tag{4.1.8}$$

$$(cF)(\mathbf{x})=F(c\mathbf{x}) \qquad \text{for all} \quad \mathbf{x}\in V, c\in R. \tag{4.1.9}$$

Then $F+G$ and cF are also linear transformations of V into R.

4.1.9. Lemma. *If $\{\mathbf{v}_1,\cdots,\mathbf{v}_n\}$ is a basis of an n-dimensional vector space V over R, and if $c_1,\cdots,c_n$ are n constants in R, there is one and only one linear transformation F on V with $F(\mathbf{v}_i)=c_i$, $i=1,\cdots,n$. This transformation F is given by the formula*

$$\sum_{i=1}^{n} F(x_i\mathbf{v}_i)=\sum_{i=1}^{n} x_ic_i, \tag{4.1.10}$$

where $\mathbf{x}=(x_1,\cdots,x_n)$ is a variable vector of V.

Proof. By induction on n, (4.1.10) follows directly from (4.1.2) for any linear transformation F with $F(\mathbf{v}_i)=c_1, i=1,\cdots,n$. Conversely, for any basis $\{\mathbf{v}_1,\cdots,\mathbf{v}_n\}$ of V, each $\mathbf{x}$ has by Theorem 3.4.7 for V a unique expression $\mathbf{x}=\Sigma_{i=1}^n x_i\mathbf{v}_i$. For any constants $c_1,\cdots,c_n$ of R, (4.1.10) therefore defines a transformation of V into R. This transformation is linear, since for any $\mathbf{x}$ and $\mathbf{y}=\Sigma_{i=1}^n y_i\mathbf{v}_i$

$$F(a\mathbf{x}+b\mathbf{y})=F\left(\sum_{i=1}^{n}(ax_i+by_i)\mathbf{v}_i\right)=\sum_{i=1}^{n}(ax_i+by_i)c_i$$

$$=a\sum_{i=1}^{n}x_ic_i+b\sum_{i=1}^{n}y_ic_i=aF(\mathbf{x})+bF(\mathbf{y}).$$

4.1.10. Theorem. *The set of all linear transformations of a vector space V over the field R of real numbers with addition and scalar multiplication defined by* (4.1.8) *and* (4.1.9) *is a vector space V^* over R, which is called the dual space of V.*

The proof of Theorem 4.1.10 is only a routine verification of conditions (3.1.3), (3.1.4), and (3.1.5), and is left as an exercise.

4.1.11. Corollary. *If a vector space V over R has a finite basis $\{\mathbf{v}_1,\cdots,\mathbf{v}_n\}$, then its dual space V^* has a basis $\{F_1,\cdots,F_n\}$ called the dual basis to the basis $\{\mathbf{v}_1,\cdots,\mathbf{v}_n\}$ of V, consisting of the n linear transformations F_i defined by $F_j(\Sigma_{i=1}^n x_i\mathbf{v}_i)=x_j$, $j=1,\cdots,n$. The n linear transformations F_i are uniquely determined by the formulas*

$$F_j(\mathbf{v}_i)=\delta_{ij}, \qquad i,j=1,\cdots,n, \tag{4.1.11}$$

where δ_{ij} is the Kronecker delta.

Proof. For n given scalars $c_1,\cdots,c_n$, the linear combination $F=\Sigma_{j=1}^n c_jF_j$ is a linear transformation; by (4.1.11), its value at any basis vector $\mathbf{v}_i$ is

$$\sum_{j=1}^{n}(c_jF_j)(\mathbf{v}_i)=\sum_{j=1}^{n}F_j(c_j\mathbf{v}_j)=c_i.$$

It follows that the transformations $F_1,\cdots,F_n$ are linearly independent in V^*, for if $F=\Sigma_{j=1}^n c_jF_j=0$, then $F(\mathbf{v}_i)=0$ for each i, hence $c_1=\cdots=c_n=0$. It also follows that the n linear transformations $F_1,\cdots,F_n$ span V^n, since by Lemma 4.1.9 any linear transformation F is determined by its value $F(\mathbf{v}_i)=c_i$, hence $F=\Sigma_{j=1}^n c_jF_j$. Q.E.D.

Corollary 4.1.12 follows immediately from Corollary 4.1.11 and Definition 3.4.5.

4.1.12. Corollary. *The dual space V^* of an n-dimensional vector space V has the same dimension n as V.*

4.2. Derivative Mappings. In E^n, let $\mathbf{p}$ be the position vector of a point, and $\mathbf{v}$ a vector at $\mathbf{p}$. Then

$$t\to\mathbf{p}+t\mathbf{v} \tag{4.2.1}$$

is a curve in E^n, and the tangent vector of the curve (4.2.1) at $t=0$, that is, at $\mathbf{p}$, is $(d/dt)(\mathbf{p}+t\mathbf{v})|_{t=0}=\mathbf{v}$. Moreover, a mapping F: $E^n\to E^m$ maps the curve (4.2.1) in E^n to the curve

$$t\to F(\mathbf{p}+t\mathbf{v}) \tag{4.2.2}$$

in E^m. Thus we can have a definition of a mapping.

4.2.1. Definition. For a mapping F: $E^n\to E^m$ we define a mapping F_* (which may also be denoted by df) such that for a vector $\mathbf{v}$ of E^n at a point $\mathbf{p}$

$$F_*(\mathbf{v})=\frac{d}{dt}F(\mathbf{p}+t\mathbf{v})|_{t=0}, \tag{4.2.3}$$

which is the tangent vector of the curve (4.2.2) in E^m at $t=0$, that is, at

$F(\mathbf{p})$. This mapping F_* (or dF) is the *derivative mapping* (or *differential*) of F.

For example, let us compute the derivative mapping F_* of the mapping F in Example 3 of Section 4.1.3. Suppose that $\mathbf{p}=(p_1,p_2)$, $\mathbf{v}=(v_1,v_2)$. Then by (4.2.3) we have

$$\begin{aligned} F_*(\mathbf{v}) &= \frac{d}{dt}\left[F(p_1+tv_1,p_2+tv_2)\right]\big|_{t=0} \\ &= \frac{d}{dt}\left((p_1+tv_1)^2-(p_2+tv_2)^2, 2(p_1+tv_1)(p_2+tv_2)\right)\big|_{t=0} \\ &= (2(p_1v_1-p_2v_2), 2(p_1v_2+p_2v_1)) \qquad \text{at } F(p). \end{aligned}$$

4.2.2. Theorem. *Let* (4.1.1) *be a mapping of E^n into E^m. If $\mathbf{v}$ is a vector of E^n at a point $\mathbf{p}$, then*

$$F_*(\mathbf{v}) = (\mathbf{v}[f_1], \mathbf{v}[f_2], \cdots, \mathbf{v}[f_m]) \qquad \text{at} \quad F(\mathbf{p}), \tag{4.2.4}$$

so that $F_*(\mathbf{v})$ is determined by the derivatives $\mathbf{v}[f_i]$ of the coordinate functions of F with respect to $\mathbf{v}$.

Proof. For definiteness we set $m=3$. By (4.1.1) we have

$$F(\mathbf{p}+t\mathbf{v}) = (f_1(\mathbf{p}+t\mathbf{v}), f_2(\mathbf{p}+t\mathbf{v}), f_3(\mathbf{p}+t\mathbf{v})).$$

Thus (4.2.4) for $m=3$ follows immediately from (4.2.3) and (3.6.1) for $f=f_1, f_2, f_3$. Q.E.D.

Let $T_p(E^n)$ be the set of all vectors in E^n at a fixed point $\mathbf{p}$, that is, the tangent space of E^n at $\mathbf{p}$ (see Definition 3.5.1). Then the derivative mapping F_* given by (4.2.3) gives rise to a mapping

$$F_{*p}: T_p(E^n) \to T_{F(p)}(E^m) \tag{4.2.5}$$

defined by

$$F_{*p}(\mathbf{v}) = F_*(\mathbf{v}) \qquad \text{at} \quad F(\mathbf{p}), \qquad \text{for} \quad \mathbf{v} \in T_p(E^n).$$

F_{*p} is called the *derivative mapping* of F at $\mathbf{p}$. In a corresponding situation in elementary calculus, a differentiable function f: $R \to R$, where R is the field of real numbers, has a derivative function f': $R \to R$, which at each point t of R gives the derivative $f'(t)$ of f at t.

4.2.3. Corollary. *At each point $\mathbf{p} \in E^n$ the derivative mapping* (4.2.5) *of a mapping F: $E^n \to E^m$ is a linear transformation.*

Proof. For $\mathbf{v}, \mathbf{w} \in T_p(E^n)$, and $a, b \in R$, we must show that

$$F_*(a\mathbf{v}+b\mathbf{w}) = aF_*(\mathbf{v}) + bF_*(\mathbf{w}),$$

which follows immediately from Theorem 3.6.3a and Theorem 4.2.2.

Q.E.D.

The linearity of F_{*p} is a generalization of the fact that the derivative $f'(t)$ of $f\colon R \to R$ is the slope of the tangent line to the graph of f at t. Indeed for each point $\mathbf{p}$, F_{*p} is the linear transformation that best approximates the behavior of F near $\mathbf{p}$. This idea could be fully developed and used to prove Theorem 4.2.7.

Since the mapping (4.2.5) is a linear transformation, it is reasonable to compute its matrix with respect to its natural basis, defined in (3.4.3):

$$\mathbf{u}_1(\mathbf{p}),\cdots,\mathbf{u}_n(\mathbf{p}) \qquad \text{for} \quad T_p(E^n),$$
$$\mathbf{u}_1(F(\mathbf{p})),\cdots,\mathbf{u}_m(F(\mathbf{p})) \qquad \text{for} \quad T_{F(p)}(E^m).$$

This matrix is called the *Jacobian matrix* of F at $\mathbf{p}$ and is denoted by $J_F(\mathbf{p})$.

4.2.4. Corollary. *For a mapping* (4.1.1) *of E^n into E^m,*

$$F_*(\mathbf{u}_j(\mathbf{p})) = \sum_{i=1}^{m} \frac{\partial f_i}{\partial x_j}(\mathbf{p})u_i(F(\mathbf{p})), \qquad 1 \leqslant j \leqslant n, \tag{4.2.6}$$

so that $J_F(\mathbf{p}) = ((\partial f_i/\partial x_j)(\mathbf{p}))_{1\leqslant i\leqslant m, i\leqslant j\leqslant n}$, *where $x_1,\cdots,x_n$ are the coordinate functions of E^n.*

Proof. Substituting $\mathbf{u}_j(\mathbf{p})$ for $\mathbf{v}$ in (4.2.4) and using (3.6.4) for $f = f_1,\cdots,f_m$ we obtain

$$F_*(\mathbf{u}_j(\mathbf{p})) = \left(\frac{\partial f_1}{\partial x_j}(\mathbf{p}),\cdots,\frac{\partial f_m}{\partial x_j}(\mathbf{p})\right)_{F(p)},$$

which is just (4.2.6) in component form.

As usual, (4.2.6) is abbreviated to

$$F_*(\mathbf{u}_j) = \sum_{i=1}^{3} \frac{\partial f_i}{\partial x_j}\bar{\mathbf{u}}_i,$$

where $\mathbf{u}_j$ and $\partial f_i/\partial x_j$ are evaluated at $\mathbf{p}$, and $\bar{\mathbf{u}}_i$ is evaluated at $F(\mathbf{p})$.

For Example 3 of Section 4.1.3, $f_1 = u^2 - v^2$, $f_2 = 2uv$; therefore J_F at $\mathbf{p} = (p_1, p_2)$ is

$$\begin{bmatrix} \dfrac{\partial f_1}{\partial u} & \dfrac{\partial f_1}{\partial v} \\ \dfrac{\partial f_2}{\partial u} & \dfrac{\partial f_2}{\partial v} \end{bmatrix}(\mathbf{p}) = \begin{pmatrix} 2u & -2v \\ 2v & 2u \end{pmatrix}(\mathbf{p}) = \begin{pmatrix} 2p_1 & -2p_2 \\ 2p_2 & 2p_1 \end{pmatrix}.$$

4.2.5. Theorem. *Let $F\colon E^n \to E^m$ be a mapping, and $\mathbf{v}$ a vector of E^n at a point $\mathbf{p}$. Set $\mathbf{x}(t) = \mathbf{p} + t\mathbf{v}$, $\mathbf{y}(t) = F(\mathbf{x}(t))$. Then*

$$F_*(\mathbf{x}'(t)) = \mathbf{y}'(t), \tag{4.2.7}$$

where the prime denotes the derivative with respect to t.

Proof. For definiteness we assume $m=3$. By (4.1.1) we have

$$\mathbf{y}(t)=F(\mathbf{x}(t))=(f_1(\mathbf{x}(t)),\ f_2(\mathbf{x}(t)),\ f_3(\mathbf{x}(t)). \tag{4.2.8}$$

From (4.1.1) and Theorem 4.9 it follows that

$$F_*(\mathbf{x}'(t))=(\mathbf{x}'(t)[f_1],\mathbf{x}'(t)[f_2],\mathbf{x}'(t)[f_3]). \tag{4.2.9}$$

Suppose $\mathbf{x}=(x_1,x_2,x_3)$. By considering $f_i=f_i(\mathbf{x})=f_i(x_1,x_2,x_3)$ and using Lemma 3.6.2 and the chain rule (Theorem 2.3.6), we obtain

$$\mathbf{x}'(t)[f_i]=\sum_{j=1}^{3}\frac{\partial f_i}{\partial x_j}(\mathbf{x}(t))\frac{dx_j}{dt}(t)=\frac{df_i(\mathbf{x})}{dt}(t). \tag{4.2.10}$$

Substitution of (4.2.10) in (4.2.9) and use of (4.2.8) thus give (4.2.7).

4.2.6. Definition. A mapping $F\colon E^n\to E^m$ is *regular* if for every point $\mathbf{p}$ of E^n the derivative mapping F_{*p} is injective.

Since each F_{*p} is a linear transformation, by applying standard results of linear algebra we have the following equivalent conditions:

(a) F_{*p} is injective.
(b) If $F_*(v)=0$, then $\mathbf{v}=0$, where $\mathbf{v}\in T_p(E^n)$.
(c) $J_F(\mathbf{p})$ has rank n (the dimension of the domain E^n of F).

For example, the mapping F in Example 3 of Section 4.1.3 is not regular. But the injective condition fails at only a single point, the origin $(0,0)$. In fact, from the computation immediately preceding Theorem 4.2.5 it follows that J_F has rank 2 at $\mathbf{p}\neq(0,0)$, and rank 0 at $(0,0)$.

A differentiable mapping that has a differentiable inverse mapping is called a *diffeomorphism*. A diffeomorphism is thus necessarily bijective, but a differentiable bijective mapping may not be a diffeomorphism (Exercise 9 of Section 4.2).

The results of this section apply equally well to mappings defined only on open sets of E^n. Thus we may speak of an injective or surjective mapping or a diffeomorphism of one open set of E^n into another, such a mapping being called a local injective or surjective mapping, or a local diffeomorphism if about every point of E^n there is such an open set. For example, consider the mapping $F\colon E^2\to E^2$ such that

$$F(u,v)=(x,y),\qquad x=u\cos v,\ y=u\sin v.$$

The Jacobian J_F of F is equal to x and is thus never zero in the right half uv-plane $D=\{\text{all }(u,v)\text{ with }u>0\}$. However, F is not a diffeomorphism in D, since points (a,b) and $(a,b+2\pi)$ of D always have the same image in the xy-plane. The effect of the mapping may be seen from the set $S\subset D$, shown together with its image $F(S)$ in Fig. 1.6. We notice that although F is not injective in S, two distinct points of S that have the same image must

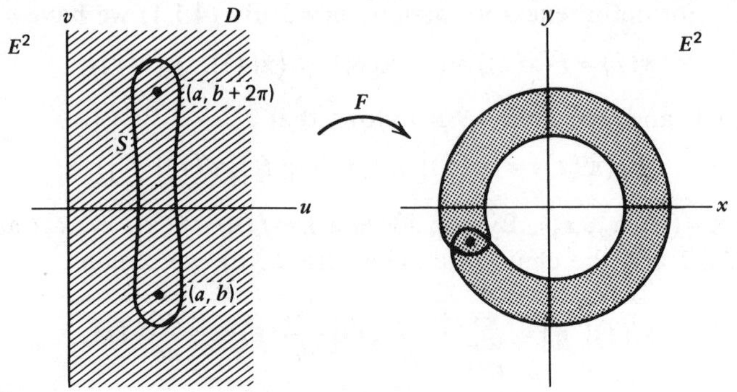

Figure 1.6

be widely separated. Thus F is a local diffeomorphism in S (and in fact in D).

For local diffeomorphisms we have the following important theorem.

4.2.7. Theorem (Inverse function theorem). *Let F: $E^n \to E^n$ be a differentiable mapping such that F_{*p} is injective at some point* $\mathbf{p}$. *Then F is a local diffeomorphism about* $\mathbf{p}$, *that is, F maps any sufficiently small open set A about* $\mathbf{p}$ *diffeomorphically onto an open set $F(A)$.*

Proof. See, for instance, T. M. Apostol [2, p. 144].

For local injective (respectively, surjective) differentiable mappings, we have the following analogous theorem.

4.2.8. Theorem (on local injective or surjective mappings). *Let U be an open neighborhood of a point $\mathbf{p} \in E^n$, and F: $U \to E^m$ a differentiable mapping. If*

$$F_{*p}\colon T_p(E^n) \to T_{F(p)}(E^m) \tag{4.2.11}$$

is injective (respectively, surjective), there is a diffeomorphism G of a neighborhood W of $F(\mathbf{p}) \in E^m$ onto a neighborhood $G(W)$ of $F(\mathbf{p}) \in E^m$ (respectively, a diffeomorphism H of a neighborhood V of $\mathbf{p} \in E^n$ onto a neighborhood $H(V)$ of $\mathbf{p} \in E^n$) such that $G \circ F$ is locally about $\mathbf{p} \in E^n$ a linear injection into E^m (respectively, $F \circ H$ is locally about $\mathbf{p} \in E^n$ a linear surjection onto E^m).

It is obvious that the converse of Theorem 4.2.8 holds.

Proof. We first consider the case in which the mapping (4.2.11) is injective. By identifying $T_p(E^n)$ and $T_{F(p)}(E^m)$ with E^n and E^m, respectively, and applying Lemma 4.1.6, we obtain a decomposition (4.1.6) of E^m

such that F_{*p}: $E^n \to E'^n$ is a bijection. Define a differentiable mapping $\tilde{G}$: $E^m = E'^n \oplus E''^{m-n} \to E^m = E'^n \oplus E''^{m-n}$ locally about $F(\mathbf{p})$ by $u=(u', u'') \to F(u') + (0, u'')$, where we have identified the first E'^n with E^n. Obviously $\tilde{G}_{*F(p)} = F_{*p} + \mathrm{id}|E''^{m-n}$ is bijective; therefore locally about $F(\mathbf{p})$ there exists the inverse mapping $\tilde{G}^{-1}$ denoted by G. Thus $G \circ F(u') = (u', 0)$; that is, $G \circ F$ is locally about $\mathbf{p}$ a linear injection: $E^n \to E'^n \subset E'^n \oplus E''^{m-n} = E^m$.

Now suppose the mapping (4.2.11) to be surjective. Then by applying Lemma 4.1.7, as above, we obtain a decomposition (4.1.7) of E^n such that F_*: $E'^m \to E^m$ is a bijection. Define a differentiable mapping $\tilde{H}$: $E^n = E'^m \oplus E''^{n-m} \to E^n = E'^m \oplus E''^{n-m}$ locally about $\mathbf{p}$ by $u = (u', u'') \to (F(u), u'')$, where we have identified the second E'^m with E^m. Since $\tilde{H}_{*p} = F_{*p}|E'^m + \mathrm{id}|E''^{n-m}$ is bijective, locally about $\mathbf{p}$ there exists the inverse mapping $\tilde{H}^{-1}$ denoted by H. Thus $H(F(u', u''), u'') = (u', u'')$, and $F \circ H(F(u', u''), u'') = F(u', u'')$; that is, $F \circ H$ is locally about $\mathbf{p}$ the projection $E^n = E'^m + E''^{n-m} \to E'^m$ and is therefore linear and surjective.

Exercises

1. For the mapping F given by (4.1.4) find all points $\mathbf{p}$, with sketches for part d, such that
   (a) $F(\mathbf{p}) = (0,0)$. (b) $F(\mathbf{p}) = (3, -4)$.
   (c) $F(\mathbf{p}) = \mathbf{p}$. (d) $F(\mathbf{p}) = (x, y)$, $1 \leqslant x \leqslant 4$, $2 \leqslant y \leqslant \frac{7}{2}$.

*2. Under the mapping F given by (4.1.4) find the image of each of the following sets:
   (a) The horizontal strip S: $1 \leqslant v \leqslant 2$.
   (b) The half-disk S: $u^2 + v^2 \leqslant 1$, $v \geqslant 0$.
   (c) The wedge S: $-u \leqslant v \leqslant u$, $u \geqslant 0$.
   In each case, show the set S and its image $F(S)$ in a single sketch.

3. Prove Theorem 4.1.10.

4. (a) Show that the derivative mapping of the mapping F in Example 1 of Section 4.1.3 is given by
$$F_*(\mathbf{v}_p) = (2v_1, v_2 - v_3, v_2 + v_3)_{F(p)}.$$
   (b) In general, if F: $E^n \to E^m$ is a linear transformation, prove that
$$F_*(\mathbf{v}_p) = F(\mathbf{v})_{F(p)}.$$

5. Find F_* for the mapping $F(x, y, z) = (x, y\cos z, y\sin z)$: $E^3 \to E^3$, and compute $F_*(\mathbf{v}_p)$ if
   (a) $\mathbf{v} = (3, -2, 1)$, $\mathbf{p} = (0,0,0)$.
   (b) $\mathbf{v} = (3, -2, 1)$, $\mathbf{p} = (4, \pi, \frac{\pi}{2})$.

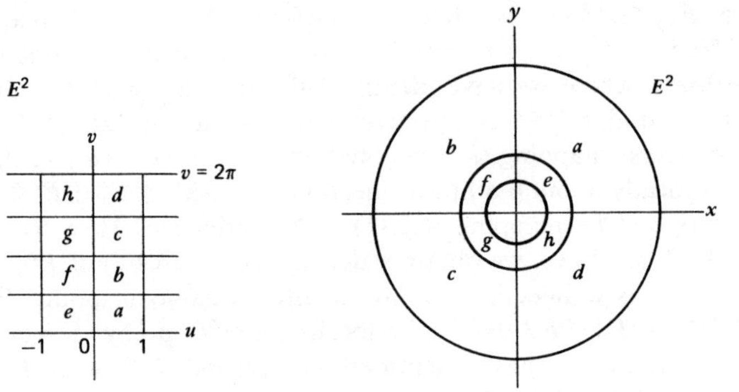

Figure 1.7

6. Is the mapping in Exercise 5 regular?

7. In Definition 4.2.1 of $F_*(\mathbf{v}_p)$, show that the straight line $\mathbf{p}+t\mathbf{v}$ can be replaced by any curve with $\mathbf{v}_p$ as its tangent at $\mathbf{p}$, that is, at $t=0$.

8. Prove that a mapping F: $E^n \to E^m$ preserves directional derivatives in the following sense: If $\mathbf{v}_p$ is a vector of E^n, and g is a differentiable function on E^m, then $F_*(\mathbf{v}_p)[g]=\mathbf{v}_p[g(F)]$.

9. In each of the following cases, show that the mapping F: $E^2 \to E^2$ is bijective, compute the inverse mapping F^{-1}, and determine whether F is a diffeomorphism (that is, whether F^{-1} is differentiable).
   (a) $F(u,v)=(v, ue^v)$.
   (b) $F(u,v)=(u-v, v^3)$.
   (c) $F(u,v)=(2u-v+1, u+3v-2)$.

10. Let F: $E^2 \to E^2$ be a mapping such that

$$F(u,v)=(x,y),\ x=e^u\cos v,\ y=e^u\sin v.$$

The nature of the mapping F is suggested by Fig. 1.7, in which certain corresponding areas are indicated by the same letters. Verify this figure and show that onto the xy-plane with the origin excluded the mapping F is not a diffeomorphism of the whole uv-plane but is a local diffeomorphism.

5. LINEAR GROUPS

5.1. Linear Transformations. In this section all definitions and results hold for E^3 as well as a Euclidean plane E^2. However, for generality we only discuss the case of E^3, and consider the case of E^2 as a special case.

A mapping of E^3 into itself is a correspondence that associates with each point $\mathbf{x}$ of E^3 a unique point $\mathbf{y}$ of E^3. By the correspondence mentioned in Section 3.1, we consider $\mathbf{x}$ and $\mathbf{y}$ here also as the position vectors of the respective points with reference to a fixed point $\mathbf{0}$ in E^3.

5.1.1. Definition. A mapping F: $E^3 \to E^3$ is a *linear transformation* of E^3 into itself if, for all vectors $\mathbf{x}$ and $\mathbf{y}$ and all scalars c and d,

$$F(c\mathbf{x}+d\mathbf{y})=c(F(\mathbf{x}))+d(F(\mathbf{y})). \tag{5.1.1}$$

Definition 5.1.1 is a special case of Definition 4.2.

Let F and G be two linear transformations of E^3. Then by defining the multiplication of F with G by

$$(GF)(\mathbf{x})=G(F(\mathbf{x})), \tag{5.1.2}$$

and introducting the identity mapping I:

$$I(\mathbf{x})=\mathbf{x}, \tag{5.1.3}$$

we can show that F has a multiplicative inverse F^{-1} if and only if F is bijective. Thus that *all the nonsingular (i.e., bijective) linear transformations of E^3 form a multiplicative group*, which is called *the general linear group* of E^3 and is denoted by GL_3. [A nonempty set G, on which there is defined a binary operation "$\circ$", is called a *group* (with respect to this operation), provided the following properties are satisfied: (i) if $a, b, c \in G$, then $(a\circ b)\circ c = a\circ(b\circ c)$; (ii) there exists an element $e\in G$ such that $e\circ a = a\circ e = a$ for every element $a\in G$; (iii) if $a\in G$, there exists an element $x\in G$ such that $a\circ x = x\circ a = e$.]

Let $\{\mathbf{x}_1,\mathbf{x}_2,\mathbf{x}_3\}$ be a basis of E^3. Then we have, uniquely,

$$F(\mathbf{x}_i)=\sum_{j=1}^{3} a_{ji}\mathbf{x}_j, \qquad i=1,2,3. \tag{5.1.4}$$

F is nonsingular if and only if the matrix

$$A=\begin{bmatrix} a_{11} & a_{12} & a_{13} \\ a_{21} & a_{22} & a_{23} \\ a_{31} & a_{32} & a_{33} \end{bmatrix} \tag{5.1.5}$$

is nonsingular, that is, if and only if $\det A \neq 0$. Let M be the set of all the nonsingular 3×3 matrices with usual row-by-column multiplication. Then the mapping $F \to A$ defined by (5.1.4) is an isomorphism of GL_3 onto M.

Now suppose $\{\mathbf{y}_1,\mathbf{y}_2,\mathbf{y}_3\}$ to be another basis of E^3 so that

$$F(\mathbf{y}_i)=\sum_{j=1}^{3} a^{*}_{ji}\mathbf{y}_j, \qquad i=1,2,3, \tag{5.1.6}$$

and denote the matrix (a^*_{ij}) by A^*. Then (see Exercise 1 of Section 5.1) there exists a unique linear transformation G of E^3 such that

$$G(\mathbf{x}_i)=\mathbf{y}_i, \qquad i=1,2,3. \tag{5.1.7}$$

Furthermore, it can be shown that

$$A^*=BAB^{-1}, \tag{5.1.8}$$

where B^{-1} is the inverse matrix of $B=(b_{ij})$ given by

$$G(\mathbf{x}_i)=\mathbf{y}_i=\sum_{j=1}^{3} b_{ji}\mathbf{x}_j, \qquad i=1,2,3. \tag{5.1.9}$$

5.1.2. Definition. A linear transformation F: $E^3 \to E^3$ is an *orthogonal transformation* if it preserves the inner product of any two vectors $\mathbf{x}$ and $\mathbf{y}$, that is, if

$$(F(\mathbf{x}))\cdot F(\mathbf{y})=\mathbf{x}\cdot\mathbf{y}. \tag{5.1.10}$$

If F and G are orthogonal transformations of E^3, then for any two vectors $\mathbf{x}$ and $\mathbf{y}$ of E^3

$$(F^{-1}(\mathbf{x}))\cdot F^{-1}(\mathbf{y})=(F(F^{-1}(\mathbf{x})))\cdot F(F^{-1}(\mathbf{y}))=\mathbf{x}\cdot\mathbf{y},$$

$$((FG)(\mathbf{x}))\cdot(FG)(\mathbf{y})=(F(G(\mathbf{x})))\cdot F(G(\mathbf{y}))=(G(\mathbf{x}))\cdot G(\mathbf{y})=\mathbf{x}\cdot\mathbf{y},$$

showing that *all the orthogonal transformations of* E^3 *form a subgroup of the linear group* GL_3, which is called the *orthogonal group* 0_3.

5.1.3. Definition. A square matrix A of order 3 is *orthogonal* if

$${}^tAA=\mathcal{I}, \tag{5.1.11}$$

where ${}^tA$ is the transpose of A, and $\mathcal{I}$ is the identity matrix.

Equation 5.1.11 implies that A is nonsingular. It may therefore be written as $A^{-1}={}^tA$ and hence also as

$$A\,{}^tA=\mathcal{I}. \tag{5.1.12}$$

From (5.1.11) or (5.1.12) it follows that if A is an orthogonal matrix, then $(\det A)^2=1$ or

$$\det A=\pm 1. \tag{5.1.13}$$

By definition 5.1.3 and direct multiplication of matrices we readily obtain Theorem 5.1.4.

5.1.4. Theorem. *The square matrix A given by (5.1.5) is orthogonal if and only if one of the following three equivalent conditions holds*:

$$\sum_{i=1}^{3} a_{ij}a_{ik}=\delta_{jk}, \qquad j,k=1,2,3, \tag{5.1.14}$$

$$\sum_{i=1}^{3} a_{ji}a_{ki}=\delta_{jk}, \qquad j,k=1,2,3, \tag{5.1.15}$$

$$\det A\neq 0,\ a_{ij}=A_{ij}, \qquad i,j=1,2,3, \tag{5.1.16}$$

where δ_{jk} is a Kronecker delta, being 1 for $j=k$ and 0 for $j\neq k$, and A_{ij} is the cofactor of a_{ij} in $\det A$.

5.1.5. Theorem. *Let $\{\mathbf{x}_1,\mathbf{x}_2,\mathbf{x}_3\}$ be an orthonormal basis of E^3. A linear transformation F: $E^3\to E^3$ given by (5.1.4) is an orthogonal transformation if and only if the matrix $A=(a_{ij})$ is orthogonal.*

Proof of Theorem 5.1.5. Suppose that F is orthogonal. Then by (5.1.4) and (5.1.10) we have

$$(F(\mathbf{x}_j))\cdot F(\mathbf{x}_k)=\left(\sum_{h=1}^{3} a_{hj}\mathbf{x}_h\right)\cdot\left(\sum_{i=1}^{3} a_{ik}\mathbf{x}_i\right)=\sum_{h=1}^{3}\sum_{i=1}^{3} a_{hj}a_{ik}\delta_{hi}$$

$$=\sum_{i=1}^{3} a_{ij}a_{ik}=\mathbf{x}_j\cdot\mathbf{x}_k=\delta_{jk}, \qquad j,k=1,2,3,$$

which shows, due to Theorem 5.1.4, that A is orthogonal. The converse of this can be proved by reversing the argument.

5.1.6. Corollary. *Let $\mathbf{x}_1,\mathbf{x}_2$ be an orthonormal basis of a plane E^2. Then the only orthogonal transformations of E^2 are F and G* given by

$$\begin{aligned} F(\mathbf{x}_1)&=\mathbf{x}_1\cos\theta+\mathbf{x}_2\sin\theta,\\ F(\mathbf{x}_2)&=-\mathbf{x}_1\sin\theta+\mathbf{x}_2\cos\theta; \end{aligned} \tag{5.1.17}$$

$$\begin{aligned} G(\mathbf{x}_1)&=\mathbf{x}_1\cos\theta+\mathbf{x}_2\sin\theta,\\ G(\mathbf{x}_2)&=\mathbf{x}_1\sin\theta-\mathbf{x}_2\cos\theta, \end{aligned} \tag{5.1.18}$$

where θ is unique such that $0\leqslant\theta<2\pi$, so that F is a rotation of E^2 through an angle θ, and G is the rotation F followed by a reflection in $\mathbf{x}_2$, which maps every point to its symmetric point with respect to $\mathbf{x}_2$ (two points in E^2 are said to be symmetric with respect to the $\mathbf{x}_2$-axis if the $\mathbf{x}_2$-axis is the perpendicular bisector of the line segment joining the two points).

Proof. Let H be an orthogonal transformation of E^2 given by

$$H(\mathbf{x}_1)=a_1\mathbf{x}_1+a_2\mathbf{x}_2, \qquad H(\mathbf{x}_2)=b_1\mathbf{x}_1+b_2\mathbf{x}_2. \tag{5.1.19}$$

Then by the special case of Theorems 5.1.5 and 5.1.4 for E^2 we have

$$a_1^2+a_2^2=1, \qquad b_1^2+b_2^2=1, \tag{5.1.20}$$

$$a_1b_1+a_2b_2=0. \tag{5.1.21}$$

By (5.1.20) there is an angle θ such that $0 \leqslant \theta < 2\pi$ and $\cos\theta = a_1$, $\sin\theta = a_2$. Then $\tan\theta = a_2/a_1 = -b_1/b_2$ by (5.1.21), whence by (5.1.20) again $b_2 = \pm\cos\theta$, $b_1 = \pm\sin\theta$. The two choices of sign give exactly the two transformations F and G. Q.E.D.

Now in E^3 we consider a right-handed rectangular trihedron $\mathbf{0x}_1\mathbf{x}_2\mathbf{x}_3$ and a rotation F that is an orthogonal transformation given by (5.1.4) with orthogonal matrix $A=(a_{ij})$ such that $\det A = +1$. Then by (5.1.14), A can be expressed as

$$A=\begin{bmatrix} \cos\alpha_1 & \cos\beta_1 & \cos\gamma_1 \\ \cos\alpha_2 & \cos\beta_2 & \cos\gamma_2 \\ \cos\alpha_3 & \cos\beta_3 & \cos\gamma_3 \end{bmatrix}, \tag{5.1.22}$$

and the columns of A are composed of the direction cosines of three mutually orthogonal lines $\mathbf{x}_1^*, \mathbf{x}_2^*, \mathbf{x}_3^*$ through the origin 0, which respectively pass through the image vectors, under F, of the unit vectors $\mathbf{u}_1 = (1,0,0)$, $\mathbf{u}_2 = (0,1,0)$, $\mathbf{u}_3 = (0,0,1)$ on the coordinate axes $\mathbf{x}_1$, $\mathbf{x}_2$, $\mathbf{x}_3$.

It is well known that any rotation in E^3 can be represented by three independent parameters called the Euler angles (Fig. 1.8). We shall determine the Euler angles for our rotation F, which takes the coordinate

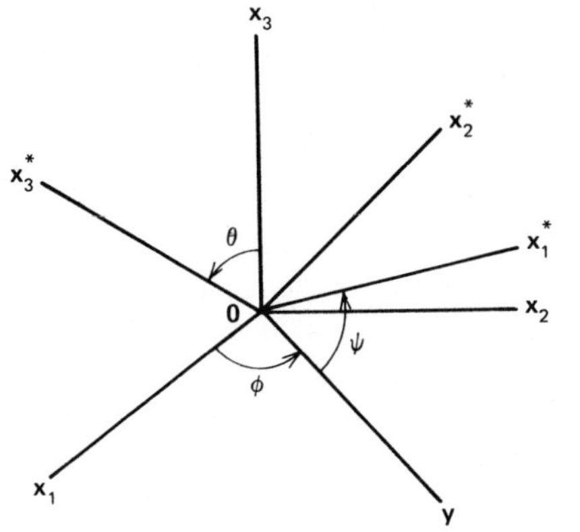

Figure 1.8

axes $\mathbf{x}_1, \mathbf{x}_2, \mathbf{x}_3$ onto three preassigned mutually orthogonal lines $\mathbf{x}_1^*, \mathbf{x}_2^*, \mathbf{x}_3^*$. To this end, first let $\mathbf{y}$ be the line of intersection of the $\mathbf{x}_1\mathbf{x}_2$- and the $\mathbf{x}_1^*\mathbf{x}_2^*$-planes, and ϕ the angle of the rotation A_ϕ about $\mathbf{x}_3$, which takes the $\mathbf{x}_1$-axis onto $\mathbf{y}$, so that

$$A_\phi = \begin{pmatrix} \cos\phi & \sin\phi & 0 \\ -\sin\phi & \cos\phi & 0 \\ 0 & 0 & 1 \end{pmatrix}.$$

The rotation

$$A_\theta = \begin{pmatrix} 1 & 0 & 0 \\ 0 & \cos\theta & \sin\theta \\ 0 & -\sin\theta & \cos\theta \end{pmatrix}$$

about $\mathbf{y}$ through an angle θ takes $\mathbf{x}_3$ onto $\mathbf{x}_3^*$, and also $\mathbf{x}_2$ into the $\mathbf{x}_1^*\mathbf{x}_2^*$-plane. Finally, the rotation

$$A_\psi = \begin{pmatrix} \cos\psi & \sin\psi & 0 \\ -\sin\psi & \cos\psi & 0 \\ 0 & 0 & 1 \end{pmatrix}$$

about $\mathbf{x}_3^*$ through an angle ψ takes $\mathbf{y}$ onto $\mathbf{x}_1^*$, and $\mathbf{x}_2$ and $\mathbf{x}_2^*$. This shows

$$A = A_\psi A_\theta A_\phi. \tag{5.1.23}$$

5.1.7. Theorem. *An orthogonal transformation F of E^3 preserves the distance between any two points, that is, F satisfies*

$$d(F(\mathbf{x}), F(\mathbf{y})) = d(\mathbf{x}, \mathbf{y}), \tag{5.1.24}$$

where $\mathbf{x}, \mathbf{y}$ are the position vectors of any two points.

Proof. From (3.2.4), (5.1.1), (3.2.3), and Definition 5.1.2 we immediately obtain

$$d(F(\mathbf{x}, F(\mathbf{y})) = \|F(\mathbf{x}-\mathbf{y})\| = \|\mathbf{x}-\mathbf{y}\| = d(\mathbf{x}, \mathbf{y}).$$

Exercises

1. If $\{\mathbf{x}_1, \mathbf{x}_2, \mathbf{x}_3\}$ is a basis of E^3, and $\mathbf{z}_1, \mathbf{z}_2, \mathbf{z}_3$ are arbitrary vectors of E^3, there exists exactly one linear transformation F of E^3 such that

$$F(\mathbf{x}_i) = \mathbf{z}_i, \qquad i = 1, 2, 3.$$

***2.** Prove Theorem 5.1.4.

3. Test the following matrices for orthogonality, and find the inverses of the orthogonal matrices:

$$\text{(a)}\begin{bmatrix} \frac{1}{2} & \frac{\sqrt{3}}{2} \\ -\frac{\sqrt{3}}{2} & \frac{1}{2} \end{bmatrix}, \quad \text{(b)}\begin{bmatrix} \frac{1}{2} & \frac{\sqrt{3}}{2} \\ \frac{\sqrt{3}}{2} & \frac{1}{2} \end{bmatrix}, \quad \text{(c)}\begin{bmatrix} \frac{3}{5} & \frac{4}{5} \\ \frac{4}{5} & -\frac{3}{5} \end{bmatrix}.$$

4. Find orthogonal 3×3 matrices with the following first rows:

$$\left(\frac{5}{13}, 0, \frac{12}{13}\right), \left(\frac{1}{3}, \frac{2}{3}, \frac{2}{3}\right).$$

5. If a rotation F has Euler angles ϕ, θ, ψ, find the Euler angles of F^{-1}.

5.2. Translations and Affine Transformations

5.2.1. *Definition.* A mapping $F: E^3 \to E^3$ is a *translation* by a fixed vector $\mathbf{c}$ if for every vector $\mathbf{x}$

$$F(\mathbf{x}) = \mathbf{x} + \mathbf{c}. \tag{5.2.1}$$

It is easy to prove Lemma 5.2.2.

5.2.2. *Lemma.* (a) *If F and G are translations of E^3, then $FG = GF$ is also a translation.*

(b) *The translation F given by* (5.2.1) *has an inverse F^{-1} that is the translation by* $-\mathbf{c}$.

(c) *Given any two points* $\mathbf{x}$ *and* $\mathbf{y}$ *of E^3, there exists a unique translation F such that* $F(\mathbf{x}) = \mathbf{y}$.

From Lemmas 5.2.2a and 5.2.2b it follows that *all the translations F of E^3 given by* (5.2.1) *form an Abelian group isomorphic to the additive groups of the vectors* $\mathbf{c}$ *of E^3.*

A useful special case of Lemma 5.2.2c is that if F is a translation such that $F(\mathbf{x}) = \mathbf{x}$ for some point $\mathbf{x}$, then F is the identity mapping I.

5.2.3. *Theorem.* *A translation F of E^3 preserves the distance between any two points, that is,* (5.1.24) *holds for a translation.*

Proof. Let F be given by (5.2.1). Then

$$d(F(\mathbf{x}), F(\mathbf{y})) = \|F(\mathbf{x}-\mathbf{y})\| = \|\mathbf{x}+\mathbf{c}-(\mathbf{y}+\mathbf{c})\| = \|\mathbf{x}-\mathbf{y}\| = d(\mathbf{x},\mathbf{y}).$$

5.2.4. *Definition.* An *affine transformation F* of E^3 is a linear transformation followed by a translation, that is, a mapping satisfying

$$F = HG, \tag{5.2.2}$$

where G is a linear transformation, and H is a translation of E^3. G is called the *linear part* of F, and H the *translation part* of F.

From (5.2.1) and (5.2.2) it follows that if H is a translation by a fixed vector $\mathbf{c}$, then for every vector $\mathbf{x}$ of E^3

$$F(\mathbf{x}) = G(\mathbf{x}) + \mathbf{c}. \tag{5.2.3}$$

The affine transformations F given by (5.2.3) include the linear transformations (with $\mathbf{c}=0$) and the translations (with $G=I$). If $\bar{F}$ is another affine transformation of E^3 given by

$$\bar{F}(\mathbf{x}) = \bar{G}(\mathbf{x}) + \bar{\mathbf{c}}, \tag{5.2.4}$$

then

$$\bar{F}F(\mathbf{x}) = \bar{G}G(\mathbf{x}) + \bar{G}(\mathbf{c}) + \bar{\mathbf{c}}. \tag{5.2.5}$$

Furthermore, if G has an inverse G^{-1}, then by solving (5.2.3) for $\mathbf{x}$ we have the inverse F^{-1} of F:

$$F^{-1}(\mathbf{x}) = G^{-1}(\mathbf{x}) - G^{-1}(\mathbf{c}). \tag{5.2.6}$$

Thus we obtain

5.2.5. Theorem. *The nonsingular affine transformations of E^3 form a group, called the affine group A_3, which contains as subgroups the general linear group GL_3 and the group T of translations.*

With respect to a right-handed trihedron $\mathbf{0}\mathbf{x}_1\mathbf{x}_2\mathbf{x}_3$ in E^3, let $A=(a_{ij})$ be the matrix representing a linear transformation G so that

$$G(\mathbf{x}_i) = \sum_{j=1}^{3} a_{ji}\mathbf{x}_j, \qquad i=1,2,3. \tag{5.2.7}$$

Then for $\mathbf{x} = \sum_{i=1}^{3} x_i\mathbf{x}_i$ we have

$$G(\mathbf{x}) = \sum_{i=1}^{3} x_i G(\mathbf{x}_i) = \sum_{j=1}^{3}\left(\sum_{i=1}^{3} a_{ji}x_i\right)\mathbf{x}_j.$$

Thus by putting $G(\mathbf{x}) = \sum_{i=1}^{3}(G(\mathbf{x}))_i\mathbf{x}_i$ we obtain

$$(G(\mathbf{x}))_i = \sum_{j=1}^{3} a_{ij}x_j, \qquad i=1,2,3. \tag{5.2.8}$$

Let $\mathbf{x}^* = F(\mathbf{x})$, $\mathbf{x}^* = (x_1^*, x_2^*, x_3^*)$, $\mathbf{c} = (c_1, c_2, c_3)$, where x_i^* and c_i, $i=1,2,3$, are the components of the respective vectors. Then in the component form, (5.2.3) can be expressed as follows:

$$x_i^* = \sum_{j=1}^{3} a_{ij}x_j + c_i, \qquad i=1,2,3. \tag{5.2.9}$$

Alternatively, using the column-vector conventions, in terms of matrix multiplication we can write $G(\mathbf{x})=A\mathbf{x}$ so that (5.2.9) becomes

$$\mathbf{x}^*=A\mathbf{x}+\mathbf{c}. \tag{5.2.10}$$

The product of the transformation (5.2.10) by $\mathbf{y}^*=B\mathbf{x}^*+\mathbf{d}$ is

$$\mathbf{y}^*=(BA)\mathbf{x}+B\mathbf{c}+\mathbf{d}, \qquad \mathbf{c},\mathbf{d}\text{: column matrices;} \tag{5.2.11}$$

the formula is parallel to (5.2.5). In E^3, of special interest are the affine transformations

$$\mathbf{x}^*=A\mathbf{x}+\mathbf{c}, \qquad \det A=1. \tag{5.2.12}$$

Since the product of two matrices of determinant 1 again has determinant 1, the affine transformations (5.2.12) form a group, which is called the *unimodular affine group*.

Exercises

1. Given the circle $x_1^2+x_2^2=1$ in the $\mathbf{x}_1\mathbf{x}_2$-plane, prove that every affine transformation of the plane carries this circle to an ellipse or a circle.
2. Prove that the group of translations of E^3 is a normal subgroup of the affine group of E^3.

5.3. Isometries or Rigid Motions

5.3.1. Definition. A mapping F: $E^3 \to E^3$ is an *isometry* or a *rigid motion* if it preserves the distance between any two points.

Lemma 5.3.2 follows from Theorems 5.1.7 and 5.2.3.

5.3.2. Lemma. *Orthogonal transformations and translations of E^3 are isometries of E^3.*

5.3.3. Lemma. *If F and $\bar{F}$ are isometries of E^3, the composite mapping $\bar{F}F$ is also an isometry of E^3.*

Proof of Lemma 5.3.3. Let $\mathbf{x}$ and $\mathbf{y}$ be the position vectors of any two points of E^3. Then

$$d((\bar{F}F)(\mathbf{x}),(\bar{F}F)(\mathbf{y}))=d(\bar{F}(F(\mathbf{x})),\bar{F}(F(\mathbf{y})))=d(F(\mathbf{x}),F(\mathbf{y}))=d(\mathbf{x},\mathbf{y}).$$

5.3.4. Lemma. *If F is an isometry of E^3 such that $F(0)=0$, then F is an orthogonal transformation.*

Proof. Let $\mathbf{x}$ be the position vector of any point with reference to the point $\mathbf{0}$. Then we have, in consequence of (3.2.4),

$$\|F(\mathbf{x})\| = d(\mathbf{0}, F(\mathbf{x})) = d(F(\mathbf{0}), F(\mathbf{x})) = d(\mathbf{0}, \mathbf{x}) = \|\mathbf{x}\|. \tag{5.3.1}$$

Let $\mathbf{y}$ be the position vector of any other point with reference to $\mathbf{0}$. Then

$$\|F(\mathbf{x}) - F(\mathbf{y})\| = d(F(\mathbf{x}), F(\mathbf{y})) = d(\mathbf{x}, \mathbf{y}) = \|\mathbf{x} - \mathbf{y}\|,$$

which, together with (3.2.3), implies

$$(F(\mathbf{x}) - F(\mathbf{y}))\cdot(F(\mathbf{x}) - F(\mathbf{y})) = (\mathbf{x} - \mathbf{y})\cdot(\mathbf{x} - \mathbf{y}),$$

or

$$\|F(\mathbf{x})\|^2 - 2(F(\mathbf{x}))\cdot F(\mathbf{y}) + \|F(\mathbf{y})\|^2 = \|\mathbf{x}\|^2 - 2\mathbf{x}\cdot\mathbf{y} + \|\mathbf{y}\|^2. \tag{5.3.2}$$

Thus from (5.3.1) and (5.3.2) we have

$$(F(\mathbf{x}))\cdot F(\mathbf{y}) = \mathbf{x}\cdot\mathbf{y}, \tag{5.3.3}$$

so that F preserves the inner product of any two vectors.

According to Definition 5.1.2 it remains to prove that F is a linear transformation. Let $\mathbf{u}_1\mathbf{u}_2\mathbf{u}_3$ be the unit natural frame with respect to a right-handed rectangular trihedron $\mathbf{0}\mathbf{x}_1\mathbf{x}_2\mathbf{x}_3$ at the point $\mathbf{0}$ so that $\mathbf{u}_1 = (1,0,0)$, $\mathbf{u}_2 = (0,1,0)$, $\mathbf{u}_3 = (0,0,1)$. Then

$$\mathbf{x} = (x_1, x_2, x_3) = \sum_{i=1}^{3} x_i \mathbf{u}_i. \tag{5.3.4}$$

Since $\{\mathbf{u}_1, \mathbf{u}_2, \mathbf{u}_3\}$ is an orthonormal basis, from (5.3.3) it follows that $\{F(\mathbf{u}_1), F(\mathbf{u}_2), F(\mathbf{u}_3)\}$ is also an orthonormal basis so that

$$F(\mathbf{x}) = \sum_{i=1}^{3} y_i F(\mathbf{u}_i). \tag{5.3.5}$$

Taking the inner product of both sides of (5.3.5) with $F(\mathbf{u}_j)$ and using the orthonormality of the basis $\{F(\mathbf{u}_1), F(\mathbf{u}_2), F(\mathbf{u}_3)\}$, we obtain

$$y_j = (F(\mathbf{x}))\cdot F(\mathbf{u}_j), \qquad j = 1,2,3. \tag{5.3.6}$$

On the other hand, from (5.3.5) and (5.3.4) we have

$$(F(\mathbf{x}))\cdot F(\mathbf{u}_j) = \mathbf{x}(\mathbf{u}_j) = x_j, \qquad j = 1,2,3. \tag{5.3.7}$$

Substituting (5.3.6) and (5.3.7) in (5.3.5) gives

$$F(\mathbf{x}) = \sum_{i=1}^{3} x_i F(\mathbf{u}_i). \tag{5.3.8}$$

Using (5.3.8) we hence obtain the linearity condition

$$F(a\mathbf{x} + b\mathbf{y}) = aF(\mathbf{x}) + bF(\mathbf{y}). \tag{5.3.9}$$

5.3.5. Theorem. *If F is an isometry of E^3, there exist a unique translation H and a unique orthogonal transformation G such that*

$$F=HG, \tag{5.3.10}$$

where G is called the orthogonal part of F, and H the translation part of F.

Proof. Let H be the translation by $F(\mathbf{0})$. By Lemma 5.2.2b, H^{-1} is the translation by $-F(\mathbf{0})$, and by Lemma 5.3.2, $H^{-1}F$ is an isometry. Furthermore

$$(H^{-1}F)(\mathbf{0})=H^{-1}(F(\mathbf{0}))=F(\mathbf{0})-F(\mathbf{0})=\mathbf{0}.$$

Thus by Lemma 5.3.4, $H^{-1}F$ is an orthogonal transformation, say $H^{-1}F=G$, from which follows immediately (5.3.10).

To prove the required uniqueness, suppose $F=\overline{H}\overline{G}$ where $\overline{H}$ is a translation and $\overline{G}$ an orthogonal transformation. Then

$$HG=\overline{H}\overline{G}, \tag{5.3.11}$$

so that $G=H^{-1}\overline{H}\overline{G}$. Since G and $\overline{G}$ are linear transformations, $G(\mathbf{0})=\overline{G}(\mathbf{0})=\mathbf{0}$. It follows that $H^{-1}\overline{H}=I$, the identity mapping, so that $\overline{H}=H$, which together with (5.3.11) implies $\overline{G}=G$. Q.E.D.

By Theorem 5.3.5 and Definition 5.2.4 we see that *an isometry of E^3 is a special affine transformation*, called an *isometric* (or a *rigid*) *affine transformation*. Since the orthogonal transformations of E^3 form a group, from (5.2.5) and (5.2.6) it follows that the totality of isometries (or rigid affine transformations) constitutes a subgroup of the affine group, called *the group of isometries* or *the Euclidean group of rigid motions* of E^3, which is the basis of Euclidean geometry.

The derivative mapping F_*, given by Definition 4.8, of an isometry F: $E^3 \to E^3$ is remarkably simple, as indicated in the following theorem.

5.3.6. Theorem. *Let F be an isometry of E^3 with orthogonal part G. Then*

$$F_*(\mathbf{v}_p)=(G\mathbf{v})_{F(p)} \tag{5.3.12}$$

for all tangent vectors $\mathbf{v}_p$ of E^3 (for a pictorial explanation see Fig. 1.9).

Proof. Write $F=HG$ as in Theorem 5.2.4. Let H be a translation by a vector $\mathbf{c}$, so that $F(\mathbf{p})=\mathbf{c}+G(\mathbf{p})$. If $\mathbf{v}_p$ is a trangent vector of E^3, then by Definition 4.2.1, $F_*(\mathbf{v}_p)$ is the tangent to the curve $t \to F(\mathbf{p}+t\mathbf{v})$ at $t=0$. But

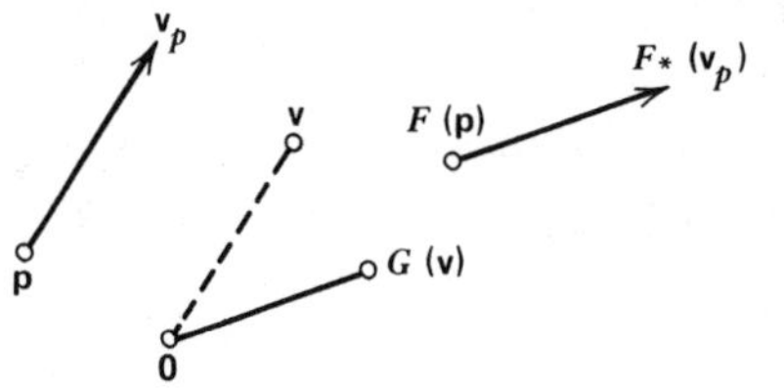

Figure 1.9

using the linearity of G we obtain

$$F(\mathbf{p}+t\mathbf{v})=HG(\mathbf{p}+t\mathbf{v})=H(G(\mathbf{p})+tG(\mathbf{v}))$$
$$=\mathbf{c}+G(\mathbf{p})+tG(\mathbf{v})=F(\mathbf{p})+tG(\mathbf{v}).$$

Hence $$F_*(\mathbf{v}_p)=\frac{d}{dt}F(\mathbf{p}+t\mathbf{v})|_{t=0}=G(\mathbf{v})|_{t=0}=G(\mathbf{v})_{F(p)}.$$ Q.E.D.

We can express (5.3.12) in terms of Euclidean coordinates as

$$F_*\left(\sum_{j=1}^{3} v_j u_j\right)=\sum_{i,j=1}^{3} a_{ij}v_j\bar{u}_i, \tag{5.3.13}$$

where $G=(a_{ij})$ is the orthogonal part of the isometry F, and $\bar{\mathbf{u}}_i$ is evaluated at $F(\mathbf{p})$ when $\mathbf{u}_i$ is evaluated at $\mathbf{p}$.

5.3.7. Corollary. *Isometries of E^3 preserve the inner products of tangent vectors. That is, if $\mathbf{v}_p$ and $\mathbf{w}_p$ are tangent vectors of E^3 at the same point $\mathbf{p}$, and F is an isometry, then*

$$F_*(\mathbf{v}_p)\cdot F_*(\mathbf{w}_p)=\mathbf{v}_p\cdot\mathbf{w}_p. \tag{5.3.14}$$

Proof. Let G be the orthogonal part of F. Then by Theorem 5.3.6 and Definitions 5.1.2 and 3.5.5 we obtain

$$F_*(\mathbf{v}_p)\cdot F_*(\mathbf{w}_p)=(G\mathbf{v})_{F(p)}\cdot(G\mathbf{w})_{F(p)}=(G\mathbf{v})\cdot G\mathbf{w}$$
$$=\mathbf{v}\cdot\mathbf{w}=\mathbf{v}_p.\mathbf{w}_p.$$ Q.E.D.

5.3.8. Remarks. 1. It should be safe to drop the point of application from the notation, and write (5.3.14) simply as $F_*(\mathbf{v})\cdot F_*(\mathbf{w})=\mathbf{v}\cdot\mathbf{w}$.

2. In other words, Corollary 5.3.7 asserts that for each point $\mathbf{p}$ of E^3 the derivative mapping F_{*p} of an isometry F of E^3 at $\mathbf{p}$ is an orthogonal transformation of tangent spaces (differing from G only by the canonical isomorphism of E^3 mentioned in the paragraph just preceding Definition 3.5.5).

From Corollary 5.3.7 it follows automatically that isometries also preserve the concepts derived from the inner product such as norm and orthogonality. Explicitly, if F is an isometry, then $\|F_*(\mathbf{v})\|=\|\mathbf{v}\|$, and if $\mathbf{v}$ and $\mathbf{w}$ are orthogonal, so are $F_*(\mathbf{v})$ and $F_*(\mathbf{w})$. Thus frames (see Definition 3.5.6) are also preserved; that is, if $\mathbf{e}_1\mathbf{e}_2\mathbf{e}_3$ is a frame at some point $\mathbf{p}$ of E^3, and F is an isometry, then $F_*(\mathbf{e}_1)F_*(\mathbf{e}_2)F_*(\mathbf{e}_3)$ is a frame at $F(\mathbf{p})$.

5.3.9. Theorem. *Given any two frames on E^3, say $\mathbf{e}_1\mathbf{e}_2\mathbf{e}_3$ at a point $\mathbf{p}$ and $\mathbf{f}_1\mathbf{f}_2\mathbf{f}_3$ at a point $\mathbf{q}$, there exists a unique isometry F of E^3 such that $F(\mathbf{p})=\mathbf{q}$ and $F_*(\mathbf{e}_i)=\mathbf{f}_i$ for $i=1,2,3$.*

Proof. First we show that there is such an isometry. Let $\bar{\mathbf{e}}_i$ and $\bar{\mathbf{f}}_i$ for $i=1,2,3$ be the points of E^3 canonically corresponding to the vectors $\mathbf{e}_i$

and $\mathbf{f}_i$, respectively. Then there is a unique linear transformation G of E^3 such that $G(\bar{\mathbf{e}}_i)=\bar{\mathbf{f}}_i$ for $i=1,2,3$ (see Exercise 1 of Section 5.1). It is easy to verify that G is orthogonal. Let H be a translation by the point $\mathbf{q}-G(\mathbf{p})$. Then the isometry $F=HG$ carries the frame $\mathbf{e}_1\mathbf{e}_2\mathbf{e}_3$ to the frame $\mathbf{f}_1\mathbf{f}_2\mathbf{f}_3$. In fact, first we have

$$F(\mathbf{q})=H(G(\mathbf{p}))=G(\mathbf{p})+\mathbf{q}-G(\mathbf{p})=\mathbf{q}.$$

Then using Theorem 5.3.6 we obtain

$$F_*(\mathbf{e}_i)=(G\bar{\mathbf{e}}_i)_{F(p)}=(\bar{\mathbf{f}}_i)_{F(p)}=(\bar{\mathbf{f}}_i)_q=\mathbf{f}_i, \qquad \text{for} \quad i=1,2,3.$$

To prove the uniqueness, we notice that by Theorem 5.3.6 only G can be the orthogonal part of the required isometry. The translation part is then completely determined also, since it must carry $G(\mathbf{p})$ to $\mathbf{q}$. Hence the isometry $F=HG$ is uniquely determined. Q.E.D.

The isometry in Theorem 5.3.6 can be computed explicitly as follows. Let $\bar{\mathbf{e}}_i=(e_{i1},e_{i2},e_{i3})$ and $\bar{\mathbf{f}}_i=(f_{i1},f_{i2},f_{i3})$ for $i=1,2,3$. Then by Theorem 5.1.4 the matrices $\mathfrak{E}=(e_{ij})$ and $\mathfrak{F}=(f_{ij})$ are orthogonal, and we claim that the matrix of G in the theorem is ${}^t\mathfrak{F}\mathfrak{E}$. It suffices to show that ${}^t\mathfrak{F}\mathfrak{E}(\bar{\mathbf{e}}_i)=\bar{\mathbf{f}}_i$, since this uniquely characterizes G. By using the column-vector convention we have

$${}^t\mathfrak{F}\mathfrak{E}\begin{pmatrix} e_{11} \\ e_{12} \\ e_{13} \end{pmatrix}={}^t\mathfrak{F}\begin{pmatrix} 1 \\ 0 \\ 0 \end{pmatrix}=\begin{pmatrix} f_{11} \\ f_{12} \\ f_{13} \end{pmatrix};$$

that is, ${}^t\mathfrak{F}\mathfrak{E}(\bar{\mathbf{e}}_1)=\bar{\mathbf{f}}_1$. Similarly, ${}^t\mathfrak{F}\mathfrak{E}(\bar{\mathbf{e}}_i)=\bar{\mathbf{f}}_i$ for $i=2,3$. Thus $G={}^t\mathfrak{F}\mathfrak{E}$. As mentioned above, H must be a translation by $\mathbf{q}-G(\mathbf{p})$.

Exercises

Throughout these exercises, G and $\bar{G}$ denote orthogonal transformations (or their matrices), $\mathbf{c},\mathbf{d},\mathbf{p},\mathbf{q}$ and $\mathbf{x}=(x_1,x_2,x_3)$ are vectors of E^3, and H_a is a translation of E^3 by a fixed vector $\mathbf{a}$.

1. Prove that $GH_c=H_{G(c)}G$.
2. Given isometries $F=H_cG$ and $\bar{F}=H_d\bar{G}$, find the translation and orthogonal parts of $F\bar{F}$ and $\bar{F}F$.
3. Let $\mathbf{c}=(2,3,-1)$ and

$$G=\begin{pmatrix} \dfrac{1}{\sqrt{2}} & 0 & -\dfrac{1}{\sqrt{2}} \\ 0 & 1 & 0 \\ \dfrac{1}{\sqrt{2}} & 0 & \dfrac{1}{\sqrt{2}} \end{pmatrix}.$$

If $\mathbf{p}=(3,-2,5)$, find the coordinates of the point $\mathbf{q}$ for which (a) $\mathbf{q}=H_cG(\mathbf{p})$; (b) $\mathbf{q}=(H_cG)^{-1}(\mathbf{p})$; (c) $\mathbf{q}=GH_c(\mathbf{p})$.

4. In each case decide whether F is an isometry of E^3. If isometry exists, find the translation and orthogonal parts.
(a) $F(\mathbf{x})=-\mathbf{x}$. (b) $F(\mathbf{x})=(\mathbf{x}\cdot\mathbf{c})\mathbf{c}$, where $\|\mathbf{c}\|=1$.
(c) $F(\mathbf{x})=(x_3-3, x_2-2, x_1+1)$. (d) $F(\mathbf{x})=(x_1, x_2, 2)$.

5. If H is a translation of E^3, then for every tangent vector $\mathbf{v}$ of E^3 show that $H_*(\mathbf{v})$ is parallel to $\mathbf{v}$.

6. Prove the general formulas $(\bar{F}F)_*=\bar{F}_*F$ and $(F^{-1})_*=(F_*)^{-1}$ for the special case where F and $\bar{F}$ are isometries of E^3.

7. (a) Let $\mathbf{e}_1\mathbf{e}_2\mathbf{e}_3$ be a frame at a point $\mathbf{p}$, and let $\mathfrak{E}=(\mathbf{e}_1,\mathbf{e}_2,\mathbf{e}_3)$ given by (3.5.7). If F is the isometry that carries the natural frame $\mathbf{u}_1\mathbf{u}_2\mathbf{u}_3$ at the point $\mathbf{0}$ to this frame $\mathbf{e}_1\mathbf{e}_2\mathbf{e}_3$, show that $F=H_p\mathfrak{E}^{-1}$.
(b) Let $\mathbf{f}_1\mathbf{f}_2\mathbf{f}_3$ be a frame at a point $\mathbf{q}$, and let $\mathfrak{F}=(\mathbf{f}_1,\mathbf{f}_2,\mathbf{f}_3)$. Use Exercise 6 to prove the result in the text that the isometry that carries the frame $\mathbf{e}_1\mathbf{e}_2\mathbf{e}_3$ to the frame $\mathbf{f}_1\mathbf{f}_2\mathbf{f}_3$ has the orthogonal part $\mathfrak{F}^{-1}\mathfrak{E}$.

8. (a) Prove that an isometry $F=HG$ carries the plane through a point $\mathbf{p}$ orthogonal to the vector $\mathbf{q}$ to the plane through $F(\mathbf{p})$ orthogonal to $G(\mathbf{q})$.
(b) If π is the plane through $(\frac{1}{2},-1,0)$ orthogonal to $(0,1,0)$, find an isometry $F=HG$ such that $F(\pi)$ is the plane through $(1,-2,1)$ orthogonal to $(1,0,-1)$.

9. Given $\mathbf{e}_1=\frac{1}{3}(2,2,1)$, $\mathbf{e}_2=\frac{1}{3}(-2,1,2)$, $\mathbf{e}_3=\frac{1}{3}(1,-2,2)$ at the point $\mathbf{p}=(0,1,0)$, and $\mathbf{f}_1=(1/\sqrt{2})(1,0,1)$, $\mathbf{f}_2=(0,1,0)$, $\mathbf{f}_3=(1/\sqrt{2})(1,0,-1)$ at the point $\mathbf{q}=(3,-1,1)$, find the isometry $F=HG$ that carries the frame $\mathbf{e}_1\mathbf{e}_2\mathbf{e}_3$ to the frame $\mathbf{f}_1\mathbf{f}_2\mathbf{f}_3$.

5.4. Orientations. In Section 1.1 we introduced a right-handed rectangular trihedron in E^3. Similarly, we can define a left-handed rectangular trihedron in E^3, so that we obtain an intuitive orientation of a trihedron in E^3. Replacing the trihedrons by frames gives the following definition of a mathematical orientation.

5.4.1. Definition. Let $\mathbf{e}_1\mathbf{e}_2\mathbf{e}_3$ be a frame in E^3 at a point $\mathbf{p}$. Then from (3.3.7) and Definition 3.5.6 it follows that

$$|\mathbf{e}_1,\mathbf{e}_2,\mathbf{e}_3|=\det\mathfrak{E}=\pm 1, \tag{5.4.1}$$

where $\mathfrak{E}$ is the matrix $(\mathbf{e}_1,\mathbf{e}_2,\mathbf{e}_3)$ given by (3.5.7). The frame $\mathbf{e}_1\mathbf{e}_2\mathbf{e}_3$ is *positively oriented* (or, *right-handed*) or *negatively oriented* (or, *left-handed*) according as det $\mathfrak{E}$ is $+1$ or -1.

The following results are obvious.

5.4.2. Remarks. 1. At each point $\mathbf{p}$ of E^3 the frame assigned by the natural frame field $\mathbf{u}_1\mathbf{u}_2\mathbf{u}_3$ given by (3.5.4) is positively oriented.

2. A frame $\mathbf{e}_1\mathbf{e}_2\mathbf{e}_3$ is positively oriented if and only if $\mathbf{e}_1 \times \mathbf{e}_2 = \mathbf{e}_3$.
3. For a positively oriented frame $\mathbf{e}_1\mathbf{e}_2\mathbf{e}_3$, the vector products are

$$\mathbf{e}_1 = \mathbf{e}_2 \times \mathbf{e}_3 = -\mathbf{e}_3 \times \mathbf{e}_2,$$
$$\mathbf{e}_2 = \mathbf{e}_3 \times \mathbf{e}_1 = -\mathbf{e}_1 \times \mathbf{e}_3,$$
$$\mathbf{e}_3 = \mathbf{e}_1 \times \mathbf{e}_2 = -\mathbf{e}_2 \times \mathbf{e}_1. \tag{5.4.2}$$

For a negatively oriented frame $\mathbf{e}_1\mathbf{e}_2\mathbf{e}_3$, reverse the vectors in each of the vector products above.

Next, we use (5.1.13) to attach a sign to each isometry of E^3.

5.4.3. Definition. The *sign* of an isometry F of E^3 is the determinant of the matrix of the orthogonal part G of F, and is denoted by

$$\operatorname{sgn} F = \det G. \tag{5.4.3}$$

The derivative mapping of an isometry carries a frame to a frame, and the following lemma gives its effect on the orientation of a frame.

5.4.4. Lemma. *If* $\mathbf{e}_1\mathbf{e}_2\mathbf{e}_3$ *is a frame at a point of* E^3, *and* F *is an isometry, then*

$$|F_*(\mathbf{e}_1), F_*(\mathbf{e}_2), F_*(\mathbf{e}_3)| = (\operatorname{sgn} F)|\mathbf{e}_1, \mathbf{e}_2, \mathbf{e}_3|. \tag{5.4.4}$$

Proof. Let $\mathbf{e}_j = \Sigma_{k=1}^3 e_{jk}\mathbf{u}_k$. Then using (5.3.13) we obtain

$$F_*(\mathbf{e}_j) = \sum_{i,k=1}^{3} a_{ik} e_{jk} \bar{\mathbf{u}}_i, \tag{5.4.5}$$

where $G = (a_{ij})$ is the orthogonal part of F. Thus the matrix of the frame $F_*(\mathbf{e}_1)F_*(\mathbf{e}_2)F_*(\mathbf{e}_3)$ is

$$\left(\sum_{k=1}^{3} a_{ik} e_{jk} \right) = \left(\sum_{k=1}^{3} a_{ik}\, {}^t e_{kj} \right) = G\,{}^t\mathfrak{E}, \tag{5.4.6}$$

where ${}^t\mathfrak{E}$ is the transpose of $\mathfrak{E} = (e_{ij})$. From (5.4.6) and (5.4.1) it follows that

$$|F_*(\mathbf{e}_1), F_*(\mathbf{e}_2), F_*(\mathbf{e}_3)| = \det(G^t\mathcal{E}) = (\det G)\det\mathcal{E}$$
$$= |\operatorname{sgn} F\,||\mathbf{e}_1, \mathbf{e}_2, \mathbf{e}_3|,$$

where the last step follows from (5.4.1 and (5.4.3).

By using Definition 5.4.1 we readily obtain a corollary.

5.4.5. Corollary. *Let F be an isometry of E^3. If* sgn $F = +1$, *then F_* carries positively (respectively, negatively) oriented frames to positively (respectively, negatively) oriented frames. On the other hand, if* sgn $F = -1$, *positive goes to negative, and negative to positive.*

5.4.6. Definition. An isometry F of E^3 is *orientation preserving* (or a *proper motion*) if sgn $F = +1$, and is *orientation reversing* (or an *improper motion*) if sgn $F = -1$.

5.4.7. Examples. 1. *Translations.* All translations are orientation preserving. In fact, from (5.2.1) the orthogonal part of a translation F is just the identity mapping I, so that sgn $F = \det I = +1$.

2. *Rotations.* Consider the orthogonal transformation F given by (5.1.17), which rotates E^3 clockwise through angle θ around the x_3-axis. The matrix of F is

$$\begin{pmatrix} \cos\theta & \sin\theta & 0 \\ -\sin\theta & \cos\theta & 0 \\ 0 & 0 & 1 \end{pmatrix}.$$

Thus sgn $F = \det F = +1$ showing that F is orientation preserving.

3. *Reflections.* A reflection in a given plane, say $\mathbf{x}_2\mathbf{x}_3$-plane, is a mapping F such that $F(\mathbf{p}) = (-p_1, p_2, p_3)$ for all points $\mathbf{p} = (p_1, p_2, p_3)$ of E^3. Obviously, F is an orthogonal transformation with matrix

$$\begin{pmatrix} -1 & 0 & 0 \\ 0 & 1 & 0 \\ 0 & 0 & 1 \end{pmatrix},$$

and is therefore an orientation-reversing isometry, as confirmed by the fact that the mirror image of a right hand is a left hand.

5.4.8. Lemma. *Let $\mathbf{e}_1\mathbf{e}_2\mathbf{e}_3$ be a frame at a point of E^3. If*

$$\mathbf{v} = \sum_{i=1}^{3} v_i\mathbf{e}_i, \qquad \mathbf{w} = \sum_{i=1}^{3} w_i\mathbf{e}_i, \tag{5.4.7}$$

then the vector product of $\mathbf{v}$ *and* $\mathbf{w}$ *is*

$$\mathbf{v}\times\mathbf{w}=\mathcal{E}\begin{vmatrix}\mathbf{e}_1 & \mathbf{e}_2 & \mathbf{e}_3\\ v_1 & v_2 & v_3\\ w_1 & w_2 & w_3\end{vmatrix}, \tag{5.4.8}$$

where $\mathcal{E}=|\mathbf{e}_1,\mathbf{e}_2,\mathbf{e}_3|=\pm 1$.

Proof. Equation 5.4.8 follows easily from (5.4.7), (3.3.5), (5.4.2), and the corresponding equations for a negatively oriented frame $\mathbf{e}_1\mathbf{e}_2\mathbf{e}_3$.

The effect of an isometry on the vector products is the content of the following theorem.

5.4.9. Theorem. *Let* $\mathbf{v}$ *and* $\mathbf{w}$ *be tangent vectors of* E^3 *at a point* $\mathbf{p}$. *If* F *is an isometry of* E^3, *then*

$$F_*(\mathbf{v}\times\mathbf{w})=(\operatorname{sgn} F)F_*(\mathbf{v})\times F_*(\mathbf{w}). \tag{5.4.9}$$

Proof. Write $\mathbf{v}=\Sigma_{i=1}^3 v_i\mathbf{u}_i(\mathbf{p})$ and $\mathbf{w}=\Sigma_{i=1}^3 w_i\mathbf{u}_i(\mathbf{p})$, and let $\mathbf{e}_i=F_*(\mathbf{u}_i(\mathbf{p}))$. Since F_* is linear, we have

$$F_*(\mathbf{v})=\sum_{i=1}^3 v_i\mathbf{e}_i,\qquad F_*(\mathbf{w})=\sum_{i=1}^3 w_i\mathbf{e}_i.$$

By means of Lemma 5.4.8, a straightforward computation shows that

$$F_*(\mathbf{v})\times F_*(\mathbf{w})=\mathcal{E}F_*(\mathbf{v}\times\mathbf{w}),$$

where

$$\mathcal{E}=|\mathbf{e}_1,\mathbf{e}_2,\mathbf{e}_3|=|F_*(\mathbf{u}_1(\mathbf{p})),\ F_*(\mathbf{u}_2(\mathbf{p})),\ F_*(\mathbf{u}_3(\mathbf{p}))|.$$

Since $\mathbf{u}_1\mathbf{u}_2\mathbf{u}_3$ is positively oriented, by Lemma 5.4.4 we obtain

$$\mathcal{E}=(\operatorname{sgn} F)|\mathbf{u}_1(\mathbf{p}),\mathbf{u}_2(\mathbf{p}),\mathbf{u}_3(\mathbf{p})|=\operatorname{sgn} F;$$

hence (5.4.9) is proved.

5.4.10. Theorem. *Let* $\mathbf{u}$, $\mathbf{v}$, *and* $\mathbf{w}$ *be tangent vectors of* E^3 *at a point* $\mathbf{p}$. *If* F *is an isometry of* E^3, *then*

$$|F_*(\mathbf{u}),\ F_*(\mathbf{v}),\ F_*(\mathbf{w})|=(\operatorname{sgn} F)|\mathbf{u},\mathbf{v},\mathbf{w}|. \tag{5.4.10}$$

Proof. Using Corollary 5.3.7 and Theorem 5.4.9, and noticing that $|\mathbf{u},\mathbf{v},\mathbf{w}|=\mathbf{u}\cdot\mathbf{v}\times\mathbf{w}$, we obtain (5.4.10) immediately.

Exercises

1. Let F and $\bar{F}$ be isometries of E^3. Prove that
$$\operatorname{sgn}(F\bar{F}) = (\operatorname{sgn} F)\operatorname{sgn} \bar{F} = \operatorname{sgn}(\bar{F}F),$$
and deduce that $\operatorname{sgn} F = \operatorname{sgn}(F^{-1})$.
2. If F_0 is an orientation-reversing isometry of E^3, show that every orientation-reversing isometry has a unique expression F_0F, where F is an orientation-preserving isometry.
3. Let $\mathbf{v} = (3, 1, -1)$ and $\mathbf{w} = (-3, -3, 1)$ be tangent vectors of E^3 at a point $\mathbf{p}$. Show that
$$G = \begin{pmatrix} -\frac{2}{3} & \frac{2}{3} & -\frac{1}{3} \\ \frac{2}{3} & \frac{1}{3} & -\frac{2}{3} \\ \frac{1}{3} & \frac{2}{3} & \frac{2}{3} \end{pmatrix}$$
is orthogonal, and verify the formula.
$$G_*(\mathbf{v} \times \mathbf{w}) = (\operatorname{sgn} G) G_*(\mathbf{v}) \times G_*(\mathbf{w}).$$

*4. Prove that a rotation F (i.e., an orthogonal transformation F with det $F = +1$) does rotate E^3 around an axis. Explicitly, given a rotation F, show that there exist a number θ and points $\mathbf{e}_1, \mathbf{e}_2, \mathbf{e}_3$ with $\mathbf{e}_i \cdot \mathbf{e}_j = \delta_{ij}$ such that
$$F(\mathbf{e}_1) = \mathbf{e}_1 \cos\theta + \mathbf{e}_2 \sin\theta,$$
$$F(\mathbf{e}_2) = -\mathbf{e}_1 \sin\theta + \mathbf{e}_2 \cos\theta,$$
$$F(\mathbf{e}_3) = \mathbf{e}_3.$$

*5. Let $\mathbf{c}$ be a point of E^3 such that $\|\mathbf{c}\| = 1$. Prove that the formula
$$G(\mathbf{p}) = \mathbf{c} \times \mathbf{p} + \mathbf{p} \cdot \mathbf{c}\mathbf{c}$$
for every point $\mathbf{p}$ of E^3 defines an orthogonal transformation. Describe its general effect on E^3.

6. Prove that all orientation-preserving isometries of E^3 form a normal subgroup of index 2 in the group of isometries of E^3.
7. Let F_1 and F_2 be two orientation-preserving isometries of E^3. Prove that there is C^0-family of orientation-preserving isometries $F(t)$, $0 \leqslant t \leqslant 1$, of E^3 such that $F(0) = F_1$ and $F(1) = F_2$.

6. DIFFERENTIAL FORMS

6.1. 1-Forms

6.1.1. Definition. A *1-form* ϕ on E^3 is a real-valued function on the set $T(E^3)$ of all tangent vectors of E^3 such that ϕ is linear at each point, that is,

$$\phi(a\mathbf{v}+b\mathbf{w})=a\phi(\mathbf{v})+b\phi(\mathbf{w}) \tag{6.1.1}$$

for any numbers a, b and tangent vectors $\mathbf{v}, \mathbf{w}$ at the same point of E^3. [Thus at each point $\mathbf{p}$, ϕ_p is an element of the dual space of $T_p(E^3)$. In this sense the notion of 1-form is dual to that of vector field.]

On E^3, let ϕ and ψ be 1-forms, and f a real-valued function. Then $\phi+\psi$ and $f\phi$ are defined as follows:

$$(\phi+\psi)(\mathbf{v})=\phi(\mathbf{v})+\psi(\mathbf{v}) \qquad \text{for all tangent vectors } \mathbf{v}, \tag{6.1.2}$$

$$(f\phi)(\mathbf{v}_p)=f(\mathbf{p})\phi(\mathbf{v}_p) \qquad \text{for all tangent vectors } \mathbf{v}_p \text{ at a point } \mathbf{p}. \tag{6.1.3}$$

For a vector field $\mathbf{v}$ on E^3, at each point $\mathbf{p}$ of E^3 the value of $\phi(\mathbf{v})$ is the number $\phi(\mathbf{v}_p)$. If $\phi(\mathbf{v})$ is *differentiable* whenever $\mathbf{v}$ is, then ϕ is said to be *differentiable*. As with vector fields, all 1-forms will be assumed to be differentiable. From the definitions it follows readily that $\phi(\mathbf{v})$ is linear in both ϕ and $\mathbf{v}$, that is,

$$\phi(f\mathbf{v}+g\mathbf{w})=f\phi(\mathbf{v})+g\phi(\mathbf{w}), \tag{6.1.4}$$

$$(f\phi+g\psi)(\mathbf{v})=f\phi(\mathbf{v})+g\psi(\mathbf{v}), \tag{6.1.5}$$

where $\mathbf{w}$ is a vector field, and f, g are functions.

In terms of directional derivatives, a most important way of assigning 1-forms to a function is given in Definition 6.1.2.

6.1.2. Definition. If f is a differentiable real-valued function on E^3, the *differential df* of f is the 1-form such that

$$df(\mathbf{v}_p)=\mathbf{v}_p[f] \qquad \text{for all tangent vectors } \mathbf{v}_p. \tag{6.1.6}$$

In fact, df is a 1-form, since by Definition 3.6.1 it is a real-valued function on tangent vectors, and by Theorem 3.6.3a it is linear at each point $\mathbf{p}$.

6.1.3. Example. *1-forms on* E^3. 1. The differentials dx_1, dx_2, dx_3 of the natural coordinate functions x_1, x_2, x_3. By Lemma 3.6.2, from (6.1.6) we obtain

$$dx_i(\mathbf{v}_p)=\mathbf{v}_p[x_i]=\sum_{j=1}^{3} v_j\frac{\partial x_i}{\partial x_j}(\mathbf{p})=\sum_{j=1}^{3} v_j\delta_{ij}=v_i. \tag{6.1.7}$$

Thus *the value of* dx_i *on an arbitrary tangent vector* $\mathbf{v}_p$ *is the ith coordinate* v_i *of its vector part*, and does not depend on the point of application $\mathbf{p}$.

2. The 1-form $\phi=\sum_{i=1}^3 f_i\,dx_i$. Since dx_i is a 1-form, from (6.1.2) and (6.1.3) it follows that ϕ is also a 1-form for any functions f_1, f_2, f_3. The value of ϕ on an arbitrary tangent vector $\mathbf{v}_p$ is

$$\phi(\mathbf{v}_p)=\left(\sum_{i=1}^{3} f_i\,dx_i\right)(\mathbf{v}_p)=\sum_{i=1}^{3} f_i(\mathbf{p})\,dx_i(\mathbf{v})=\sum_{i=1}^{3} f_i(\mathbf{p})v_i. \tag{6.1.8}$$

Example 1 shows that the 1-forms dx_1, dx_2, dx_3 are the analogues of the tangent vectors of the natural coordinate functions x_1, x_2, x_3 for points. Actually, dx_1, dx_2, dx_3 are the *duals* of the natural unit vector fields $\mathbf{u}_1, \mathbf{u}_2, \mathbf{u}_3$ since from (6.1.7) we have (see Corollary 4.1.11)

$$dx_i(\mathbf{u}_j)=\delta_{ij}. \tag{6.1.9}$$

6.1.4. Lemma. *Let* x_1, x_2, x_3 *be the natural coordinate functions for points in* E^3. *Then* dx_1, dx_2, dx_3 *constitute a basis of the dual space of the tangent space* $T(E^3)$, *that is,*

(a) dx_1, dx_2, dx_3 *are linearly independent, and*
(b) *if* ϕ *is a 1-form on* E^3, *then*

$$\phi=\sum_{i=1}^{3} f_i\,dx_i, \tag{6.1.10}$$

where

$$f_i=\phi(\mathbf{u}_i). \tag{6.1.11}$$

These functions f_1, f_2, f_3 are called the *Euclidean coordinate functions of* ϕ.

Proof. (a) Suppose that there are numbers c_1, c_2, c_3 such that $\sum_{i=1}^3 c_i\,dx_i=0$. Then $\sum_{i=1}^3 c_i\,dx_i(\mathbf{v}_p)=0$ for every tangent vector $\mathbf{v}_p$, and, in particular, $\sum_{i=1}^3 c_i\,dx_i(u_j)=0$, for $j=1,2,3$. From (6.1.9) it follows that $c_j=0$ for $j=1,2,3$. Thus dx_1, dx_2, dx_3 are linearly independent.

(b) Since by definition a 1-form is a function on tangent vectors, ϕ and $\sum_{i=1}^3 f_i\,dx_i$ are equal if and only if they have the same value on every tangent vector

$$\mathbf{v}_p=\sum_{i=1}^{3} v_i\mathbf{u}_i(\mathbf{p}). \tag{6.1.12}$$

For a 1-form ϕ, from (6.1.12), (6.1.1), and (6.1.11) it follows readily that

$$\phi(\mathbf{v}_p)=\phi\left(\sum_{i=1}^{3} v_i\mathbf{u}_i(\mathbf{p})\right)=\sum_{i=1}^{3} v_i\phi(\mathbf{u}_i(\mathbf{p}))=\sum_{i=1}^{3} v_i f_i(\mathbf{p}),$$

which is equal to $(\sum_{i=1}^3 f_i\,dx_i)(\mathbf{v}_p)$ by (6.1.8). Hence (6.1.10) is established.

6.1.5. **Corollary.** *If f is a differentiable function on E^3, then*

$$df=\sum_{i=1}^{3}\frac{\partial f}{\partial x_i}dx_i. \qquad (6.1.13)$$

Proof. The value of the right-hand side of (6.1.13) on an arbitrary tangent vector $\mathbf{v}_p$ given by (6.1.12) is $\Sigma_{i=1}^{3}(\partial f/\partial x_i)(\mathbf{p})v_i$, which is equal to $df(\mathbf{v}_p)$ by Lemma 3.6.2 and Definition 6.1.2.

Either Corollary 6.1.5 or Definition 6.1.2 gives immediately that for differentiable functions f and g

$$d(f+g)=df+dg.$$

Substituting fg and $h(f)$ in turn for f in Corollary 6.1.5 and using the chain rule (Theorem 2.3.6) for the latter case, we can easily obtain the following two lemmas about the effect of d on products and compositions of functions.

6.1.6. **Lemma.** *Let fg be the product of differentiable functions f and g on E^3. Then*

$$d(fg)=g\,df+f\,dg. \qquad (6.1.14)$$

6.1.7. **Lemma.** *Let f: $E^3\to E^1$ and h: $E^1\to E^1$ be differentiable functions, so that the composite function $h(f)$: $E^3\to E^1$ is also differentiable. Then*

$$d(h(f))=h'(f)\,df, \qquad (6.1.15)$$

where the prime denotes the ordinary derivative.

To compute df for a given function f we can also use (6.1.14) and (6.1.15) rather than substitute in (6.1.13). Then by means of (6.1.6) we can obtain all the directional derivatives of f. For example, suppose

$$f=(1-y^2)x+(z^2+2)y.$$

From (6.1.14) and (6.1.15) it follows immediately that

$$df=(1-y^2)\,dx+(-2xy+z^2+2)\,dy+2yz\,dz.$$

Thus for $\mathbf{v}=(v_1,v_2,v_3)$ and $\mathbf{p}=(p_1,p_2,p_3)$ we obtain, in consequence of (6.1.8),

$$\mathbf{v}[f]=df(\mathbf{v}_p)=(1-p_2^2)v_1+(-2p_1p_2+p_3^2+2)v_2+2p_2p_3v_3.$$

Exercises

1. Let $\mathbf{v}=(3,2,-1)$ and $\mathbf{p}=(0,-5,2)$. Evaluate the following 1-forms on the tangent vector $\mathbf{v}_p$: (a) $z^2\,dy$; (b) $y\,dx-z\,dy$; (c) $y\,dx+(1-z^2)\,dy-x^2\,dz$.

2. If $\phi=\Sigma_{i=1}^{3} f_i\,dx_i$ and $\mathbf{v}=\Sigma_{i=1}^{3} v_i\mathbf{u}_i$, show that the 1-form ϕ evaluated on the vector field $\mathbf{v}$ is the function $\phi(\mathbf{v})=\Sigma_{i=1}^{3} f_i v_i$.

3. Evaluate the 1-form $\phi=yz\,dx-xz\,dy$ on the vector fields:

(a) $\mathbf{v}=y\mathbf{u}_1+z\mathbf{u}_2+x\mathbf{u}_3$.
(b) $\mathbf{w}=y(\mathbf{u}_1-\mathbf{u}_3)+z(\mathbf{u}_1-\mathbf{u}_2)$.
(c) $(1/x)\mathbf{v}+(1/y)\mathbf{w}$.

4. Express the following differentials in terms of df: (a) $d(f^5)$, and (b) $d(\sqrt{f})$, where $f>0$.

5. In each of the following cases, compute the differential of f and find the directional derivative $\mathbf{v}_p[f]$ for $\mathbf{v}=(3,2,-1)$ and $\mathbf{p}=(1,5,2)$: (a) $f=xz^2-y^2z$; (b) $f=ye^{xz}$; (c) $f=\sin(xy)+\cos(xz)$.

6. Which of the following are 1-forms? In each case ϕ is the function on tangent vectors such that the value of ϕ on $\mathbf{v}_p=(v_1,v_2,v_3)_p$ for $\mathbf{p}=(p_1,p_2,p_3)$ is (a) v_2-v_3; (b) p_2-p_3; (c) $v_1p_2+v_3p_1$; (d) $\mathbf{v}_p[x_1^2-x_2^2+x_3^2]$; (e) 0; (f) $(p_2)^2$. In case ϕ is a 1-form, express it as $\Sigma_{i=1}^{3} f_i\,dx_i$.

7. Prove Lemma 6.1.6 directly from Definition 6.1.2.

8. A 1-form ϕ is *zero* at a point $\mathbf{p}$ if $\phi(\mathbf{v}_p)=0$ for all tangent vectors $\mathbf{v}$ at $\mathbf{p}$. Use Lemma 3.6.2 to prove that $\mathbf{p}$ is a critical point (see Definition 2.3.3) of a function f if and only if $df=0$ at $\mathbf{p}$.

6.2. Exterior Multiplication and Differentiation. To study differential forms of higher degrees on E^3, we first have

6.2.1. Definition. The *exterior* or *wedge multiplication* $\wedge$ on the differentials dx_1, dx_2, dx_3 of the natural coordinate functions x_1, x_2, x_3 of E^3 is associative, distributive, and anticommutative, that is, for all $i, j, k=1,2,3$, it satisfies

$$(dx_i\wedge dx_j)\wedge dx_k=dx_i\wedge(dx_j\wedge dx_k), \tag{6.2.1}$$

$$(dx_i+dx_j)\wedge dx_k=dx_i\wedge dx_k+dx_j\wedge dx_k, \tag{6.2.2}$$

$$dx_i\wedge dx_j=-dx_j\wedge dx_i. \tag{6.2.3}$$

From (6.2.3) we immediately have

$$dx_i\wedge dx_i=0. \tag{6.2.4}$$

We shall further restrict our definition by assuming

$$dx_i\wedge dx_j\neq 0 \qquad \text{for} \quad i\neq j. \tag{6.2.5}$$

Using (6.2.3) and (6.2.4) we can easily see that for $k\geqslant 2$

$$dx_{i_1}\wedge\cdots\wedge dx_{i_k}$$

cannot be zero unless at least one index i_j is repeated. Moreover, for any permutation σ of $(1,\cdots,k)$, $k\geq 2$, we have

$$dx_{i_{\sigma(1)}}\wedge\cdots\wedge dx_{i_{\sigma(k)}}=\varepsilon_\sigma dx_{i1}\wedge\cdots\wedge dx_{i_k}, \tag{6.2.6}$$

where ε_σ is the sign of the permutation σ, which is $+1$ or -1 depending on whether σ is even or odd.

To the exterior multiplication of differentials above we also adjoin

$$f\wedge dx_i=dx_i\wedge f=f\,dx_i,$$
$$dx_i\wedge(f\,dx_j)=f\,dx_i\wedge dx_j, \tag{6.2.7}$$

where f is a real function on E^3.

Including 1-forms defined in Section 6.1, a p-form or a differential form of degree p, $p=0,1,2,3$, on E^3 is defined as follows.

6.2.2. Definition. Let f, f_1, f_2, f_3 be differentiable functions on E^3. On E^3, a *0-form* is just an f, a *1-form* is an expression $\Sigma_{i=1}^3 f_i\,dx_i$ (according to Lemma 6.1.4), a 2-form is an expression $f_3\,dx_1\wedge dx_2+f_2\,dx_1\wedge dx_3+f_1\,dx_2\wedge dx_3$, and a *3-form* is an expression $f\,dx_1\wedge dx_2\wedge dx_3$. A form is *zero* (respectively, *of class* C^r) if all the coefficient functions of its expression are zero (respectively, of class C^r).

Every p-form with $p>3$ *on* E^3 *is zero*, since in each term of its expression some dx_i must be repeated.

From the definition of a vanishing form we can easily prove Lemma 6.2.3.

6.2.3. Lemma. *Two forms of the same degree with respect to the same variables are equal if and only if all their corresponding coefficient functions are equal.*

Section 6.1 defined the sum of two 1-forms

$$\phi=\sum_{i=1}^{3} f_i\,dx_i, \qquad \psi=\sum_{i=1}^{3} g_i\,dx_i \tag{6.2.8}$$

to be

$$\phi+\psi=\sum_{i=1}^{3}(f_i+g_i)\,dx_i. \tag{6.2.9}$$

Similarly, we have definitions of the sums of two 2-forms and of two 3-forms.

6.2.4. Definition. The *sum of two 2-forms*

$$\omega^2=\sum_{i,j=1}^{3} f_{ij}\,dx_i\wedge dx_j, \qquad \bar{\omega}^2=\sum_{i,j=1}^{3} \bar{f}_{ij}\,dx_i\wedge dx_j \tag{6.2.10}$$

is

$$\omega^2+\bar{\omega}^2=\sum_{i,j=1}^{3}\left(f_{ij}+\bar{f}_{ij}\right)dx_i\wedge dx_j, \tag{6.2.11}$$

and the *sum of two 3-forms*

$$\omega^3=f\,dx_1\wedge dx_2\wedge dx_3, \qquad \bar{\omega}^3=\bar{f}\,dx_1\wedge dx_2\wedge dx_3 \tag{6.2.12}$$

is

$$\omega^3+\bar{\omega}^3=\left(f+\bar{f}\,\right)dx_1\wedge dx_2\wedge dx_3. \tag{6.2.13}$$

By using Definition 6.2.1 and (6.2.4) and (6.2.7), we can easily compute the exterior product of two forms, and in general we obtain Lemma 6.2.5.

6.2.5. Lemma. *On E^3 the exterior product of a p-form and a q-form is a $(p+q)$-form, so that such a product is automatically zero whenever $p+q>3$.*

6.2.6. Lemma. *If ϕ and ψ are 1-forms, then*

$$\phi\wedge\psi=-\psi\wedge\phi. \tag{6.2.14}$$

Proof of Lemma 6.2.6. Let ϕ and ψ be expressed by (6.2.8). Then by (6.2.7) and Definition 6.2.1 we have

$$\phi\wedge\psi=\sum_{i,j=1}^{3} f_i g_j\,dx_i\wedge dx_j=-\sum_{i,j=1}^{3} g_j f_i\,dx_j\wedge dx_i=-\psi\wedge\phi. \tag{6.2.15}$$

Q.E.D.

The operator d of Definition 6.1.2 assigns a 1-form df to a 0-form f of class C^1. We can generalize d to an operator, also denoted by d, which assigns a $(p+1)$-form $d\omega^p$ to a p-form ω^p of class C^1 by simply applying d to the coefficient functions of the expression of ω^p. More precisely we have Definition 6.2.7.

6.2.7. Definition. The *exterior derivatives* of forms ϕ and ω^2 of class C^1 expressed by (6.2.8) and (6.2.10) are

$$d\phi=\sum_{i=1}^{3} df_i\wedge dx_i, \qquad d\omega^2=\sum_{i,j=1}^{3} df_{ij}\wedge dx_i\wedge dx_j. \tag{6.2.16}$$

From Definition 6.2.7 it follows at once that the exterior derivative of any form of degree greater than 3 is zero. By using Corollary 6.1.5 we can easily obtain the exterior derivative of the 1-form expressed by (6.2.8):

$$d\phi=\left(\frac{\partial f_2}{\partial x_1}-\frac{\partial f_1}{\partial x_2}\right)dx_1\wedge dx_2+\left(\frac{\partial f_3}{\partial x_1}-\frac{\partial f_1}{\partial x_3}\right)dx_1\wedge dx_3$$
$$+\left(\frac{\partial f_3}{\partial x_2}-\frac{\partial f_2}{\partial x_3}\right)dx_2\wedge dx_3. \tag{6.2.17}$$

It is also easy to verify that the exterior differentiation is linear (this is true for the particular case in Definition 6.1.2), that is,

$$d(a\phi+b\psi)=ad\phi+bd\psi, \tag{6.2.18}$$

where ϕ and ψ are arbitrary forms, and a and b are numbers.

The following theorem shows that the exterior differentiation does not always satisfy the ordinary product formula.

6.2.8. Theorem. *Let f be a C^1 function, ϕ and ψ 1-forms of class C^1, and ω an arbitrary C^1 form. Then*

$$d(\phi\wedge\psi)=d\phi\wedge\psi-\phi\wedge d\psi, \tag{6.2.19}$$

$$d(f\omega)=df\wedge\omega+fd\omega. \tag{6.2.20}$$

Proof. Let ϕ and ψ be expressed by (6.2.8). Then by (6.2.15), (6.2.16), and (6.2.14) we obtain

$$\begin{aligned} d(\phi\wedge\psi)&=\sum_{i,j=1}^{3} d(f_i g_j)\wedge dx_i\wedge dx_j\\ &=\sum_{i,j=1}^{3}\left[(df_i)g_j+f_i dg_j\right]\wedge dx_i\wedge dx_j\\ &=\sum_{i,j=1}^{3} df_i\wedge dx_i\wedge g_j dx_j-\sum_{i,j=1}^{3} f_i dx_i\wedge dg_j\wedge dx_j\\ &=d\phi\wedge\psi-\phi\wedge d\psi. \end{aligned}$$

The proof of (6.2.20) will be left as an exercise.

6.2.9. Theorem (H. Poincaré). *If ϕ is an arbitrary C^2 form, then*

$$d^2\phi=0. \tag{6.2.21}$$

Proof. Since every p-form with $p>3$ on E^3 is zero, we need only consider the cases of ϕ a 0-form and ϕ a 1-form. When ϕ is a 0-form f, by (6.1.13), (6.2.4), and Definition 6.2.7 we obtain

$$\begin{aligned} d^2f&=\sum_{i=1}^{3} d\left(\frac{\partial f}{\partial x_i}\right)dx_i=\sum_{i,j=1}^{3}\frac{\partial^2 f}{\partial x_j\partial x_i}dx_j\wedge dx_i\\ &=\sum_{i,j=1}^{3}\left(\frac{\partial^2 f}{\partial x_j\partial x_i}-\frac{\partial^2 f}{\partial x_i\partial x_j}\right)dx_j\wedge dx_i. \end{aligned}$$

Since f is of class C^2, $\partial^2 f/(\partial x_j\partial x_i)-\partial^2 f/(\partial x_i\partial x_j)=0$ for all distinct i and j; hence $d^2f=0$.

When ϕ is a 1-form, let ϕ be expressed by (6.2.8). Then from (6.2.16) it follows that

$$d^2\phi = d\left(\sum_{i,j=1}^{3} \frac{\partial f_i}{\partial x_j} dx_j \wedge dx_i\right) = \sum_{i,j,k=1}^{3} \frac{\partial^2 f_i}{\partial x_k \partial x_j} dx_k \wedge dx_j \wedge dx_i$$

$$= \sum_{\substack{i,j,k=1\\k<j}}^{3} \left(\frac{\partial^2 f_i}{\partial x_k \partial x_j} - \frac{\partial^2 f_i}{\partial x_j \partial x_k}\right) dx_k \wedge dx_j \wedge dx_i,$$

which is zero, since f_i is of class C^2 for every i.

6.2.10. *Definition.* A C^1 form ω is *closed* if $d\omega = 0$. A form ω is *exact* if there is a C^1 form $\bar{\omega}$ such that $\omega = d\bar{\omega}$.

From Theorem 6.2.9 it follows that *every exact form is closed.*

Let $\mathbf{v} = (v_1, v_2, v_3)$ be a vector field of class C^1 on E^3. Then we define $d\mathbf{v}$ to be a vector field in the componentwise, that is,

$$d\mathbf{v} = (dv_1, dv_2, dv_3). \tag{6.2.22}$$

By (6.1.13) we thus obtain

$$d\mathbf{v} = \left(\sum_{i=1}^{3} \frac{\partial v_i}{\partial x_i} dx_i, \sum_{i=1}^{3} \frac{\partial v_2}{\partial x_i} dx_i, \sum_{i=1}^{3} \frac{\partial v_3}{\partial x_i} dx_i\right) = \sum_{i=1}^{3} \frac{\partial \mathbf{v}}{\partial x_i} dx_i, \tag{6.2.23}$$

where as in Section 3.3, $\partial \mathbf{v}/\partial x_i$ for each i is a vector field defined by

$$\frac{\partial \mathbf{v}}{\partial x_i} = \left(\frac{\partial v_1}{\partial x_i}, \frac{\partial v_2}{\partial x_i}, \frac{\partial v_3}{\partial x_i}\right). \tag{6.2.24}$$

Now we can define the following combined operator $\hat{\times}$ of the vector product $\times$ and the exterior product $\wedge$, which is very useful as we shall see in some of its important applications in Chapter 4.

6.2.11. *Definition.* Let $\mathbf{v} = (v_1, v_2, v_3)$ and $\mathbf{w} = (w_1, w_2, w_3)$ be two vector fields of class C^1 on E^3. Then

$$\begin{aligned} \mathbf{v} \hat{\times} \mathbf{w} &= \mathbf{v} \times \mathbf{w}, \\ \mathbf{v} \hat{\times} d\mathbf{w} &= \sum_{i=1}^{3} \left(\mathbf{v} \times \frac{\partial \mathbf{w}}{\partial x_i}\right) dx_i, \\ d\mathbf{v} \hat{\times} d\mathbf{w} &= \sum_{i,j=1}^{3} \left(\frac{\partial \mathbf{v}}{\partial x_i} \times \frac{\partial \mathbf{w}}{\partial x_j}\right) dx_i \wedge dx_j. \end{aligned} \tag{6.2.25}$$

From (6.2.25) it follows that $\mathbf{v} \hat{\times} d\mathbf{w}$ is a vector field and, at the same time, is also a 1-form. Similarly, $d\mathbf{v} \hat{\times} d\mathbf{w}$ is a vector field and also a 2-form.

Moreover, we can have

$$d\mathbf{v}\hat{\times}d\mathbf{w}=d\mathbf{w}\hat{\times}d\mathbf{v},$$
$$d\mathbf{v}\hat{\times}(d\mathbf{w}+d\overline{\mathbf{w}})=d\mathbf{v}\hat{\times}d\mathbf{w}+d\mathbf{v}\hat{\times}d\overline{\mathbf{w}}, \qquad (6.2.26)$$

where $\overline{\mathbf{w}}$ is another vector field of class C^1 on E^3.

Exercises

1. Let $\phi=yz\,dx+dz$ and $\psi=x\,dx-y\,dy$. Find the standard expressions for (a) $\phi\wedge\psi$ and (b) $d\phi$.
2. Let f and g be real C^1 functions. Simplify the following forms: (a) $d(f-g)\wedge(df+dg)$; (b) $d(f\,dg\wedge g\,df)$; (c) $d(gf\,df)+d(f\,dg)$.
3. Prove (6.2.20).
4. Prove that the 1-form $x\,dx+y\,dy+z\,dz$ is both closed and exact on E^3.
5. Prove that the 2-form

$$(3y^2z-3xz^2)dy\wedge dz+x^2y\,dz\wedge dx+(z^3-x^2z)dx\wedge dy$$

is exact on E^3.

6. For any three 1-forms $\phi_i=\Sigma_{j=1}^3 f_{ij}dx_j$, $i=1,2,3$, prove that

$$\phi_1\wedge\phi_2\wedge\phi_3=\begin{vmatrix} f_{11} & f_{12} & f_{13} \\ f_{21} & f_{22} & f_{23} \\ f_{31} & f_{32} & f_{33} \end{vmatrix} dx_1\wedge dx_2\wedge dx_3.$$

7. Let f and g be real C^1 functions on E. Prove that

$$df\wedge dg=\begin{vmatrix} f_u & f_v \\ g_u & g_v \end{vmatrix} du\wedge dv,$$

and deduce that $df\wedge dg=-dg\wedge df$, where the subscripts u and v denote the partial derivatives with respect to u and v, respectively.

8. On E^3 between the system of differential forms and the system of vector analysis there are one-to-one correspondences (1) and (2) defined as follows:

$$\sum_{i=1}^{3} f_i dx_i \overset{(1)}{\leftrightarrow} \sum_{i=1}^{3} f_i\mathbf{u}_i \overset{(2)}{\leftrightarrow} f_3\,dx_1\wedge dx_2-f_2\,dx_1\wedge dx_3+f_1\,dx_2\wedge dx_3.$$

In vector analysis of E^3 the following operations are basic:
Gradient of a function f:

$$\operatorname{grad} f=\sum_{i=1}^{3}\frac{\partial f}{\partial x_i}\mathbf{u}_i.$$

Curl of a vector field $\mathbf{v}=\Sigma_{i=1}^{3} f_i \mathbf{u}_i$:

$$\operatorname{curl} \mathbf{v}=\left(\frac{\partial f_3}{\partial x_2}-\frac{\partial f_2}{\partial x_3}\right)\mathbf{u}_1+\left(\frac{\partial f_1}{\partial x_3}-\frac{\partial f_3}{\partial x_1}\right)\mathbf{u}_2+\left(\frac{\partial f_2}{\partial x_1}-\frac{\partial f_1}{\partial x_2}\right)\mathbf{u}_3.$$

Divergence of a vector field $\mathbf{v}=\Sigma_{i=1}^{3} f_i \mathbf{u}_i$:

$$\operatorname{div} \mathbf{v}=\sum_{i=1}^{3} \frac{\partial f_i}{\partial x_i}.$$

Prove that all three operations may be expressed by exterior derivatives as follows:

(a) $df \overset{(1)}{\leftrightarrow} \operatorname{grad} f$.

(b) If $\phi \overset{(1)}{\leftrightarrow} \mathbf{v}$, then $d\phi \overset{(2)}{\leftrightarrow} \operatorname{curl} \mathbf{v}$.

(c) If $\omega \overset{(2)}{\leftrightarrow} \mathbf{v}$, then $d\omega=(\operatorname{div} \mathbf{v})\, dx_1 \wedge dx_2 \wedge dx_3$.

6.3. Structural Equations. Let $\mathbf{e}_1\mathbf{e}_2\mathbf{e}_3$ be a positively oriented frame at a point $\mathbf{x}$ on E^3. Then by Definition 3.5.6

$$\mathbf{e}_i \cdot \mathbf{e}_j=\delta_{ij}, \qquad i, j=1,2,3. \tag{6.3.1}$$

Since, according to (6.2.22), $d\mathbf{x}$ is a tangent vector of E^3 at $\mathbf{x}$, we have

$$d\mathbf{x}=\sum_{i=1}^{3} \omega_i \mathbf{e}_i, \tag{6.3.2}$$

where $\omega_1, \omega_2, \omega_3$ are 1-forms on E^3. Similarly,

$$d\mathbf{e}_i=\sum_{j=1}^{3} \omega_{ij} \mathbf{e}_j, \qquad i=1,2,3, \tag{6.3.3}$$

where ω_{ij} are 1-forms on E^3.

Now consider a curve $t \to \mathbf{x}(t)$ in E^3. Then by the chain rule (Theorem 2.3.6) we can write the tangent vector of this curve at the point $\mathbf{x}$ as

$$\frac{d\mathbf{x}}{dt}=\sum_{i=1}^{3} \frac{\partial \mathbf{x}}{\partial x_i} \frac{dx_i}{dt}. \tag{6.3.4}$$

Thus the vectors

$$\frac{\partial \mathbf{x}}{\partial x_1}, \frac{\partial \mathbf{x}}{\partial x_2}, \frac{\partial \mathbf{x}}{\partial x_3} \tag{6.3.5}$$

span the tangent space $T_x(E^3)$ of E^3 at the point $\mathbf{x}$, that is, the space that contains the tangent vectors of all the curves in E^3 through $\mathbf{x}$. Since $T_x(E^3)$ is isomorphic to E^3, tangent vectors (6.3.5) are linearly independent, so that

$$\frac{\partial \mathbf{x}}{\partial x_i} = \sum_{j=1}^{3} a_{ij}\mathbf{e}_j, \qquad i=1,2,3, \tag{6.3.6}$$

where (see Exercise 4 of Section 3.4)

$$\det(a_{ij}) \neq 0. \tag{6.3.7}$$

On the other hand, replacing $\mathbf{v}$ by $\mathbf{x}$ in (6.2.23) we have

$$d\mathbf{x} = \sum_{i=1}^{3} \frac{\partial \mathbf{x}}{\partial x_i} dx_i. \tag{6.3.8}$$

Comparison of (6.3.2) with (6.3.8) and use of (6.3.6) yield

$$\omega_i = \sum_{j=1}^{3} a_{ji}\, dx_j, \qquad i=1,2,3. \tag{6.3.9}$$

Thus from (6.3.9), (6.3.7), Lemma 6.1.4, and Exercise 4 of Section 3.4 again follows easily a lemma for ω_i.

6.3.1. Lemma. *$\omega_1, \omega_2, \omega_3$ defined by* (6.3.2) *constitute a basis of the dual space of the tangent space $T(E^3)$ at the point* $\mathbf{x}$.

Taking the exterior derivatives of both sides of (6.3.1) we obtain

$$\mathbf{e}_i \cdot d\mathbf{e}_j + d\mathbf{e}_i \cdot \mathbf{e}_j = 0, \qquad i,j=1,2,3. \tag{6.3.10}$$

Substituting (6.3.3) in (6.3.10) and using (6.3.1) we have

$$\omega_{ij} + \omega_{ji} = 0, \qquad i,j=1,2,3, \tag{6.3.11}$$

which imply

$$\omega_{ii} = 0, \qquad i=1,2,3. \tag{6.3.12}$$

Exterior differentiation of (6.3.2) and use of (6.2.20), (6.3.3), and Theorem 6.2.9 give

$$0 = \sum_{i=1}^{3} (d\mathbf{e}_i \wedge \omega_i + \mathbf{e}_i\, d\omega_i) = \sum_{i=1}^{3} \mathbf{e}_i\, d\omega_i + \sum_{i,j=1}^{3} \mathbf{e}_j \omega_{ij} \wedge \omega_i.$$

Since $\mathbf{e}_1, \mathbf{e}_2, \mathbf{e}_3$ are linearly independent, the equation above implies

$$d\omega_i = \sum_{j=1}^{3} \omega_j \wedge \omega_{ji}, \qquad i=1,2,3. \tag{6.3.13}$$

Similarly, by exterior differentiation of (6.3.3) we obtain

$$d\omega_{ij} = \sum_{k=1}^{3} \omega_{ik} \wedge \omega_{kj}, \qquad i,j=1,2,3. \tag{6.3.14}$$

We call (6.3.13) and (6.3.14) the *structural equations* for the forms ω_i, ω_{ij}, where $i, j = 1,2,3$.

Exercises

1. Prove (6.3.14).

7. THE CALCULUS OF VARIATIONS

To study the Euler equations of the calculus of variations on plane curves, consider the problem of finding a C^1 curve α given by $y=y(x)$ passing through two points $\mathbf{p}=(x_0, y_0)$, $\mathbf{q}=(x_1, y_1)$ such that the integral

$$I=\int_{x_0}^{x_1} f(x, y, y')dx, \tag{7.1}$$

where $y'=dy/dx$ and f is a C^2 function in all three variables x, y, y', attains an extremal value. The equations

$$\bar{y}=y+\varepsilon\omega, \tag{7.2}$$

where ε is small and ω is a C^1 function of x such that

$$\omega=0 \qquad \text{for} \quad x=x_0, x_1 \tag{7.3}$$

defines a neighboring curve $\bar{\alpha}$ of C passing through $\mathbf{p}$ and $\mathbf{q}$. Let $I(\varepsilon)$ be the integral (7.1) corresponding to $\bar{\alpha}$, that is,

$$I(\varepsilon)=\int_{x_0}^{x_1} f(x, y+\varepsilon\omega, y'+\varepsilon\omega')dx, \tag{7.4}$$

where $\omega'=d\omega/dx$. Then on expanding f in Taylor's series (Theorem 2.2.2) we have

$$I(\varepsilon)-I=\varepsilon\int_{x_0}^{x_1}(f_y\omega+f_{y'}\omega')dx+\cdots, \tag{7.5}$$

where $f_y=\partial f/\partial y'$, $f_{y'}=\partial f/\partial y'$, and the unwritten terms are of degree $\geqslant 2$ in ε, and we call

$$\delta I=\varepsilon\int_{x_0}^{x_1}(f_y\omega+f_{y'}\omega')dx, \tag{7.6}$$

the *first variation* of the integral I. Integrating the second term of the integrand of (7.6) by parts and making use of (7.3) we obtain

$$\delta I=\varepsilon\int_{x_0}^{x_1}\left(f_y-\frac{d}{dx}f_{y'}\right)\omega\, dx. \tag{7.7}$$

The integral I is said to be *stationary* and α the corresponding *extremal*, if the first variation δI is zero for every C^1 function ω satisfying (7.3).

7.1. Lemma. *If $\int_{x_0}^{x_1}\rho(x)F(x)dx=0$ for continuous $F(x)$ and all C^1 (or continuous) $\rho(x)$ on the interval $x_0 \leqslant x \leqslant x_1$, then $F(x)=0$ for $x\in[x_0, x_1]$.*

Proof. If there is a ξ, $x_0<\xi<x_1$, such that $F(\xi)\neq 0$, say $F(\xi)>0$. Then by continuity there exists a whole interval $\xi_0<\xi<\xi_1$, in which $F(\xi)>0$. Choose now

$$\rho(x)=0, \qquad x<\xi_0 \quad \text{or} \quad x>\xi_1,$$

$$\rho(x)=(x-\xi_0)^2(\xi_1-x)^2, \qquad \xi_0 \leqslant x \leqslant \xi_1.$$

Then clearly $\int_{x_0}^{x_1}\rho(x)F(x)dx>0$, which contradicts the assumption. Hence no such ξ may exist. Q.E.D.

Theorem 7.2 follows immediately from (7.7) and Lemma 7.1.

7.2. Theorem. *α is an extremal curve of the integral* (7.1) *if and only if it satisfies*

$$f_y - \frac{d}{dx}f_{y'}=0, \tag{7.8}$$

which is called the Euler equation for the extremal α.

The Euler equations for surface functionals can be derived in the same way. Moreover, Theorem 7.2 can be extended to piecewise C^1 curves (see Definition 1.2.2) as follows.

7.3. Theorem (G. Erdmann). *For integral* (7.1) *for a piecewise C^1 curve α given by $y=y(x)$ passing through two points $\mathbf{p}=(x_0, y_0)$, $\mathbf{q}=(x_1, y_1)$ with a vertex $\mathbf{r}=(x_2, y_2)$ on the line $x=x_2$ (see Fig.* 1.10) *to attain an extremal value, the curve α must satisfy, in addition to the Euler equations for the two arcs* **pr** *and* **rq**,

$$f_{+y'}=f_{-y'}, \tag{7.9}$$

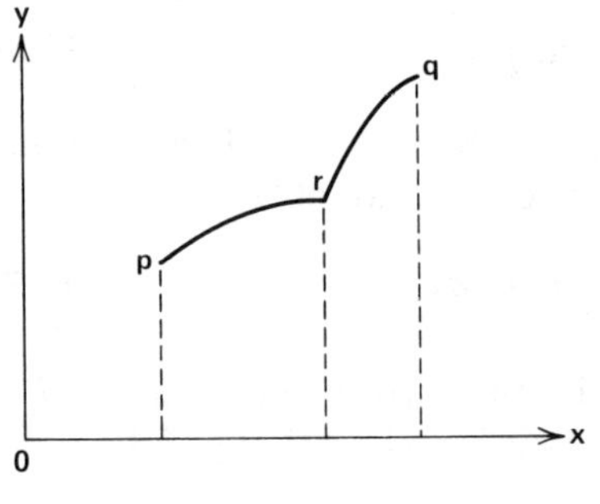

Figure 1.10

where

$$f_{+y'}=f_{y'}(x_2,y_2,y'_+),\qquad f_{-y'}=f_{y'}(x_2,y_2,y'_-),$$
$$y'_+=\lim_{\delta\to 0} y'(x_2+\delta),\qquad y'_-=\lim_{\delta\to 0} y'(x_2-\delta). \tag{7.10}$$

Proof. In this case, since we assume that the vertices of piecewise neighboring curves of α still move along the line $x=x_2$, the integral (7.4) and therefore the integral (7.5) split into two parts, respectively,

$$I(\varepsilon)=\int_{x_0}^{x_2} f(x,y+\varepsilon\omega,y'+\varepsilon\omega')\,dx$$
$$+\int_{x_2}^{x_1} f(x,y+\varepsilon\omega,y'+\varepsilon\omega')\,dx, \tag{7.11}$$

$$\delta I=\varepsilon\int_{x_0}^{x_2}(f_y\omega+f_{y'}\omega')\,dx+\varepsilon\int_{x_2}^{x_1}(f_y\omega+f_{y'}\omega')\,dx. \tag{7.12}$$

Integrating the second term of the integrand of the first integral of (7.12) and making use of (7.3), we obtain

$$\int_{x_0}^{x_2} f_{y'}\omega'\,dx=f_{y'}\omega\Big|_{x_0}^{x_2}-\int_{x_0}^{x_2}\omega\frac{d}{dx}(f_{y'})\,dx$$
$$=\omega_2 f_{-y'}-\int_{x_0}^{x_2}\omega\frac{d}{dx}(f_{y'})\,dx, \tag{7.13}$$

where $\omega_2=\omega(x_2)$. Substituting (7.13) and a similar equation for the second integral of (7.12) in (7.12), we have

$$\delta I=\varepsilon\int_{x_0}^{x_2}\left(f_y-\frac{d}{dx}f_{y'}\right)\omega\,dx+\varepsilon\int_{x_2}^{x_1}\left(f_y-\frac{d}{dx}f_{y'}\right)\omega\,dx+\varepsilon\omega_2(f_{-y'}-f_{+y'}).$$

From Lemma 7.1 and Theorem 7.2 for the arcs **pr** and **rq** it follows that $\delta I=0$ gives the required conditions, and Theorem 7.3 is therefore proved.

Exercises

1. Complete the argument of the proof of Lemma 7.1 by considering the cases $\xi=x_0$ and $\xi=x_1$.

2. Find extremals for the following integrals:

(a) $\int_{x_0}^{x_1} y'(1+x^2y')\,dx.$ (b) $\int_{x_0}^{x_1}(y'^2+2yy'-16y^2)\,dx.$

(c) $\int_{x_0}^{x_1}(xy'+y'^2)\,dx.$ (d) $\int_{x_0}^{x_1}(y^2+y'^2-2y\sin x)\,dx.$

2

Curves

In this chapter we study curves in a Euclidean 3-space E^3 and also in a Euclidean plane E^2 independently. We give local properties of the curves first, then global properties.

1. GENERAL LOCAL THEORY

1.1. Parametric Representations. Let $I=[a,b]$ be an interval on the real line E^1. Throughout this chapter we shall assume this interval I to be in a generalized sense, which allows $a=-\infty$, $b=+\infty$, or both.

There are various notions of a curve in E^3. However, this chapter deals only with the curves given by the following definition.

1.1.1 Definition. A curve in E^3 is a differentiable mapping $\mathbf{x}$: $I \rightarrow E^3$ of I into E^3.

For each $t \in I$ we have

$$\mathbf{x}(t)=(x_1(t), x_2(t), x_3(t)), \tag{1.1.1}$$

where x_1, x_2, x_3 are the Euclidean coordinate functions of $\mathbf{x}$ (by Definition 4.1.1, Chapter 1), and t is called the *parameter* of the curve $\mathbf{x}$. Thus $\mathbf{x}(t)$ can be considered as the position vector of a moving point on the image set $\mathbf{x}(I)$ of the curve $\mathbf{x}$, which can also be expressed by $t \rightarrow (x_1(t), x_2(t), x_3(t))$.

In the following we give several examples of curves that are used later. A curve $\mathbf{x}$: $I \rightarrow E^3$ should be distinguished from its image set $\mathbf{x}(I)$. Two curves may have the same image set as illustrated by item 6 of Examples 1.1.2.

1.1.2. Examples. 1. *Straight line*. It is the simplest type of curve in E^3. Explicitly the curve $\mathbf{x}$: $E^1 \rightarrow E^3$ given by

$$\mathbf{x}(t)=\mathbf{p}+t\mathbf{q}=(p_1+tq_1, p_2+tq_2, p_3+tq_3), \tag{1.1.2}$$

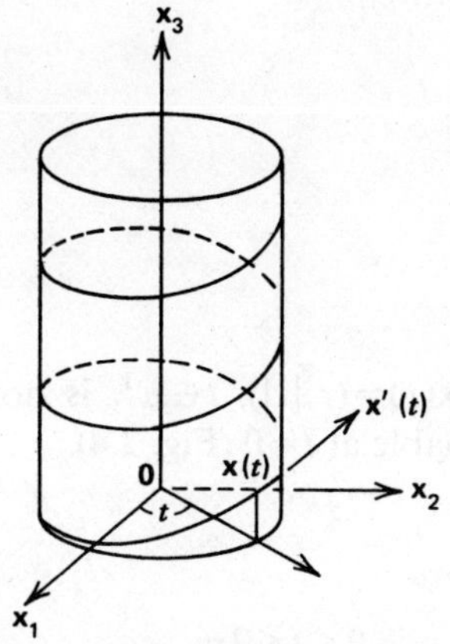

Figure 2.1

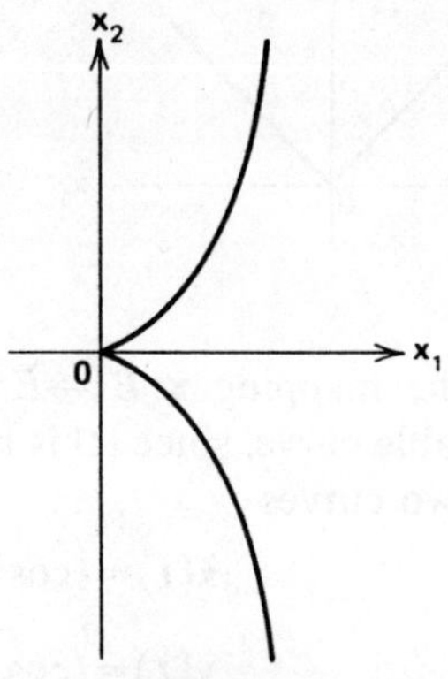

Figure 2.2

where $\mathbf{p}=(p_1,p_2,p_3)$ and $\mathbf{q}=(q_1,q_2,q_3)$, is the straight line through the point $\mathbf{p}=\mathbf{x}(0)$ in the $\mathbf{q}$ direction.

2. *Helix* (Fig. 2.1). It is a curve $\mathbf{x}\colon E^1\to E^3$ given by

$$\mathbf{x}(t)=(a\cos t, a\sin t, bt), \qquad a>0, b>0, \tag{1.1.3}$$

rising at a constant rate on the cylinder $x_1^2+x_2^2=a^2$, and is said to be right winding. If b is negative instead of positive, the helix is said to be left winding. The $\mathbf{x}_3$-axis is called the *axis*, and $2\pi b$ the *pitch* of the helix.

3. *Curve with cusp* (Fig. 2.2). The curve $\mathbf{x}\colon E^1\to E^2$ such that

$$\mathbf{x}(t)=(t^2,t^3). \tag{1.1.4}$$

Notice that $\mathbf{x}'(0)=(0,0)$. The origin $(0,0)$ is called a *cusp*.

4. The curve $\mathbf{x}\colon E^1\to E^2$ such that

$$\mathbf{x}(t)=(t^3-3t-2, t^2-t-2), \qquad t\in E^1. \tag{1.1.5}$$

Elimination of t from (1.1.5) gives $x_2^3=x_1^2-3x_1x_2$, which would not make the sketch of the graph of the curve (Fig. 2.3) easier. Notice that $\mathbf{x}(-1)=\mathbf{x}(2)=(0,0)$, so that the mapping $\mathbf{x}$ is not injective.

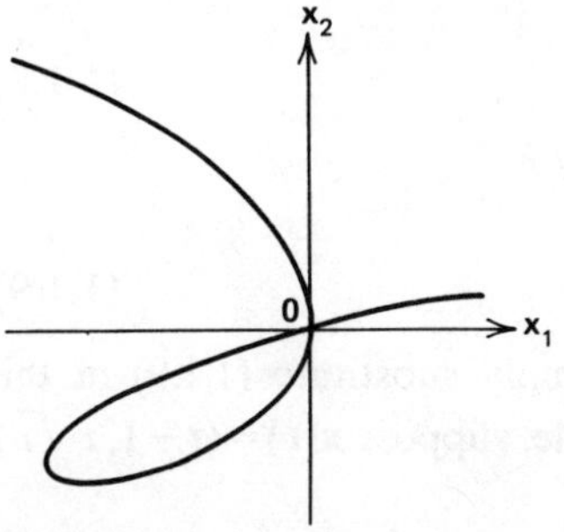

Figure 2.3

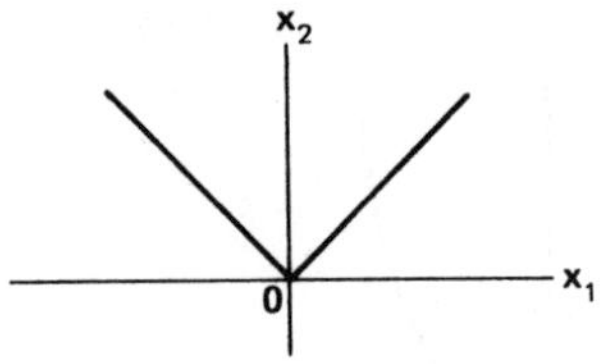

Figure 2.4

5. The mapping $\mathbf{x}: E^1 \to E^2$ given by $\mathbf{x}(t) = (t, |t|)$, $t \in E^1$, is not a differentiable curve, since $|t|$ is not differentiable at $t=0$ (Fig. 2.4).

6. Two curves

$$\mathbf{x}(t) = (\cos t, \sin t, 0),$$
$$\mathbf{y}(t) = (\cos 2t, \sin 2t, 0), \qquad 0 \leq t \leq 2\pi, \tag{1.1.6}$$

have the same image set, namely, the unit circle $x_1^2 + x_2^2 = 1$, $x_3 = 0$, and both describe the motion of a point on the unit circle counterclockwise. But the circle is traversed once in the first case and twice in the second.

1.1.3. Definition. By (3.3.10) of Chapter 1 and Definition 1.1.1, for a curve $\mathbf{x}$: $I \to E^3$ given by (1.1.1) we have

$$\mathbf{x}'(t) = (x_1'(t), x_2'(t)\, x_3'(t)), \tag{1.1.7}$$

where the prime denotes the derivative with respect to t. The vector (1.1.7) is called the *tangent vector* of the curve $\mathbf{x}$ at the point $\mathbf{x}(t)$; cf. (4.10) of Chapter 1.

Since a curve $\mathbf{x}$ can be considered as a motion of a moving particle $\mathbf{x}$ that is located at the point $\mathbf{x}(t)$ at each "time" t, the tangent vector $\mathbf{x}'(t)$ may also be called the *velocity vector* of $\mathbf{x}$ at t.

Given a curve $\mathbf{x}$, we can construct many new curves that follow the same path as $\mathbf{x}$ but move at different speeds.

1.1.4. Definition. Let I and J be generalized intervals on the real line E^1. Let $\mathbf{x}$: $I \to E^3$ be a curve and let h: $J \to I$ be a differentiable (real-valued) function. Then the composite function

$$\mathbf{y} = \mathbf{x}(h): J \to E^3 \tag{1.1.8}$$

is a curve called the *reparametrization* of $\mathbf{x}$ by h.

Let $\tau \in J$, $t \in I$, and

$$t = h(\tau). \tag{1.1.9}$$

To compute the coordinates of $\mathbf{y}$ we can simply substitute (1.1.9) in the coordinates $x_1(t)$, $x_2(t)$, $x_3(t)$ of $\mathbf{x}$. For example, suppose $\mathbf{x}(t) = (t-1, t\sqrt{t}\,,$

$\sqrt{t}$ on I: $0 \leqslant t \leqslant 9$. If $h(\tau)=\tau^2$ on J: $0 \leqslant \tau \leqslant 3$, then

$$\mathbf{y}(\tau)=\mathbf{x}(h(\tau))=\mathbf{x}(\tau^2)=(\tau^2-1,\tau^3,\tau).$$

Thus the curve $\mathbf{x}$: $I \to E^3$ has been reparametrized by h to give the curve $\mathbf{y}$: $J \to E^3$.

Using the chain rule (Theorem 2.3.6, Chapter 1) and (1.1.7), (1.1.8), and (1.1.9), we immediately obtain the following relationship between the tangent vector of a curve and the corresponding tangent vector of a reparametrization.

1.1.5. Lemma. *Let* $\mathbf{y}(\tau)$, $\tau \in J$, *be the reparametrization of a curve* $\mathbf{x}(t)$, $t \in I$, *by* h *so that* (1.1.9) *holds. Then*

$$\mathbf{y}'(\tau)=\frac{dh}{d\tau}(\tau)\mathbf{x}'(h(\tau)), \tag{1.1.10}$$

and $\mathbf{y}'(\tau) \neq 0$ *if*

$$\frac{dh}{d\tau}(\tau) \neq 0 \quad \textit{for all} \quad \tau \in J. \tag{1.1.11}$$

Moreover, under condition (1.1.11) *if* $h(\tau)$ *and* $\mathbf{x}(t)$ *have continuous second derivatives* $h''(\tau)$ *and* $\mathbf{x}''(t)$ *in* J *and* I, *respectively*, $\mathbf{y}(\tau)$ *also has continuous* $\mathbf{y}''(\tau)$ *in* J.

The last part of Lemma 1.1.5 can be proved as follows. Condition (1.1.11) implies that h is a monotone function of τ, so that I and J are in a one-to-one correspondence. Thus the inverse function $\tau = k(t)$ exists, and $k(t)$ would, as $h(\tau)$, have a continuous second derivative. Hence from (1.1.10) it follows that $\mathbf{y}(\tau)$ has continuous $\mathbf{y}''(\tau)$ in J.

By the chain rule and Lemma 3.6.2 of Chapter 1 we can obtain the following lemma concerning the directional tangent of a function with respect to a tangent vector of a curve.

1.1.6. Lemma. *Let* $\mathbf{x}$ *be a curve in* E^3, *and* f *a differentiable function on* E^3. *Then*

$$\mathbf{x}'(t)[f]=\frac{d(f(\mathbf{x}))}{dt}(t). \tag{1.1.12}$$

This lemma shows that the rate of change of f along the line through the point $\mathbf{x}(t)$ in the direction $\mathbf{x}'(t)$ is the same as that along the curve $\mathbf{x}$ itself.

Exercises

1. Sketch the graphs of the following curves in E^2 and express each curve in terms of a parameter. (a) $2x_1+3x_2=5$; (b) $x_2=e^{x_1}$; (c) $2x_1^2+$

$x_2=1$; (d) The folium of Descartes $x_1^3+x_2^3=3x_1x_2$, with slope t of a line through the origin as the parameter.

2. Find a curve $\mathbf{x}(t)$ whose image set is the circle $x_1^2+x_2^2=1$, $x_3=0$ such that $\mathbf{x}(t)$ moves clockwise around the circle with $\mathbf{x}(0)=(0,1,0)$.
3. Let $\mathbf{x}(t)$ be a curve in E^3 not passing through the origin. If $\mathbf{x}(t_0)$ is the point on the image set of $\mathbf{x}$ closest to the origin, and $\mathbf{x}'(t_0)\neq 0$, show that the position vector $\mathbf{x}(t_0)$ is orthogonal to the vector $\mathbf{x}'(t_0)$.
4. Find the (unique) curve $\mathbf{x}$ such that $\mathbf{x}(0)=(2,3,0)$ and $\mathbf{x}'(t)=(e^t,-2t,t^2)$.
5. What is a curve $\mathbf{x}(t)$ whose second derivative $\mathbf{x}''(t)$ is identically zero?
6. Let $\mathbf{x}$: $I\to E^3$ be a curve with $\mathbf{x}'(t)\neq 0$ for all $t\in I$. Show that the norm $\|\mathbf{x}(t)\|$ [for the definition see (3.2.3) of Chapter 1] of the vector $\mathbf{x}(t)$ is a nonzero constant if and only if $\mathbf{x}(t)$ is orthogonal to $\mathbf{x}'(t)$ for all $t\in I$.
7. For a fixed t, the *tangent line* of a curve at $\mathbf{x}(t)$, where $\mathbf{x}'(t)\neq 0$, is the straight line $u\to\mathbf{x}(t)+u\mathbf{x}'(t)$. Find the tangent lines of the helix $\mathbf{x}(t)=(2\cos t, 2\sin t, t)$ at the points $\mathbf{x}(0)$ and $\mathbf{x}(\frac{1}{4}\pi)$.
8. Suppose that
$$\mathbf{x}(t)=(2\cos^2 t,\sin 2t,2\sin t)\qquad\text{for}\qquad 0\leqslant t\leqslant \tfrac{1}{2}\pi.$$
Find the coordinate functions of the curve $\mathbf{y}=\mathbf{x}(h)$, where h is the function on J: $0\leqslant\tau\leqslant 1$ such that $h(\tau)=\sin^{-1}\tau$.
9. Prove Lemma 1.1.6.

1.2. Arc Length, Vector Fields, and Knots. If a curve $\mathbf{x}$: $I\to E^3$ has a zero tangent vector at a point $\mathbf{x}(t)$, then from (1.1.7) we have $\mathbf{x}'(t)=0$, and this point is called a *singular point* of $\mathbf{x}$. The point $t=0$ in item 3 of Examples 1.1.2 is such a singular point.

1.2.1. Definition. A curve $\mathbf{x}$: $I\to E^3$ is *regular* if $\mathbf{x}'(t)\neq 0$ for all $t\in I$.

Thus a regular curve has no singular points. On the other hand, we have

1.2.2. Definition. Let $\mathbf{x}$: $I=[a,b]\to E^3$ be a mapping of I into E^3. $\mathbf{x}$ is a *piecewise* (or *sectionally*) *regular* (*smooth*, i.e., C^k for some $k\geqslant 1$) *curve* if there exists a finite partition
$$\mathcal{P}: a=t_0<t_1<\cdots<t_n=b$$
of $[a,b]$ such that $\mathbf{x}$ is differentiable and regular (smooth) in each $[t_i,t_{i+1}]$ for $i=0,1,\cdots,n-1$. The points $\mathbf{x}(t_i)$, $i=0,1,\cdots,n$, are the *vertices* of $\mathbf{x}$, and the image sets of $\mathbf{x}([t_i,t_{i+1}])$ are the *regular* (*smooth*) *arcs* of $\mathbf{x}$.

Intuitively, the tangents of the two arcs $\mathbf{x}([t_{i-1}, t_i])$ and $\mathbf{x}([t_i, t_{i+1}])$ through the vertex $\mathbf{x}(t_i)$ may be different. Thus a piecewise regular (smooth) curve fails to have a well-defined tangent line only at a finite number of vertices.

The curve in item 3 of Examples 1.1.2 is a piecewise regular curve.

Most of our discussions on curves are restricted to regular curves; therefore from now on all curves are assumed to be regular unless stated otherwise.

Let $\mathbf{x}$: $I=[a, b]\to E^3$ be a curve. For every partition $\mathcal{P}$ of $[a, b]$ mentioned above, we consider $\sum_{i=1}^{n}\|\mathbf{x}(t_i)-\mathbf{x}(t_{i-1})\|=L(\mathbf{x}, \mathcal{P})$, which is the sum of the lengths of the straight segments $\mathbf{x}(t_i)\mathbf{x}(t_{i-1})$, $i=1,\cdots,n$, or the length of the polygon $\mathbf{x}(t_0)\mathbf{x}(t_1)\cdots\mathbf{x}(t_n)$ inscribed in $\mathbf{x}(I)$. Let $\|\mathcal{P}\|$ be the norm of the partition $\mathcal{P}$ defined by

$$\|\mathcal{P}\|=\max(t_i-t_{i-1}), \qquad i=1,\cdots,n.$$

Then

$$\lim_{\substack{n\to\infty \\ \|\mathcal{P}\|\to 0}} L(x, \mathcal{P})$$

is called the *arc length*, or shorter, the *length* of $\mathbf{x}(I)$, and $\mathbf{x}(I)$ is said to be *rectifiable*, provided such a limit does exist and is finite.

1.2.3. Theorem. *A curve* $\mathbf{x}$: $I\to E^3$ *with continuous derivative* $\mathbf{x}'(t)$ *is rectifiable, and its arc length from a fixed point* t_0 *to a variable point* t *is given by*

$$s(t)=\int_{t_0}^{t}\|\mathbf{x}'(t)\|\,dt. \tag{1.2.1}$$

The proof of Theorem 1.2.3 can be found in calculus books, and is therefore omitted here. Theorem 1.2.3 is also true for a piecewise regular curve $\mathbf{x}$: $I\to E^3$ with continuous derivative $\mathbf{x}'(t)$ except at the vertices.

For a regular curve $\mathbf{x}$: $I\to E^3$, since $\mathbf{x}'(t)\neq 0$ for all t, from (1.2.1) it follows that $s(t)$ is a differentiable function of t and

$$\frac{ds}{dt}=\|\mathbf{x}'(t)\|. \tag{1.2.2}$$

Thus if the original parameter t is the arc length s, then

$$\|\mathbf{x}'(s)\|=1, \tag{1.2.3}$$

and (1.2.1) gives that the arc length of the curve $\mathbf{x}$ from $s=a$ to $s=b$ with $a<b$ is just $b-a$.

1.2.4. Corollary. *The arc lengths of a curve* $\mathbf{x}$ *and a regular reparametrization of* $\mathbf{x}$ *are equal.*

Proof. Let $\mathbf{x}(t)$, $a \leqslant t \leqslant b$, be a curve in E^3, and let $\mathbf{x}(h(\tau))$, $\alpha \leqslant \tau \leqslant \beta$, be a regular reparametrization of $\mathbf{x}(t)$ so that $h'(\tau) \neq 0$, $t = h(\tau)$. Then by (1.2.1) the length of the curve $\mathbf{x}(h(\tau))$ is

$$\int_\alpha^\beta \left\| \frac{d\mathbf{x}(h(\tau))}{d\tau} \right\| d\tau = \int_\alpha^\beta \left\| \frac{d\mathbf{x}(h(\tau))}{dt} \right\| \frac{dt}{d\tau} d\tau = \int_a^b \left\| \frac{d\mathbf{x}(t)}{dt} \right\| dt,$$

for $dt/d\tau > 0$ (for $dt/d\tau < 0$ we use $h(\alpha) = b$, $h(\beta) = a$), the last integral being the arc length of the curve $\mathbf{x}(t)$.

1.2.5. Corollary. *The arc length of a curve* $\mathbf{x}$: $I \to E^3$ *is invariant under isometries of* E^3.

Proof. This corollary follows from the fact that the integrand $\|\mathbf{x}'(t)\|$ of (1.2.1) is an inner product of vectors that is invariant under isometries of E^3 (Corollary 5.3.7 of Chapter 1).

1.2.6. Theorem. *There exists a reparametrization* $\mathbf{y}$ *of a curve* $\mathbf{x}$: $I \to E^3$ *such that* $\mathbf{y}$ *has unit tangent vectors.*

Proof. Fix a number t_0 in I, and consider the arc-length function $s(t)$ given by (1.2.1).

Since $\mathbf{x}'(t) \neq 0$ for all t, from (1.2.2) it follows that $ds/dt > 0$. By a standard theorem of calculus, the function s has an inverse function $t = t(s)$, whose derivative dt/ds at $s = s(t)$ is the reciprocal of ds/dt at $t = t(s)$. In particular, $dt/ds > 0$.

Now let $\mathbf{y}$ be the reparametrization $\mathbf{y}(s) = \mathbf{x}(t(s))$ of $\mathbf{x}$. Then by Lemma 1.1.5 we have

$$\mathbf{y}'(s) = \frac{dt}{ds}(s)\mathbf{x}'(t(s)),$$

which together with (1.2.2) thus implies that

$$\|\mathbf{y}'(s)\| = \frac{dt}{ds}(s)\|\mathbf{x}'(t(s))\| = \frac{dt}{ds}(s)\frac{ds}{dt}(s) = 1.$$

1.2.7. Definition. A reparametrization $\mathbf{x}(s)$ of a curve $\mathbf{x}(t)$ by h so that $t = h(s)$ is *orientation preserving* if $h' \geqslant 0$, that is, if h is monotone increasing, and is *orientation reversing* if $h' \leqslant 0$, that is, if h is monotone decreasing.

In the latter case of the definition above, $\mathbf{x}(t)$ and $\mathbf{x}(s)$ have the same image set but are described in opposite directions. The reparametrization above by the arc length s is always orientation preserving, since $ds/dt > 0$ for a curve $\mathbf{x}(t)$.

In Definition 3.5.2 of Chapter 1 we defined a vector field on E^3. Now we use a little variation to define a vector field on a curve $\mathbf{x}$: $I \to E^3$, and then follow Definitions 3.5.6 and 5.4.1 of Chapter 1 to define oriented frame fields on the curve **x**.

1.2.8. Definition. A *vector field* **a** *on a curve x*: $I \to E^3$ is a function that assigns to each number t in I a vector $\mathbf{a}(t)$ of E^3 at the point $\mathbf{x}(t)$. A vector field **a** on a curve **x** is a *unit vector field* if each vector $\mathbf{a}(t)$ is a unit vector.

We have already known such a vector field, namely, the vector field formed by all the tangent vectors $\mathbf{x}'(t)$ of the curve **x**. However, for a general vector field **a** each vector $\mathbf{a}(t)$ need not be tangent to the curve **x**.

1.2.9. Definition. A set of three unit vector fields $\mathbf{e}_1, \mathbf{e}_2, \mathbf{e}_3$, on a curve **x**: $I \to E^3$ is a *frame field* on the curve **x**, denoted by $\mathbf{x}\mathbf{e}_1\mathbf{e}_2\mathbf{e}_3$, if $\mathbf{e}_1(t), \mathbf{e}_2(t), \mathbf{e}_3(t)$ are mutually orthogonal at each point $\mathbf{x}(t)$. A frame field $\mathbf{x}\mathbf{e}_1\mathbf{e}_2\mathbf{e}_3$ is *positively oriented* (or, *right-handed*) or *negatively oriented* (or, *left-handed*) depending on whether

$$|\mathbf{e}_1(t), \mathbf{e}_2(t), \mathbf{e}_3(t)| = +1 \text{ or } -1 \tag{1.2.4}$$

at each point $\mathbf{x}(t)$.

The properties of vector fields on curves in E^3 are analogous to those of vector fields on E^3. For example, if **a** is a vector field on a curve **x**: $I \to E^3$, then for each t in I we can write (cf. Lemma 3.5.4 of Chapter 1)

$$\mathbf{a}(t) = (a_1(t), a_2(t), a_3(t))_{\mathbf{x}(t)} = \sum_{i=1}^{3} a_i(t)\mathbf{u}_i(x(t)), \tag{1.2.5}$$

where $\mathbf{u}_i$'s are defined by (3.5.4) of Chapter 1, and the real-valued functions a_1, a_2, a_3 on I are called the *Euclidean coordinate functions* of **a**.

The operations of addition, scalar multiplication, inner product, and vector product of vector fields on the same curve are all defined in the usual pointwise fashion (cf. Sections 3.1–3.3 of Chapter 1).

To differentiate a vector field **a** on a curve **x**: $I \to E^3$, we simply differentiate its Euclidean coordinate functions a_1, a_2, a_3 given by (1.2.5) and thus obtain a new vector field on **x**: $\mathbf{a}' = \sum_{i=1}^{3}(da_i/dt)\mathbf{u}_i$.

In E^3, corresponding to different properties of a mapping, we may have different types of curve.

1.2.10. Definition. A mapping **x**: $I \to E^3$ is a *curve of class* C^k if each of the coordinate functions in the expression $\mathbf{x}(t) = (x_1(t), x_2(t), x_3(t))$ has continuous derivatives of orders up to and including k. A curve of class C^0 is merely a continuous curve.

1.2.11. Definition. A curve **x** in E^3 is *simple* if **x** is injective, that is, if $\mathbf{x}(t) = \mathbf{x}(t_1)$ only when $t = t_1$. A curve **x**: $[a, b] \to E^3$ is closed if $\mathbf{x}(a) = \mathbf{x}(b)$, and is closed of class C^k if it is closed and $\mathbf{x}'(a)=\mathbf{x}'(b),\cdots,\mathbf{x}^{(k)}(a)=\mathbf{x}^{(k)}(b)$. For a curve $\mathbf{x}$: $E^1\to E^3$ if there is a number $p>0$ such that $\mathbf{x}(t+p)=\mathbf{x}(t)$ for all t, the curve $\mathbf{x}$ is *periodic*, and the least such number p is the *period* of $\mathbf{x}$.

A differentiable function that has arbitrarily small periods must be constant (Exercise 8 of Section 1.2).

The curve in item 4 of Examples 1.1.2 is not simple, and the period of a closed periodic curve $\mathbf{x}(s)$ with arc length s is the arc length of the curve.

1.2.12. Definition. A piecewise (or sectionally) regular or smooth curve (see Definition 1.2.2) $\mathbf{x}$: $I=[a,b]\to E^3$ with vertices $\mathbf{x}(t_i)$, $i=0,1,\cdots,n$ and $a=t_0<t_1<\cdots<t_n=b$, is *closed* if $\mathbf{x}(a)=\mathbf{x}(b)$.

The simplest example of a closed piecewise smooth curve is a rectilinear polygon. It should be remarked that many interesting theorems in differential geometry are valid for piecewise smooth curves.

1.2.13. Definition. A curve C in E^3 is a *knot* if there is a homeomorphism of the unit circle

$$x_1^2+x_2^2=1, \qquad x_3=0 \tag{1.2.6}$$

into E^3, whose image is C. Two knots C_1 and C_2 are *equivalent* if there is a homeomorphism of E^3 onto itself, which maps C_1 onto C_2.

The simplest of the common knots are the overhand knot (often called the cloverleaf knot or the trefoil) in Fig. 2.5 and the figure-eight knot (also called the four-knot or Listing's knot) in Fig. 2.6.

It is obvious that the relation of knot equivalence is an equivalence relation, that is, it is reflexive, symmetric, and transitive.

1.2.14. Definition. Equivalent knots are of the same *type*, and each equivalence class of knots is a *knot type*. Those knots equivalent to the circle (1.2.6) are *trivial*.

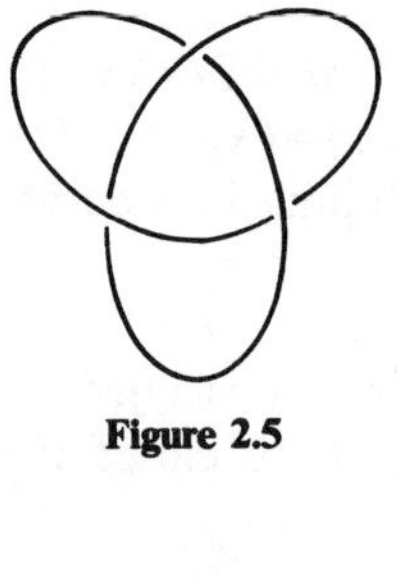

Figure 2.5

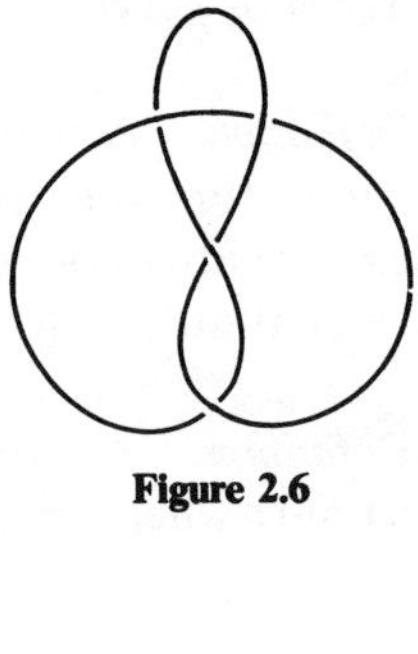

Figure 2.6

1.2.15. Theorem (for reference). *Any knot that lies in a plane is necessarily trivial.*

This is a well-known and deep theorem of plane topology. For a proof see A. H. Newman [42, p. 173].

1.2.16. Definition. A polygonal knot is the union of a finite number of closed straight-line segments called *edges*, whose end points are the *vertices* of the knot. A knot is *tame* if it is equivalent to a polygonal knot; otherwise it is *wild*.

1.2.17. Theorem (for reference). *If a knot in E^3 parametrized by the arc length is of class C^1, it is tame.*

For a proof see R. H. Crowell and R. H. Fox [18, Appendix I].

Exercises

1. For the helix (1.1.3), (a) express the coordinate functions in terms of its arc length s; (b) find the unit tangent vector at a general point; (c) find its arc length of one turn.
2. Show that all the tangent lines of the curve $\mathbf{x}(t)=(2t,3t^2,3t^3)$ make a constant angle with the line $x_2=0, x_1=x_3$.
3. Find the arc length of a complete arch of the cycloid
$$\mathbf{x}(t)=(a(t-\sin t), a(1-\cos t), 0).$$
4. Let $\mathbf{a}$ be a vector field on the curve $\mathbf{x}(t)=(e^t, e^{-t}, \sqrt{2}\, t)$. In each of the following cases, express $\mathbf{a}$ in the form $\Sigma_{i=1}^3 a_i\mathbf{u}_i$ given that:
(a) $\mathbf{a}(t)$ is the vector from the point $\mathbf{x}(t)$ to the origin of E^3;
(b) $\mathbf{a}(t)=\mathbf{x}'(t)-\mathbf{x}''(t)$;
(c) $\mathbf{a}(t)$ has unit norm and is orthogonal to both $\mathbf{x}'(t)$ and $\mathbf{x}''(t)$.
5. Show that the curve given by
$$\mathbf{x}(t)=\begin{cases}(t^2,t^2), & t\geqslant 0,\\ (t^2,-t^2), & t\leqslant 0\end{cases}$$
is of class C^1 but not of class C^2. Make a sketch of the curve and its tangent vectors.
6. Use the following scheme to prove that in E^3 a straight line is the shortest distance between two points. Let $\mathbf{x}$: $[a,b]\to E^3$ be a curve, and set $\mathbf{x}(a)=\mathbf{p}, \mathbf{x}(b)=\mathbf{q}$.

(a) Let **u** be any fixed unit vector in E^3. Use (3.2.5) of Chapter 1 to show that

$$(\mathbf{q}-\mathbf{p})\cdot\mathbf{u}=\int_a^b \mathbf{x}'(t)\cdot\mathbf{u}\,dt \leqslant \int_a^b \|\mathbf{x}'(t)\|\,dt.$$

(b) Use

$$\mathbf{u}=\frac{\mathbf{q}-\mathbf{p}}{\|\mathbf{q}-\mathbf{p}\|}$$

to show that

$$\|\mathbf{x}(b)-\mathbf{x}(a)\| \leqslant \int_a^b \|\mathbf{x}'(t)\|\,dt.$$

7. The following example shows that the arc length of a continuous curve in a closed interval may be unbounded. Let $\mathbf{x}$: $[a,b]\to E^2$ be given by

$$\mathbf{x}(t)=\begin{cases}\left(t, t\sin\dfrac{\pi}{t}\right), & \text{for } t\neq 0,\\ (0,0), & \text{for } t=0.\end{cases}$$

(a) Sketch the graph of the curve **x** and, considering the point $t=1/(n+\frac{1}{2})$, use Exercise 6 to show geometrically that the arc length of the portion of the curve **x** corresponding to $1/(n+1)\leqslant t\leqslant 1/n$ is at least $2/(n+\frac{1}{2})$.

(b) Use part a to show that the arc length of the curve **x** in the interval $1/(N+1)\leqslant t\leqslant 1$ is greater than $2\Sigma_{n=1}^{N}1/(n+\frac{1}{2})$, therefore tends to infinity as $N\to\infty$.

*8. Show that a differentiable function that has arbitrarily small periods must be constant.

1.3. The Frenet Formulas. In this section we shall use the method of frame fields (or moving frames) to derive the Frenet formulas for a curve in E^3, which are fundamental in the study of differential geometry of curves in E^3.

Let $\mathbf{x}$: $I\to E^3$ be a regular curve, and let $\mathbf{x}\mathbf{e}_1\mathbf{e}_2\mathbf{e}_3(t)$ be a positively oriented frame field on the curve **x**. Suppose **x** to be of class C^4, and $\mathbf{e}_1,\mathbf{e}_2,\mathbf{e}_3$ of class C^k, $k\geqslant 1$. (Although in most of our discussions in this chapter the order of the class of the curve **x** need not be so high, for simplicity we assume this order to cover all the cases.) Then from Definition 1.2.9 we have

$$\mathbf{e}_i(t)\cdot\mathbf{e}_j(t)=\delta_{ij}=\begin{cases}1, & i=j,\\ 0, & i\neq j,\end{cases} \tag{1.3.1}$$

$$|\mathbf{e}_1(t),\mathbf{e}_2(t),\mathbf{e}_3(t)|=1. \tag{1.3.2}$$

Since $\mathbf{e}_1, \mathbf{e}_2, \mathbf{e}_3$ are linearly independent, we can write

$$\frac{d\mathbf{x}}{dt} = \sum_{j=1}^{3} p_j(t)\mathbf{e}_j, \qquad \frac{d\mathbf{e}_i}{dt} = \sum_{j=1}^{3} q_{ij}(t)\mathbf{e}_j, \quad i=1,2,3, \tag{1.3.3}$$

where

$$p_j(t) = \frac{d\mathbf{x}}{dt}\cdot\mathbf{e}_j, \qquad q_{ij}(t) = \frac{d\mathbf{e}_i}{dt}\cdot\mathbf{e}_j, \qquad i, j = 1,2,3. \tag{1.3.4}$$

Differentiating (1.3.1) we obtain

$$\frac{d\mathbf{e}_i}{dt}\cdot\mathbf{e}_j + \mathbf{e}_i\cdot\frac{d\mathbf{e}_j}{dt} = 0.$$

Substitution of (1.3.3) in the equation above gives

$$q_{ij} + q_{ji} = 0. \tag{1.3.5}$$

In other words, the functions q_{ij} are skew symmetric in i, j; therefore $q_{ii} = 0$, $i = 1,2,3$.

For the theory of curves we take t to be the arc length s, and $\mathbf{e}_1$ the unit tangent vector. Then

$$p_1 = 1, \qquad p_2 = p_3 = 0, \tag{1.3.6}$$

and we shall write s for t. Notice that both $\mathbf{e}_1$ and s are determined up to a sign; geometrically they are determined by the orientation of the curve. We shall say that the curve is *oriented* when a definite choice of $\mathbf{e}_1$ and s has been made.

The method of frame fields is to find a frame field $\mathbf{x}\mathbf{e}_1\mathbf{e}_2\mathbf{e}_3$ on the curve $\mathbf{x}$ such that equations 1.3.3 take a form as simple as possible. This method is quite commonly used in analytic geometry to reduce the equation of a geometric figure such as a quadric surface to a canonical form by a proper choice of the coordinate system. In differential geometry the figures under consideration are more general; thus a frame field must be used.

To keep $\mathbf{e}_1$ unchanged, the most general possible change of the frame $\mathbf{x}\mathbf{e}_1\mathbf{e}_2\mathbf{e}_3$ is given by

$$\mathbf{e}_2 = \mathbf{e}_2^* \cos\theta + \mathbf{e}_3^* \sin\theta, \qquad \mathbf{e}_3 = -\mathbf{e}_2^* \sin\theta + \mathbf{e}_3^* \cos\theta, \tag{1.3.7}$$

where θ is a function of s; this is because there is a rotation in the $\mathbf{e}_2\mathbf{e}_3$-plane (see Fig. 2.7). The frame $\mathbf{x}\mathbf{e}_1\mathbf{e}_2^*\mathbf{e}_3^*(s)$ satisfies a system of differential equations of the same form as (1.3.3), whose coefficients are

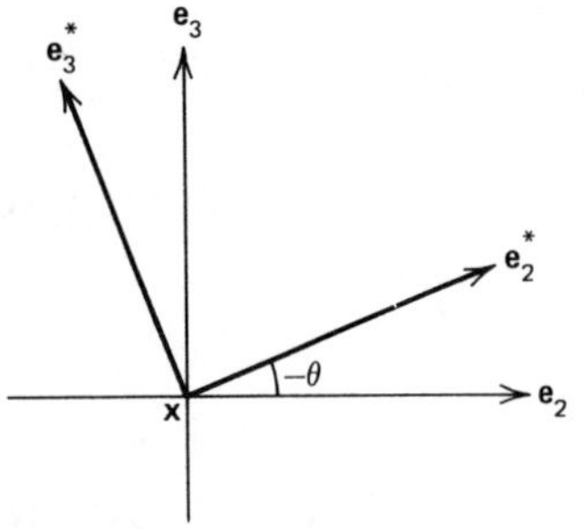

Figure 2.7

denoted by the same symbols with asterisks. Then we find

$$q_{12}^* = q_{12}\cos\theta - q_{13}\sin\theta,$$
$$q_{13}^* = q_{12}\sin\theta + q_{13}\cos\theta, \tag{1.3.8}$$

from which it follows that

$$q_{12}^{*2} + q_{13}^{*2} = q_{12}^2 + q_{13}^2. \tag{1.3.9}$$

The quantity $\kappa(s) = \sqrt{q_{12}^2 + q_{13}^2}$ is therefore independent of the choice of the vectors $\mathbf{e}_2, \mathbf{e}_3$, provided of course that $\mathbf{e}_1$ remains the tangent vector. We call $\kappa(s)$ the *curvature* of the curve at the point $\mathbf{x}(s)$.

In a neighborhood of the point $\mathbf{x}(s)$ suppose $\kappa(s) \neq 0$. Then $\theta(s)$ can be so determined that

$$q_{12}\sin\theta + q_{13}\cos\theta = 0,$$

that is, $q_{13}^* = 0$. This defines $\theta(s)$ up to a multiple of π, and $\mathbf{e}_2^*$ up to a sign. Thus we have Theorem 1.3.1.

1.3.1. Theorem. *In a neighborhood of a point* $\mathbf{x}(s)$ *on a curve in* E^3 *with arc length s and nonzero curvature there exist four frame fields*

$$\mathbf{x}\mathbf{e}_1\mathbf{e}_2\mathbf{e}_3, \qquad \mathbf{x}\mathbf{e}_1(-\mathbf{e}_2)(-\mathbf{e}_3),$$
$$\mathbf{x}(-\mathbf{e}_1)\mathbf{e}_2(-\mathbf{e}_3), \qquad \mathbf{x}(-\mathbf{e}_1)(-\mathbf{e}_2)\mathbf{e}_3, \tag{1.3.10}$$

for each of which $\mathbf{e}_1$ *is a unit tangent vector and* $\mathbf{e}_3 \cdot d\mathbf{e}_1/ds = 0$.

1.3.2. Definition. The frame fields (1.3.10) are called *Frenet frame fields*. With respect to a Frenet frame field, equations 1.3.3 become

$$\frac{d\mathbf{x}}{ds} = \mathbf{e}_1, \qquad \frac{d\mathbf{e}_1}{ds} = \kappa\mathbf{e}_2,$$
$$\frac{d\mathbf{e}_2}{ds} = -\kappa\mathbf{e}_1 + \tau\mathbf{e}_3, \qquad \frac{d\mathbf{e}_3}{ds} = -\tau\mathbf{e}_2, \tag{1.3.11}$$

and are called the *Frenet formulas* for the curve $\mathbf{x}$. $\kappa(s)$ and $\tau(s)$ are called the *curvature* and the *torsion*, and their reciprocals the *radius of curvature* and the *radius of torsion* of the curve at the point $\mathbf{x}(s)$, respectively. The lines $\mathbf{x}\mathbf{e}_2, \mathbf{x}\mathbf{e}_3$ are called the *principal normal* and the *binormal*, and the planes $\mathbf{x}\mathbf{e}_1\mathbf{e}_2$, $\mathbf{x}\mathbf{e}_2\mathbf{e}_3$, and $\mathbf{x}\mathbf{e}_1\mathbf{e}_3$, the *osculating plane*, the *normal plane*, and the *rectifying plane* of the curve at the point $\mathbf{x}(s)$, *respectively*. Any line through the point $\mathbf{x}(s)$ in the normal plane is called a *normal line* of the curve at the point $\mathbf{x}(s)$.

It is clear that the principal normal, the binormal, the osculating plane, the normal plane, and the rectifying plane are independent of the type of Frenet frame selected from among the four possible choices. The same is true of the torsion $\tau(s)$, which is therefore an invariant of the curve under orientation-preserving isometries. The curvature κ is not so; only the square of its value is determined by the curve. Among the four possible Frenet frames we can determine a definite one by the following conventions: (i) a choice of $\mathbf{e}_1$, which can be described geometrically by saying that an orientation is given on the curve and (ii) a choice of $\mathbf{e}_2$, so as to make $\kappa(s)>0$. In fact, the second choice is an orientation of the curve in a generalized sense.

The curvature of a curve has the following geometric interpretation.

1.3.3. Theorem. *The curvature of a curve C at a point P is the limit of the ratio of the angle between a tangent vector at P and a tangent vector at a neighboring point P_1 of C, to the arc PP_1, as the point P_1 approaches P along C.*

Proof. Let C be given by $\mathbf{x}(s)$ with arc length s, and let $\Delta\theta$ be the angle between the unit tangent vectors $\mathbf{e}_1(s+\Delta s)$ and $\mathbf{e}_1(s)$ at the points $\mathbf{x}(s+\Delta s)$ and $\mathbf{x}(s)$, respectively. Then from the isosceles triangle (Fig. 2.8) it follows immediately that

$$\|\mathbf{e}_1(s+\Delta s)-\mathbf{e}_1(s)\| = 2\sin\frac{1}{2}\Delta\theta.$$

Thus

$$\lim_{\Delta s\to 0}\frac{\Delta\theta}{\Delta s} = \lim_{\Delta\theta\to 0}\frac{\frac{1}{2}\Delta\theta}{\sin\frac{1}{2}\Delta\theta}\lim_{\Delta s\to 0}\frac{\|\mathbf{e}_1(s+\Delta s)-\mathbf{e}_1(s)\|}{\Delta s} = \left\|\frac{d\mathbf{e}_1}{ds}\right\| = \kappa.$$

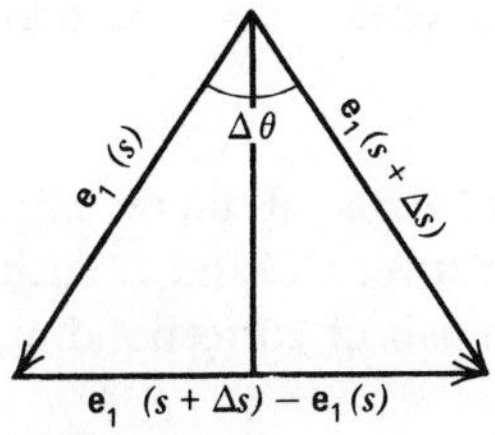

Figure 2.8

Similarly we can have the following geometric interpretation of the torsion of a curve.

1.3.4. Theorem. *The torsion of a curve C at a point P is the limit of the ratio of the angle between a binormal vector at P and a binormal vector at a neighboring point P_1 of C, to the arc PP_1, as the point P_1 approaches P along C.*

Identical vanishing of the curvature or torsion of a curve has the following significance.

1.3.5. Lemma. *A curve is a straight line if its $\kappa(s)\equiv 0$. A curve lies in a plane if its $\tau(s)\equiv 0$.*

Proof. In the first case, from the second equation of (1.3.11) it follows that $\mathbf{e}_1$ is a constant vector. Integrating the first equation of (1.3.11), we obtain

$$\mathbf{x}(s)=\mathbf{e}_1 s+\mathbf{c}, \qquad \mathbf{c}=\text{a constant vector},$$

which is the equation of a straight line [cf. (1.1.2)].

In the second case, from the last equation of (1.3.11) it follows that $\mathbf{e}_3$ is a constant vector. Since $\mathbf{e}_1\cdot\mathbf{e}_3=0$, we have

$$\frac{d}{ds}(\mathbf{x}\cdot\mathbf{e}_3)=\mathbf{e}_1\cdot\mathbf{e}_3=0.$$

Hence $\mathbf{x}\cdot\mathbf{e}_3=c=\text{constant}$, and the curve lies in a plane.

1.3.6. Lemma. *If all the tangents to a curve pass through a fixed point, the curve is a straight line.*

Proof. Let $\mathbf{a}$ be the fixed point, and $\mathbf{x}(s)$ the curve with arc length s (Fig. 2.9). Then

$$\mathbf{x}=\mathbf{a}-c\mathbf{x}', \tag{1.3.12}$$

where the prime denotes the derivative with respect to s, and c is not zero. Differentiation of (1.3.12) with respect to s and use of the first two equations of (1.3.11) give

$$(1+c')\mathbf{x}'=-c\kappa\mathbf{e}_2. \tag{1.3.13}$$

Taking the inner product of (1.3.13) with $\mathbf{e}_2$, we thus obtain $\kappa=0$, so that the curve is a straight line by Lemma 1.3.5.

1.3.7. Definition. A curve C is called a *Bertrand curve* if there exist another curve $\bar{C}$, distinct from C, and a bijection f between C and $\bar{C}$ such that C and $\bar{C}$ have the same principal normal at each pair of corresponding

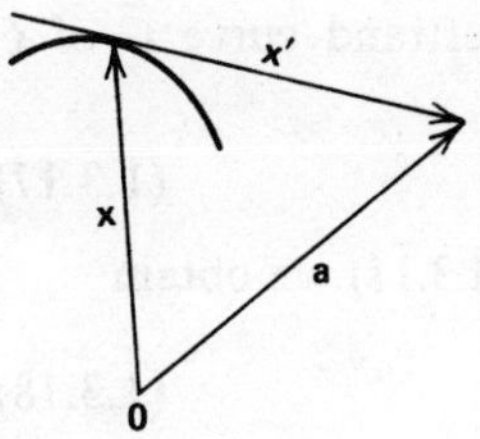

Figure 2.9

points under f, and then both C and $\bar{C}$ are called *associated Bertrand curves*.

The existence of Bertrand curves is at once evident from the following theorem.

1.3.8. Theorem. *Every plane curve is a Bertrand curve.*

Proof. The theorem is obvious for a straight line. Now let C_0 be the locus of the centers of curvature of a nonrectilinear plane curve C given by $\mathbf{x}(s)$ with arc length s. Then the center of curvature (for the definition see Definition 1.4.2.5) of C at the point $\mathbf{x}(s)$ is

$$\mathbf{x}_0 = \mathbf{x} + \frac{1}{\kappa}\mathbf{e}_2, \tag{1.3.14}$$

where κ and $\mathbf{e}_2$ are, respectively, the curvature and the unit principal normal vector of C at $\mathbf{x}(s)$. Differentiating (1.3.14) with respect to s and using the first and the third equations of (1.3.11) with $\tau = 0$, we obtain

$$\frac{d\mathbf{x}_0}{ds} = -\frac{\kappa'}{\kappa^2}\mathbf{e}_2, \tag{1.3.15}$$

where the prime denotes the derivative with respect to s. Thus the curve C has the tangents of C_0 for principal normals, and the same is true of any other orthogonal trajectory $\bar{C}$ of the tangents of C_0 (for the case $\kappa' \equiv 0$, see Exercise 19 of Section 1.3). Hence C and $\bar{C}$ have the same principal normals.

1.3.9. Theorem. *A curve C with nonzero curvature κ and nonzero torsion τ is a Bertrand curve if and only if it satisfies the condition*

$$a\kappa + b\tau = 1, \tag{1.3.16}$$

where a and b are constants.

Proof. Let the curve C be given by $\mathbf{x}(s)$ with arc length s, and let $\mathbf{e}_1(s)\mathbf{e}_2(s)\mathbf{e}_3(s)$ be a Frenet frame of C at $\mathbf{x}(s)$. At first suppose that C is a

Bertrand curve. Then the point of an associated Bertrand curve $\bar{C}$ of C corresponding to $\mathbf{x}(s)$ is of the form

$$\bar{\mathbf{x}}(s)=\mathbf{x}(s)+a(s)\mathbf{e}_2(s). \tag{1.3.17}$$

Differentiating (1.3.17) with respect to s and using (1.3.11) we obtain

$$\frac{d\bar{\mathbf{x}}}{ds}=(1-a\kappa)\mathbf{e}_1+a'\mathbf{e}_2+a\tau\mathbf{e}_3, \tag{1.3.18}$$

where the prime denotes the derivative with respect to s. Since $d\bar{x}/ds$ is orthogonal to $\mathbf{e}_2$, $a'=0$ or $a=\text{const}$. Then differentiation of (1.3.18) with respect to s and use of the condition $a'=0$ give

$$\frac{d^2\bar{\mathbf{x}}}{ds^2}=-a\kappa'\mathbf{e}_1+\left[(1-a\kappa)\kappa-a\tau^2\right]\mathbf{e}_2+a\tau'\mathbf{e}_3. \tag{1.3.19}$$

The vector $d^2\bar{\mathbf{x}}(s)/ds^2$ lies in the osculating plane of $\bar{C}$ and is therefore a linear combination of the vectors $d\bar{\mathbf{x}}/ds$ and $\mathbf{e}_2$. Thus the vector $-a\kappa'\mathbf{e}_1+a\tau'\mathbf{e}_3$ is collinear with the vector $d\bar{\mathbf{x}}/ds$. From (1.3.18) it follows that when $a\kappa\neq 1$ we have, because of $\tau\neq 0$,

$$\frac{-a\kappa'}{1-a\kappa}=\frac{\tau'}{\tau}$$

and therefore (1.3.16) by solving this differential equation.

Now let us discuss the excluded case $a\kappa=1$ in which $\kappa=1/a=\text{const}$, so that C is of constant curvature and (1.3.17) becomes (1.3.14) with $\mathbf{x}_0$ replaced by $\mathbf{x}$, which is the equation of the locus of the centers of curvature of C. To show that the $\bar{\mathbf{x}}$ in (1.3.14) generates a Bertrand curve $\bar{C}$ associated with C, we shall find a Frenet frame of $\bar{C}$. From (1.3.18) we readily obtain the tangent vector of $\bar{C}$:

$$\frac{d\bar{\mathbf{x}}}{ds}=\frac{\tau}{\kappa}\mathbf{e}_3, \tag{1.3.20}$$

which is parallel to the binormal vector $\mathbf{e}_3$ of C, since we have assumed $\tau\neq 0$. Differentiation of (1.3.20) with respect to s and use of (1.3.11) give

$$\frac{d^2\bar{\mathbf{x}}}{ds^2}=\frac{\tau'}{\kappa}\mathbf{e}_3-\frac{\tau^2}{\kappa}\mathbf{e}_2.$$

Thus

$$\frac{d\bar{\mathbf{x}}}{ds}\times\frac{d^2\bar{\mathbf{x}}}{ds^2}=-\frac{\tau^3}{\kappa^2}\mathbf{e}_3\times\mathbf{e}_2=\frac{\tau^3}{\kappa^2}\mathbf{e}_1, \tag{1.3.21}$$

which means that the binormal vector of $\bar{C}$ is parallel to the tangent vector of C. Hence the corresponding rectifying planes of C and $\bar{C}$ are parallel, so that the corresponding principal normals of C and $\bar{C}$ are parallel, and $\bar{C}$ is a Bertrand curve associated with C. Obviously, (1.3.16) is also satisfied in this case with $a=1/\kappa$ and $b=0$.

Now for the converse, suppose that C satisfies (1.3.16), and define a curve $\bar{C}$ by

$$\bar{\mathbf{x}}=\mathbf{x}+a\mathbf{e}_2. \tag{1.3.22}$$

Differentiating (1.3.22) with respect to s and using (1.3.11) and (1.3.16) we obtain

$$\frac{d\bar{\mathbf{x}}}{ds}=\tau(b\mathbf{e}_1+a\mathbf{e}_3)$$

and therefore a unit tangent vector $\bar{\mathbf{e}}_1$ of $\bar{C}$:

$$\bar{\mathbf{e}}_1=\frac{(b\mathbf{e}_1+a\mathbf{e}_3)}{\sqrt{a^2+b^2}}. \tag{1.3.23}$$

Form (1.3.23) it follows that

$$\frac{d\bar{\mathbf{e}}_1}{ds}=\frac{b\kappa-a\tau}{\sqrt{a^2+b^2}}\mathbf{e}_2,$$

which implies $\bar{\mathbf{e}}_2(s)=\pm\mathbf{e}_2(s)$. Hence C is a Bertrand curve.

1.3.10. Corollary. *A curve C of constant curvature κ is a Bertrand curve. Furthermore, if the torsion of C is not constant, the associated curve of C is the locus of the centers of curvature of C and has the same curvature κ as C.*

Proof. If κ is nonzero, then in the proof of Theorem 1.3.9 we have shown that C is a Bertrand curve with, as its associated curve, the locus of the centers of curvature of C, and (1.3.16) is reduced to $a=1/\kappa$, $b=0$.

Furthermore, using Exercise 3 and (1.3.21) and (1.3.20), we obtain the curvature of $\bar{C}$:

$$\bar{\kappa}=\left\|\frac{d\bar{\mathbf{x}}}{ds}\times\frac{d^2\bar{\mathbf{x}}}{ds^2}\right\|\left\|\frac{d\bar{\mathbf{x}}}{ds}\right\|^{-3}=\frac{|\tau|^3}{\kappa^2}\left(\frac{|\tau|}{\kappa}\right)^{-3}=\kappa.$$

Thus $\bar{C}$ is a curve of the curvature κ. Since being associated in the sense of Bertrand is a symmetric relation, C must be the locus of the centers of curvature of $\bar{C}$.

1.3.11. Corollary. *A circular helix C ($\kappa=$const and $\tau=$const) has infinitely many associated Bertrand curves, all of which are circular helices with the same axis and pitch as C.*

Proof. By Corollary 1.3.10 a circular helix C is a Bertrand curve, so that (1.3.16) holds for C. An associated curve of such a circular helix C is given by (1.3.17) with $a(s)$ constant. Since, in general, condition (1.3.16) determines the constants a and b unless both κ and τ are constant, and since in our case a can be chosen arbitrarily and b matched subsequently,

C has infinitely many associated curves. Using Exercise 1b it is easy to check that all the associated curves are circular helices with the same axis and pitch as C.

1.3.12. Corollary. *If a curve C has more than one associated Bertrand curve, it has infinitely many associated Bertrand curves, and this case occurs if and only if C is a circular helix.*

Proof. Suppose that a curve C has two distinct associated Bertrand curves given by

$$\bar{\mathbf{x}}_1(s)=\mathbf{x}(s)+a_1\mathbf{e}_2(s), \qquad \bar{\mathbf{x}}_2(s)=\mathbf{x}(s)+a_2\mathbf{e}_2(s),$$

where a_1 and a_2 are two distinct constants. Then by Theorem 1.3.9 there are two constants b_1 and b_2 such that

$$1-a_1\kappa=b_1\tau, \qquad 1-a_2\kappa=b_2\tau. \tag{1.3.24}$$

By eliminating τ from (1.3.24) we obtain $(a_2b_1-a_1b_2)\kappa=b_1-b_2$, which implies that $a_2b_1\neq a_1b_2$, since $b_1\neq b_2$. Thus $\kappa=\text{const}$. Similarly, $\tau=\text{const}$. Using uniqueness theorem 1.5.2, we hence see that the circular helix is the only such curve.

Exercises

1. For the helix (1.1.3), find (cf. Exercise 1 of Section 1.2) (a) the curvature and torsion (note that their ratio is constant); (b) the unit principal normal and binormal vectors; (c) the equations of the osculating, normal, and rectifying planes.
2. Find the curvature and torsion for each of the following curves:
   (a) $\mathbf{x}(s)=(\frac{5}{13}\cos s, \frac{8}{13}-\sin s, -\frac{12}{13}\cos s)$.
   (b) $\mathbf{x}(t)=(3t-t^3, 3t^2, 3t+t^3)$.
   (c) $\mathbf{x}(t)=(1+t^2, t, t^3)$.
3. Prove that for the curvature κ and the torsion τ of a curve $\mathbf{x}(t)$,
   (a) $\kappa=\|\mathbf{x}'\times\mathbf{x}''\|\|\mathbf{x}'\|^{-3}$,
   (b) $\kappa^2\tau=|\mathbf{x}',\mathbf{x}'',\mathbf{x}'''|\|\mathbf{x}'\|^{-6}$,
   where the primes denote the derivative with respect to t.
4. Prove Theorem 1.3.4.
5. Consider the mapping

$$\mathbf{x}(t)=\begin{cases}(t, e^{-1/t^2}, 0), & t<0,\\ (0,0,0), & t=0,\\ (t, 0, e^{-1/t^2}), & t>0.\end{cases}$$

(a) Prove that $\mathbf{x}(t)$ is regular for all t.
(b) Prove that for the curvature $\kappa(t)$ of the curve $\mathbf{x}(t)$, $\kappa(0)=0$.

6. Prove that if all the tangents to a curve are parallel, the curve is a straight line.

7. Prove that all the normal planes of the curve $\mathbf{x}(t)=(a\sin^2 t, a\sin t\cos t, a\cos t)$ pass through the origin $(0,0,0)$.

8. Prove that if all the osculating planes of a curve pass through a fixed point, the curve is a plane curve.

***9.** Prove that if all the normal planes of a curve $\mathbf{x}(t)$ pass through a fixed point $\mathbf{a}$, the curve is spherical.

***10.** Prove that the curve $\mathbf{x}(\theta)=(-\cos 2\theta, -2\cos\theta, \sin 2\theta)$ is spherical.

11. Prove that if all the principal normals of a curve pass through a fixed point, the curve is either a circular arc or a whole circle.

12. A *cylindrical helix* is defined to be a curve that intersects a given direction at a constant angle, and the direction is called the *axis* of the helix. Prove that a necessary and sufficient condition for a curve to be a cylindrical helix is that the ratio of its curvature to its torsion be constant.

***13.** (a) Prove that the curve $\mathbf{x}(t)=(at, bt^2, t^3)$ with constants a and b is a cylindrical helix if and only if $4b^4=9a^2$.
(b) What is the axis of the helix in this case?

14. Prove that the curve

$$\mathbf{x}(s)=\left(\frac{a}{c}\int \sin\theta(s)\,ds,\ \frac{a}{c}\int \cos\theta(s)\,ds,\ \frac{b}{c}s\right), \qquad a^2=b^2+c^2,$$

is a cylindrical helix, and $\kappa/\tau=b/a$.

15. Prove that if the principal normals of a curve are all parallel to a fixed plane, the curve is a cylindrical helix.

16. Find a function $f(t)$ such that all the principal normals of the curve $\mathbf{x}(t)=(x_1,x_2,x_3)$ where $x_1=a\cos t$, $x_2=a\sin t$, $x_3=f(t)$ are parallel to the $\mathbf{x}_1\mathbf{x}_2$-plane.

17. Prove that a nonplanar curve is a cylindrical helix if and only if its tangent indicatrix (or binormal indicatrix) given by Definition 3.1.2 (or a similar definition) is a plane curve, hence a circular arc or a whole circle.

18. Prove that associated Bertrand curves intercept a constant distance on their common principal normals.

19. Prove that two concentric circles in the same plane are associated Bertrand curves.

***20.** Prove that the angle ω between the tangents at corresponding points of a Bertrand curve C and an associated Bertrand curve $\bar{C}$ is constant.

***21.** Let C be a curve with nonzero curvatures κ and nonzero torsion τ. Prove that if C is a Bertrand curve, it satisfies the condition

$$a\kappa \sin\omega + a\tau \cos\omega = \sin\omega,$$

where a is a constant, and ω is the angle between the tangents at corresponding points of C and an associated Bertrand curve $\bar{C}$. Conversely, prove that if C satisfies the condition above with constant ω, C is a Bertrand curve.

***22.** Prove that the product of the torsions at corresponding points of two associated Bertrand curves is constant and positive.

23. Prove that if a curve C is such that there exists another curve, distinct from C, with the same binormals, the curve C is a plane curve, and the two curves intercept a constant distance on their common binormals.

***24.** A curve C is called a *Mannheim curve* if there exist another curve C, distinct from C, and a bijection f between C and $\bar{C}$ such that the principal normal at each point of C is the binormal of $\bar{C}$ at the corresponding point under f. Prove that a curve C with nonzero curvature κ and nonzero torsion τ is a Mannheim curve if and only if it satisfies the condition

$$(1 - a\kappa)\kappa - a\tau^2 = 0,$$

where a is a constant.

25. Let C and $\bar{C}$ be two curves such that $\bar{C}$ intersects the tangents of C orthogonally. Then $\bar{C}$ is called an *involute* of C, and C an *evolute* of $\bar{C}$ (Fig. 2.10).

In other words, an involute of a curve is an *orthogonal trajectory* of

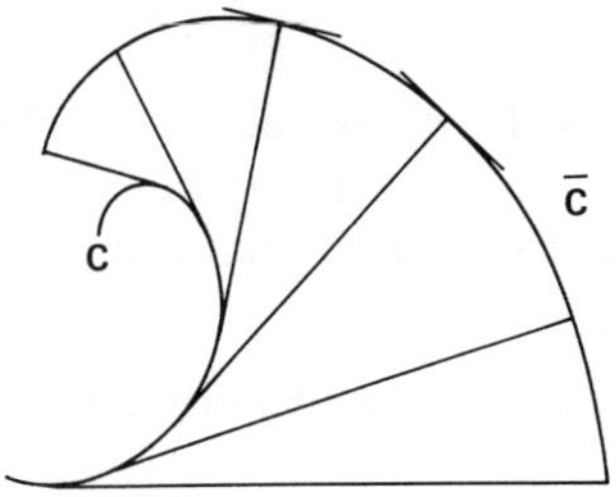

Figure 2.10

the tangents of the curve. Prove each of the following:

(a) An involute of a curve $\mathbf{x}(s)$ with arc length s is given by

$$\bar{\mathbf{x}}(s)=\mathbf{x}(s)+(c-s)\mathbf{e}_1(s),$$

where c is a constant, and $\mathbf{e}_1(s)$ is the unit tangent vector of the curve $\mathbf{x}(s)$.

(b) The tangent at a point of an involute of a curve C is orthogonal to the rectifying plane at the corresponding point of C.

(c) The distance between two involutes of a curve C, measured along the tangents of C, is constant; and the tangents of the two involutes at points on each tangent of C are parallel.

(d) If one end of a string of constant length l is fastened at the point where $s=0$ on a curve C, and if the string is held taut and wound along C so as to remain tangent to C, the locus of the other end of the string is an involute of C.

(e) An evolute of a curve $\mathbf{x}(s)$ with arc length s, nonzero curvature κ, and torsion τ is given by

$$\bar{\mathbf{x}}(s)=\mathbf{x}(s)+a(s)\mathbf{e}_2(s)+b(s)\mathbf{e}_3(s),$$

where $\mathbf{e}_2(s)$, $\mathbf{e}_3(s)$ are unit principal normal and binormal vectors of $\mathbf{x}(s)$, respectively, and

$$a=\frac{1}{\kappa}, \quad \frac{a'-b\tau}{a}=\frac{b'+a\tau}{b},$$

the prime denoting the derivative with respect to s.

(f) Any two involutes of a plane curve are associated Bertrand curves.

26. Two curves are said to be *parallel* if between them there is a bijection under which corresponding points are equally distant and the tangents of the curves at corresponding points are parallel. By Exercise 25c, two involutes of a curve C are parallel curves, for which corresponding points are on a tangent of C.

*(a) Prove that the curves parallel to a nonrectlinear curve C are the curves obtainable from C by a translation and the orthogonal trajectories of the normal planes of C.

(b) Find all curves parallel to the circle

$$\mathbf{x}(s)=\left(r\cos\frac{s}{r}, r\sin\frac{s}{r}, k\right),$$

where r, k are constants, and s is the arc length of the circle.

***27.** Let C be a closed C^3 curve given by (1.3.3) with $d\mathbf{x}/dt=\mathbf{e}_1$ and t replaced by the arc length s. Show that $\frac{1}{2\pi}\int_C q_{23}\,ds$, mod 1, is invariant under a rotation field given by (1.3.7) along C. $\frac{1}{2\pi}\int_C q_{23}\,ds$ is called the *total twist number* of the curve C [F. B. Fuller, 22]. For an interesting application of this number to molecular biology see F. H. C. Crick [17].

1.4. Local Canonical Form and Osculants

1.4.1. Local Canonical Form. To study the properties of a curve C in a neighborhood of a point O, we take a Frenet frame $O\mathbf{e}_1^0\mathbf{e}_2^0\mathbf{e}_3^0$ at O as the coordinate frame with respect to which the equations of the curve will have a simple form. Let s be the arc length of C, and for simplicity we assume that $s=0$ at O. Then the coordinates $x_1(s)$, $x_2(s)$, $x_3(s)$ of a point $\mathbf{x}(s)$ on the curve with respect to $O\mathbf{e}_1^0\mathbf{e}_2^0\mathbf{e}_3^0$ are defined by

$$\mathbf{x}(s)=\mathbf{x}(0)+x_1(s)\mathbf{e}_1^0+x_2(s)\mathbf{e}_2^0+x_3(s)\mathbf{e}_3^0. \tag{1.4.1}$$

It is obvious that the osculating, normal, and rectifying planes of the curve C at the point O are given by $x_3=0$, $x_1=0$, and $x_2=0$, respectively.

By successive differentiation from (1.3.11) we find

$$\mathbf{x}'=\mathbf{e}_1,\qquad \mathbf{x}''=\kappa\mathbf{e}_2,\qquad \mathbf{x}'''=-\kappa^2\mathbf{e}_1+\kappa'\mathbf{e}_2+\kappa\tau\mathbf{e}_3, \tag{1.4.2}$$

where the primes denote the derivatives with respect to s. Substitution of (1.4.2) in the Maclaurin's formula

$$\mathbf{x}(s)=\mathbf{x}(0)+x_0's+\frac{1}{2}\mathbf{x}_0''s^2+\frac{1}{3!}\mathbf{x}'''s^3+(4)$$

gives immediately

$$x_1(s)=s-\frac{\kappa_0^2}{6}s^3+(4),$$

$$x_2(s)=\frac{\kappa_0}{2}s^2+\frac{\kappa_0}{6}s^3+(4), \tag{1.4.3}$$

$$x_3(s)=\frac{\kappa_0\tau_0}{6}s^3+(4),$$

where the subscript 0 denotes the value at the point O, and (4) denotes terms of degree $\geqslant 4$ in s. The representation (1.4.3) is called the *local canonical form* of the curve C in a neighborhood of the point where $s=0$.

Much information about the shape of a curve C in a neighborhood of a point O could be acquired by studying the orthogonal projections of C onto the $\mathbf{e}_1^0\mathbf{e}_2^0$-, $\mathbf{e}_2^0\mathbf{e}_3^0$-, and $\mathbf{e}_1^0\mathbf{e}_3^0$-planes. By eliminating s between equations

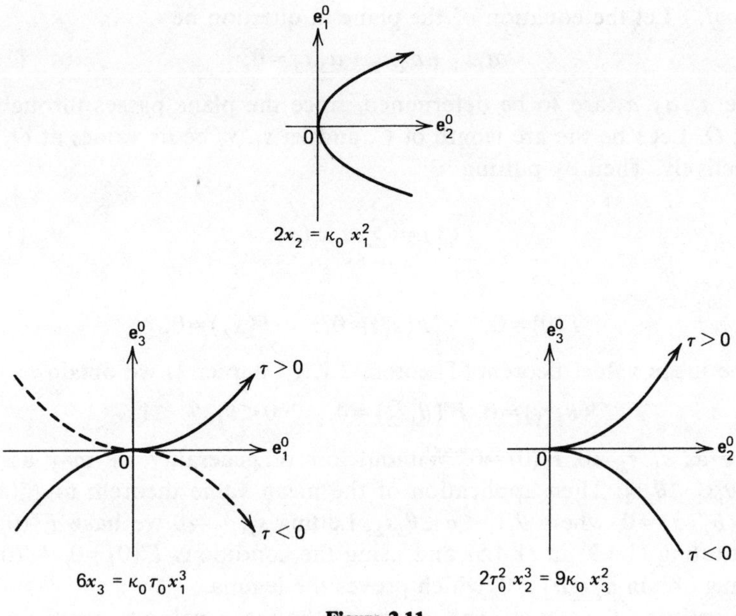

Figure 2.11

(1.4.3), taken in pairs, with all terms after the first in the right-hand member of each equation dropped, we can obtain an approximation to these orthogonal projections, as indicated in Fig. 2.11.

1.4.2. Osculants.

1.4.2.1. Definition. An *osculant of order k* of a curve C at a point O is the limiting position of a geometric figure (such as a straight line, a plane, a circle, or a sphere) through O and k neighboring points on C, as each of these neighboring points independently approaches O along C. In this case, the osculant is said to have *contact of order k* with C at O.

The tangent of a curve at a point is an osculant of order 1. The osculating plane of a curve C at a point O has the equation $x_3 = 0$ in our coordinate system in Section 1.4.1, and is an osculant of order 2 by the following theorem.

1.4.2.2 Theorem. *The osculating plane of a curve C at a point O is the limiting position of the plane through O and two neighboring points O_1, O_2 on C as each of O_1, O_2 independently approaches O along C.*

Proof. Let the equation of the plane in question be

$$a_1x_1 + a_2x_2 + a_3x_3 = 0, \tag{1.4.4}$$

where a_1, a_2, a_3 are to be determined, since the plane passes through the point O. Let s be the arc length of C, and let s_1, s_2 be its values at O_1, O_2, respectively. Then by putting

$$F(s) = \sum_{i=1}^{3} a_i x_i(s), \tag{1.4.5}$$

we have

$$F(0)=0, \qquad F(s_1)=0, \qquad F(s_2)=0.$$

By the mean value theorem (Theorem 2.1.1, Chapter 1) we obtain

$$F'(\theta_1 s_1)=0,\ F'(\theta_2 s_2)=0, \qquad 0<\theta_1,\theta_2<1.$$

Hence as $s_1, s_2 \to 0$, $F'(0)=0$. Without loss of generality we may assume that $\theta_1 s_1 < \theta_2 s_2$. Then application of the mean value theorem to $F'(\theta_2 s_2)$ gives $F''(\sigma)=0$, where $\theta_1 s_1 < \sigma < \theta_2 s_2$. Letting $s_1, s_2 \to 0$ we have $F''(0)=0$. Substituting (1.4.3) in (1.4.5) and using the conditions $F'(0)=0$, $F''(0)=0$ we thus obtain $a_1 = a_2 = 0$, which proves the lemma.

By omitting the direct applications of the mean value theorem we can replace the proof above by the following simple equivalence.

Demanding that (1.4.4) be satisfied by the power series (1.4.3) for x_1, x_2, x_3 identically in s as far as the terms in s^2, we obtain $a_1 = a_2 = 0$ again.

1.4.2.3. Remark. From now on we shall use the simple method just described to find the equation or equations of any osculant of a curve at a point.

The osculating plane has also the following geometric interpretation.

1.4.2.4. Lemma. *The osculating plane of a curve C at a point O where $\kappa\tau \neq 0$ is the only plane through the tangent of C at O, which cuts the curve into two parts.*

Proof. A plane through the tangent of C at O has the equation

$$ax_2 + bx_3 = 0,$$

where a and b are to be determined. Substituting (1.4.3) in the left-hand side of this equation we obtain

$$ax_2(s) + bx_3(s) = \frac{1}{2}a\kappa_0 s^2 + \frac{1}{6}\kappa_0(a + b\tau_0)s^3 + (4),$$

which keeps the same sign for all sufficiently small positive and negative values of s unless $a=0$. Thus the proof is complete.

The tangent and the osculating plane are the simplest geometric figures among all the osculants of a curve at a point. In the following we shall define more osculants of higher orders.

1.4.2.5. Definition. The *osculating circle* of a curve C at a point O is the limiting position of the circle through O and two neighboring points O_1, O_2 on C, as each of O_1, O_2 independently approaches O along C.

Clearly, the osculating circle lies in the osculating plane, and therefore has equations of the form

$$x_3 = 0, \qquad x_1^2 + x_2^2 - 2a_1x_1 - 2a_2x_2 = 0.$$

Let

$$F(s) = x_1^2(s) + x_2^2(s) - 2a_1x_1(s) - 2a_2x_2(s).$$

Demanding that $F(s) = 0$ be satisfied by the power series (1.4.3) for x_1, x_2, x_3 identically in s as far as the terms in s^2, we can easily obtain

$$a_1 = 0, \qquad a_2 = \frac{1}{\kappa_0}.$$

Hence the osculating circle of the curve C at O has radius $1/\kappa_0$ and its center at the point

$$\mathbf{x}(0) + \frac{1}{\kappa_0}\mathbf{e}_2^0, \tag{1.4.6}$$

which is on the principal normal $\mathbf{e}_2^0$ and called the *center of curvature* of the curve C at the point O.

1.4.2.6. Definition. The *osculating sphere* of a curve C at a point O is the limiting position of a sphere through O and three neighboring points O_1, O_2, O_3 on C, as each of O_1, O_2, O_3 independently approaches O along C.

The equation of a sphere through O can be written as

$$x_1^2 + x_2^2 + x_3^2 - 2a_1x_1 - 2a_2x_2 - 2a_3x_3 = 0. \tag{1.4.7}$$

Demanding that (1.4.7) be satisfied by (1.4.3) identically in s as far as the terms in s^3, we obtain

$$a_1 = 0, \qquad a_2 = \frac{1}{\kappa_0}, \qquad a_3 = \frac{1}{\tau_0}\left(\frac{1}{\kappa}\right)_0'.$$

Thus the center of the osculating sphere of C at O is the point

$$\mathbf{x}(s) + \frac{1}{\kappa_0}\mathbf{e}_2 + \frac{1}{\tau_0}\left(\frac{1}{\kappa}\right)_0'\mathbf{e}_3, \tag{1.4.8}$$

and the radius R is given by

$$R^2 = \frac{1}{\kappa_0^2} + \left[\frac{1}{\tau_0}\left(\frac{1}{\kappa}\right)_0'\right]^2. \tag{1.4.9}$$

Exercises

1. Use Theorem 1.4.2.2 and the method in Section 1.4.2 to prove that the equation of the osculating plane of a curve C at a point $\mathbf{x}(t)$ is
$$|\mathbf{X}-\mathbf{x}(t),\mathbf{x}',\mathbf{x}''|=0,$$
where the left-hand side is a determinant, $\mathbf{X}$ is the vector of the natural coordinate functions on E^3, and the primes denote the derivatives with respect to the arc length s of C.

*2. Let C be the cubic parabola $\mathbf{x}(t)=(t,t^2,t^3)$. Show that:
   (a) Through a point Q in space there pass three osculating planes of C at three points P_1, P_2, P_3, respectively.
   (b) Q and P_1, P_2, P_3 are coplanar.

3. Let C be a curve given by $\mathbf{x}(s)$ with arc length s, $\mathbf{c}(s)$ the center of curvature of C at the point $\mathbf{x}(s)$, and K the locus of the center $\mathbf{c}(s)$ for all s. Show that:
   (a) The tangent of K at $\mathbf{c}(s)$ is orthogonal to the tangent of C at $\mathbf{x}(s)$.
   (b) If C is of constant curvature, then the tangent of K at $\mathbf{c}(s)$ is orthogonal to the osculating plane of C at $\mathbf{x}(s)$, and K is also of constant curvature.

4. Prove that the locus of the centers of curvature of a circular helix is a circular helix (cf. Exercise 1 of Section 1.3).

5. Prove that a necessary and sufficient condition that a nonplanar curve $\mathbf{x}(s)$ be a spherical curve is
$$\frac{\tau}{\kappa} + \left[\frac{1}{\tau}\left(\frac{1}{\kappa}\right)'\right]' = 0,$$
where the prime denotes the derivative with respect to the arc length s of the curve.

6. Prove that a curve $\mathbf{x}(s)$ with arc length s is a spherical curve if and only if $\kappa>0$ and there exists a differentiable function $f(s)$ such that $f\tau=(1/\kappa)'$, $f'+\tau/\kappa=0$ [Y. C. Wong, 62].

7. Prove that the only spherical curves of constant curvature are circles.

1.5. Existence and Uniqueness Theorems. This section gives the existence and uniqueness theorems for curves, which are fundamental in the local theory of curves.

1.5.1. Existence Theorem. *Given a C^1 function $\kappa(s)>0$ and a continuous function $\tau(s)$, $s\in I=[a,b]$, there exists a C^3 regular curve* $\mathbf{x}$: $I\to E^3$ *such that s is the arc length, $\kappa(s)$ is the curvature, and $\tau(s)$ is the torsion of the curve* $\mathbf{x}$.

Proof. See Appendix 1.

From Definition 1.3.2 of the Frenet frame fields and Corollary 5.4.5 of Chapter 1, it follows that if an orientation-preserving isometry T carries a curve C into another curve C', the derivative mapping T_* of T carries the Frenet frame fields of C into those of C'; hence C and C' have the same values of $|\kappa|$ and τ at corresponding points, respectively. To this important fact there is a converse.

1.5.2. Uniqueness Theorem. *If there is a bijection between two curves such that the two curves have the same values of s, κ, and τ at corresponding points, respectively, the bijection is an orientation-preserving isometry, and in this case the two curves are said to be congruent.*

It may be useful to establish this theorem in the following more general form.

1.5.3. Generalized Uniqueness Theorem. *Let* $\mathbf{x}\mathbf{e}_1\mathbf{e}_2\mathbf{e}_3(t)$, $t\in I=[a,b]$, *be a positively oriented frame field (Definitions* 3.5.6 *and* 5.4.1, *Chapter* 1) *satisfying the differential equations* (1.3.3), *and* $\mathbf{y}\mathbf{f}_1\mathbf{f}_2\mathbf{f}_3(t)$ *another such frame field satisfying the differential equations*

$$\begin{aligned}\frac{d\mathbf{y}}{dt}&=\sum_j p_j(t)\mathbf{f}_j(t),\\ \frac{d\mathbf{f}_i}{dt}&=\sum_j q_{ij}(t)\mathbf{f}_j(t),\qquad i=1,2,3.\end{aligned}\tag{1.5.1}$$

Then there exists an orientation-preserving isometry whose derivative mapping carries $\mathbf{x}\mathbf{e}_1\mathbf{e}_2\mathbf{e}_3(t)$ *into* $\mathbf{y}\mathbf{f}_1\mathbf{f}_2\mathbf{f}_3(t)$ *for all t.*

Proof. Since the coefficients in (1.3.3) and (1.5.1) are invariant under an orientation-preserving isometry, after performing such an isometry we can assume that

$$\mathbf{x}\mathbf{e}_1\mathbf{e}_2\mathbf{e}_3(t_0)=\mathbf{y}\mathbf{f}_1\mathbf{f}_2\mathbf{f}_3(t_0)\tag{1.5.2}$$

for a certain t_0. It remains to prove that under this isometry $\mathbf{x}\mathbf{e}_1\mathbf{e}_2\mathbf{e}_3(t)$ is identical with $\mathbf{y}\mathbf{f}_1\mathbf{f}_2\mathbf{f}_3(t)$ for all t in the interval I. We suppose for the moment that all indices i, j, l, m run from 1 to 3.

Let (e_{i1}, e_{i2}, e_{i3}) and (f_{i1}, f_{i2}, f_{i3}) be, respectively, the components of the vectors $\mathbf{e}_i$ and $\mathbf{f}_i$ with respect to a fixed positively oriented frame in the space E^3. Written in terms of the components, the differential systems (1.3.3) and (1.5.1) give

$$\frac{de_{il}}{dt} = \sum_j q_{ij} e_{jl}, \qquad \frac{df_{il}}{dt} = \sum_j q_{ij} f_{jl}. \tag{1.5.3}$$

Since the functions q_{ij} are skew symmetric in i and j, using (1.5.3) we obtain

$$\frac{d}{dt}\left(\sum_i e_{im} f_{il}\right) = \sum_{i,j} q_{ij} e_{jm} f_{il} + \sum_{i,j} q_{ij} f_{jl} e_{im} = 0.$$

Hence $\Sigma_i e_{im} f_{il}$ is a constant and is equal to its value for $t = t_0$, that is,

$$\sum_i e_{im}(t) f_{il}(t) = \sum_i e_{im}(t_0) f_{il}(t_0) = \sum_i e_{im}(t_0) e_{il}(t_0) = \delta_{ml}. \tag{1.5.4}$$

For simplicity of notation, we introduce the matrices

$$E = (e_{ij}), \qquad F = (f_{ij}).$$

Then (1.5.4) can be written as

$${}^tFE = \mathcal{I},$$

where ${}^tF$ is the transpose of F, and $\mathcal{I}$ is the identity matrix. Thus ${}^tF$ is the inverse of E. Since by Theorem 5.1.4, Chapter 1, E is an orthogonal matrix, from Definition 5.1.3, Chapter 1, it follows that ${}^tF = {}^tE$ or $F = E$. In other words, $e_i = f_i$, which together with the first equations of (1.3.3) and (1.5.1) gives

$$\frac{d}{dt}(\mathbf{x} - \mathbf{y}) = 0,$$

or, in consequence of (1.5.2),

$$\mathbf{x} - \mathbf{y} = \mathbf{x}(t_0) - \mathbf{y}(t_0) = 0.$$

Hence we have proved that $\mathbf{x}\mathbf{e}_1\mathbf{e}_2\mathbf{e}_3(t) = \mathbf{y}\mathbf{f}_1\mathbf{f}_2\mathbf{f}_3(t)$ for all values of t.

1.5.4. Remark. Generalized uniqueness theorem 1.5.3 is reduced to uniqueness theorem 1.5.2 if the differential systems (1.3.3) and (1.5.1) are, in particular, replaced by the Frenet formulas (1.3.11) and similar ones for $\mathbf{y}\mathbf{f}_1\mathbf{f}_2\mathbf{f}_3(s)$, respectively.

In general, given the curvature κ and the torsion τ, it is very difficult to solve the Frenet equations (1.3.11) and to find the curve $\mathbf{x}(s)$. However, it can (almost) be done in the case of a cylindrical helix.

1.5.5. Example. Determine $\mathbf{x}(s)$ for a cylindrical helix with $\kappa>0$ and $\tau=c\kappa$ for some constant c.

For this purpose it will be useful to reparametrize $\mathbf{x}$ by a parameter t given by

$$t(s)=\int_0^s \kappa(r)\,dr. \tag{1.5.5}$$

Note that this is an allowable change of parameter, since $t'=\kappa>0$, where the prime denotes the derivative with respect to s, implies that $t(s)$ is injective, and that both $t(s)$ and $s(t)$ are differentiable. Substituting $\tau=c\kappa$ in (1.3.11) and using $t'=\kappa$ we have

$$\frac{d\mathbf{e}_1}{dt}=\mathbf{e}_2, \qquad \frac{d\mathbf{e}_2}{dt}=-\mathbf{e}_1+c\mathbf{e}_3, \qquad \frac{d\mathbf{e}_3}{dt}=-c\mathbf{e}_2. \tag{1.5.6}$$

Thus

$$\frac{d^2\mathbf{e}_2}{dt^2}=-\alpha^2\mathbf{e}_2, \tag{1.5.7}$$

where $\alpha=\sqrt{1+c^2}$. Solving (1.5.7) gives

$$\mathbf{e}_2=\mathbf{a}\cos\alpha t+\mathbf{b}\sin\alpha t, \tag{1.5.8}$$

where $\mathbf{a}$ and $\mathbf{b}$ are constant vectors. Substituting (1.5.8) in the first equation of (1.5.6) and integrating the resulting equation we obtain

$$\alpha\mathbf{e}_1=\mathbf{a}\sin\alpha t-\mathbf{b}\cos\alpha t+\mathbf{c}, \tag{1.5.9}$$

where $\mathbf{c}$ is a constant vector. Hence by further integration,

$$\mathbf{x}(s)=\frac{1}{\alpha}\left(\int_0^s \mathbf{a}\sin\alpha t(r)\,dr-\int_0^s \mathbf{b}\cos\alpha t(r)\,dr+s\mathbf{c}+\mathbf{d}\right), \tag{1.5.10}$$

where $\mathbf{d}$ is a constant vector. However the integration constants $\mathbf{a},\mathbf{b},\mathbf{c},\mathbf{d}$ are not arbitrary.

Since $\mathbf{e}_2$ is a unit vector, we have $\mathbf{e}_2\cdot(d\mathbf{e}_2/dt)=0$, which together with (1.5.8) implies

$$\alpha(-\|\mathbf{a}\|^2+\|\mathbf{b}\|^2)\sin\alpha t\cos\alpha t+\alpha(\mathbf{a}\cdot\mathbf{b})(\cos^2\alpha t-\sin^2\alpha t)=0 \tag{1.5.11}$$

for every t. In particular, when $t=0$, we have $\mathbf{a}\cdot\mathbf{b}=0$. Thus (1.5.11) becomes $\frac{1}{2}\alpha(-\|\mathbf{a}\|^2+\|\mathbf{b}\|^2)\sin 2\alpha t=0$ for every t, so that $\|\mathbf{a}\|^2=\|\mathbf{b}\|^2$.

On the other hand, from (1.5.8) we obtain

$$1=\|\mathbf{e}_2\|^2=\|\mathbf{a}\|^2\cos^2\alpha t+\|\mathbf{b}\|^2\sin^2\alpha t,$$

so that $\|\mathbf{a}\|=\|\mathbf{b}\|=1$. Hence $\mathbf{a}$ and $\mathbf{b}$ are orthogonal unit vectors.

Similarly, $\mathbf{e}_1\cdot\mathbf{e}_2=0$ yields $\mathbf{a}\cdot\mathbf{c}=0$ and then $\mathbf{b}\cdot\mathbf{c}=0$. Moreover, $1=\|\mathbf{e}_1\|^2$ gives $\|\mathbf{c}\|=|c|$, and so $\mathbf{c}=\pm c(\mathbf{a}\times\mathbf{b})$. On the other hand, from the second equation of (1.5.6) it follows that

$$c\mathbf{e}_1=\mathbf{e}_2\times c\mathbf{e}_3=\mathbf{e}_2\times\left(\frac{d\mathbf{e}_2}{dt}+\mathbf{e}_1\right). \tag{1.5.12}$$

Substituting (1.5.8) and (1.5.9) in (1.5.12) we can easily obtain

$$\frac{c}{\alpha}(\mathbf{a}\sin\alpha t-\mathbf{b}\cos\alpha t+\mathbf{c})=\frac{1}{\alpha}\left[(\mathbf{a}\times\mathbf{c})\cos\alpha t+(\mathbf{b}\times\mathbf{c})\sin\alpha t+c^2(\mathbf{a}\times\mathbf{b})\right],$$

which gives that $\mathbf{c}=c(\mathbf{a}\times\mathbf{b})$. Hence with respect to the orthonormal basis $\{\mathbf{a},\mathbf{b},\mathbf{a}\times\mathbf{b}\}$ we have

$$\mathbf{x}(s)=\frac{1}{\alpha}\left(\int_0^s\sin\alpha t(r)\,dr,\ -\int_0^s\cos\alpha t(r)\,dr,\ cs\right)+\mathbf{d}_1, \tag{1.5.13}$$

where $t(r)$ is given by (1.5.5), and $\mathbf{d}_1=\mathbf{x}(0)$.

1.5.6. Example. Using Theorem 1.5.2 we can give an analytic proof of the second statement of Lemma 1.3.5.

By putting $c=0$ in Example 1.5.5, (1.5.13) yields the curve

$$\mathbf{y}(s)=\left(\int_0^s\sin t(r)\,dr,\ -\int_0^s\cos t(r)\,dr, 0\right), \tag{1.5.14}$$

where $t(r)=\int_0^r\kappa(s)\,ds$. This curve $\mathbf{y}(s)$ lies in the plane spaned by $\mathbf{a}$ and $\mathbf{b}$ and is therefore a plane curve. Since $\mathbf{y}(s)$ has arc length s, curvature $\kappa(s)$, and torsion 0, by Theorem 1.5.2, $\mathbf{x}(s)$ is congruent to $\mathbf{y}(s)$, hence is a plane curve.

Although this proof is not as simple as the original one, (1.5.14) has some applications; for instance, we can obtain a plane curve, expressed in arc length, from its curvature by three integrations; see Exercise 1 below.

Exercises

1. Find a curve $\mathbf{x}(s)$ in arc length s with torsion $\tau\equiv 0$ and curvature $\kappa(s)$ given in each of the following:

(a) $\kappa(s)=\text{constant}>0$.

(b) $\kappa(s)=\dfrac{1}{s+m}$, where m is constant.

(c) $\kappa(s)=\dfrac{1}{1+s^2}$.

2. Let $\mathbf{x}(s)$ be a curve with arc length s, curvature $\kappa>0$, torsion $\tau>0$, and a Frenet frame $\mathbf{e}_1\mathbf{e}_2\mathbf{e}_3$. Suppose that

$$\mathbf{y}(s)=\int_0^s\mathbf{e}_3(t)\,dt.$$

(a) Prove that s is also the arc length of the curve $\mathbf{y}(s)$.
(b) Let $\bar{\kappa}, \bar{\tau}, \bar{\mathbf{e}}_1\bar{\mathbf{e}}_2\bar{\mathbf{e}}_3$ be the curvature, the torsion, and a Frenet frame of the curve $\mathbf{y}(s)$. Show that $\bar{\kappa}=\tau$, $\bar{\tau}=\kappa$, $\bar{\mathbf{e}}_1=\mathbf{e}_3$, $\bar{\mathbf{e}}_2=-\mathbf{e}_2$, and $\bar{\mathbf{e}}_3=\mathbf{e}_1$.
(c) Show that if $\kappa=\tau$, the curves $\mathbf{x}(s)$ and $\mathbf{y}(s)$ are congruent.

2. PLANE CURVES

The developments in the last section include the theory of plane curves as a special case. We can also easily give an independent treatment of plane curves, since it has the advantage that the plane can be taken to be oriented; an orientation of a plane may be given by a fixed oriented frame $\mathbf{0x}_1\mathbf{x}_2$ at a point $\mathbf{0}$, called the origin, in the plane, which now consists of only two orthogonal unit vectors.

Most theorems in this section are global theorems on plane curves, which may be contrasted with local theorems. The latter theorems are concerned with the invariant properties in a neighborhood of a general point of a curve, whereas the former ones view a curve as a whole. For instance, the length of a closed curve and the area enclosed by a closed plane curve are simple global properties, since each requires the entire curve. Similarly, the simple theorem that the number of points of intersection of a straight line with a conic in the same plane is at most two is global.

2.1. Frenet Formulas and the Jordan Curve Theorem

2.1.1. Frenet Formulas. In a plane E^2 a vector will have two components. We associate with every point of an oriented curve $\mathbf{x}(s)$ in E^2, s being the arc length, two orthogonal unit vectors $\mathbf{e}_1(s), \mathbf{e}_2(s)$ such that they form an oriented frame with the same orientation as that of E^2 and that $\mathbf{e}_1(s)$ is the tangent vector of the curve. Then the Frenet formulas become

$$\frac{d\mathbf{x}}{ds}=\mathbf{e}_1, \qquad \frac{d\mathbf{e}_1}{ds}=\kappa\mathbf{e}_2, \qquad \frac{d\mathbf{e}_2}{ds}=-\kappa\mathbf{e}_1, \tag{2.1}$$

where the function κ is the *curvature* of the curve. It is important to remark that contrary to the case of space curves, $\kappa(s)$ together with sign is well defined, and changes its sign if the orientation of the curve or the plane is reversed.

2.1.2. Lemma. *Let a curve C be given by $\mathbf{x}(s)=(x_1(s), x_2(s))$ with arc length s. Then*

$$\mathbf{e}_1(s)=(\mathbf{x}_1'(s),\mathbf{x}_2'(s)), \qquad \mathbf{e}_2(s)=(-\mathbf{x}_2'(s),\mathbf{x}_1'(s)), \tag{2.2}$$

where the prime denotes the derivative with respect to s.

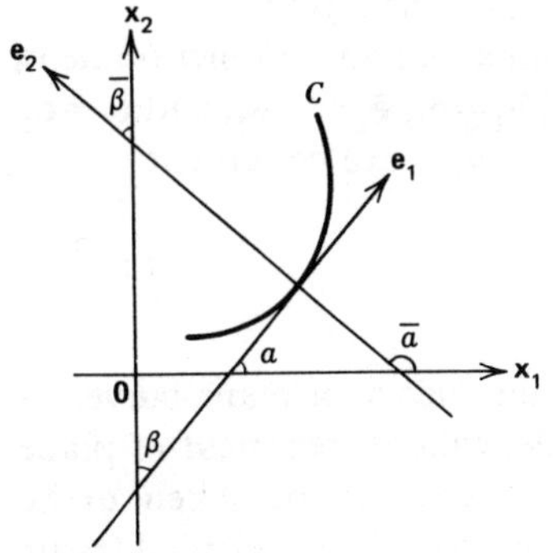

Figure 2.12

Proof. The first equation of (2.2) is obvious, and the second follows immediately from (see Fig. 2.12): $\mathbf{e}_2(\cos\bar{\alpha}, \cos\bar{\beta})$, $\cos\alpha = x_1'$, $\cos\beta = x_2'$, $\cos\bar{\alpha} = \cos(\alpha + \pi/2)$, $\cos\bar{\beta} = \cos\alpha$.

One of the most important global theorems for plane curves is without doubt the Jordan curve theorem.

2.1.3. Jordan Curve Theorem. *Let C be a simple closed curve in the plane E^2. Then $E^2 - C$ has exactly two connected components whose common boundary is C.*

One of the two components in the theorem is bounded and the other is not. The bounded component is called the *interior* of C, and the other the *exterior* of C. Whenever we speak of the area bounded by a simple closed curve C, we mean the area of the interior of C.

2.2. Winding Number and Rotation Index. In the plane let $\mathbf{0}$ be a fixed point, which we take as the origin of our fixed frame. Denote by Γ the unit circle with center at $\mathbf{0}$.

2.2.1. Definition. A simple closed curve is *positively oriented* if one goes along the curve in the direction of increasing parameter, then the interior of the curve remains to one's left-hand side. Tangents of a positively oriented simple closed curve in the positive direction are *positively oriented tangents*.

2.2.2. Definition. Let an oriented closed curve C be given by $\mathbf{x}(t), t \in [a, b]$, and $\mathbf{p}$ a point of the plane not on the curve. Then the mapping f: $C \to \Gamma$, given by

$$f(\mathbf{x}(t)) = \frac{\mathbf{x}(t) - \mathbf{p}}{\|\mathbf{x}(t) - \mathbf{p}\|}, \qquad t \in [a, b],$$

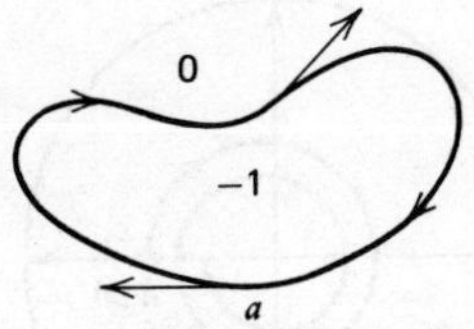

Figure 2.13

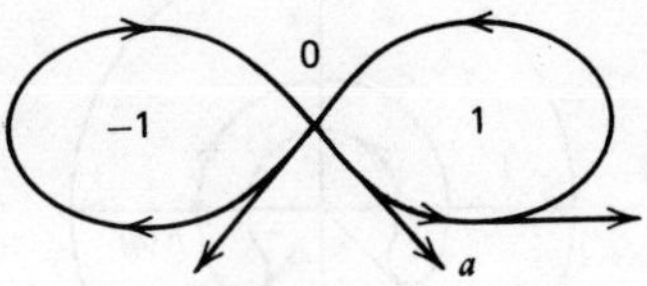

Figure 2.14

is defined to be the *position mapping* of the curve C relative to $\mathbf{p}$. Intuitively it is clear that when the point $\mathbf{x}(t)$ goes around the curve C once, its image point $f(\mathbf{x}(t))$ will go around Γ a number of times. This number is called the *winding number* of the curve C relative to the point $\mathbf{p}$.

Notice that the winding number remains unchanged if $\mathbf{p}$ moves along an arc that does not meet the curve, and it is the same relative to points in the same connected region. Furthermore, the winding number changes sign when the orientation of the curve changes. By convention, the winding number of a positively oriented simple closed curve is positive (see Figs. 2.13–2.15).

Similarly, we have a definition of the tangential mapping of a curve C.

2.2.3. *Definition.* Let an oriented closed curve C of length L be given by $\mathbf{x}(s)$ with arc length s. Then the mapping h: $C \to \Gamma$ given by

$$h(\mathbf{x}(s)) = \mathbf{x}'(s), \qquad s \in [0, L],$$

is defined to be the *tangential mapping* of the curve C. Clearly h is a continuous mapping. Intuitively it is obvious that when the point $\mathbf{x}(s)$ goes around the curve C once, its image point $h(\mathbf{x}(s))$ will go around Γ a number of times. This number is called the *rotation index* of the curve C.

Like the winding number, the rotation index changes sign when the orientation of the curve changes, and the rotation index of a positively oriented simple closed curve is positive. The rotation indices of the curves

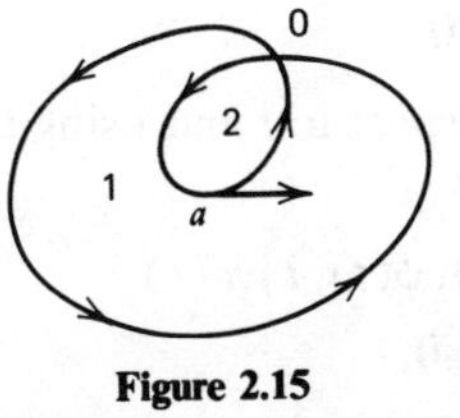

Figure 2.15

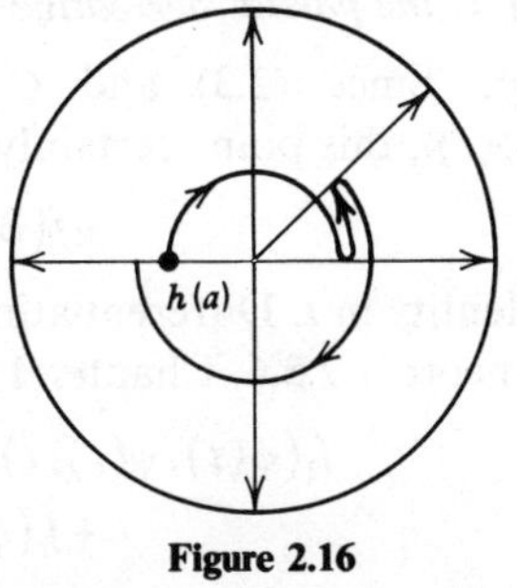

Figure 2.16

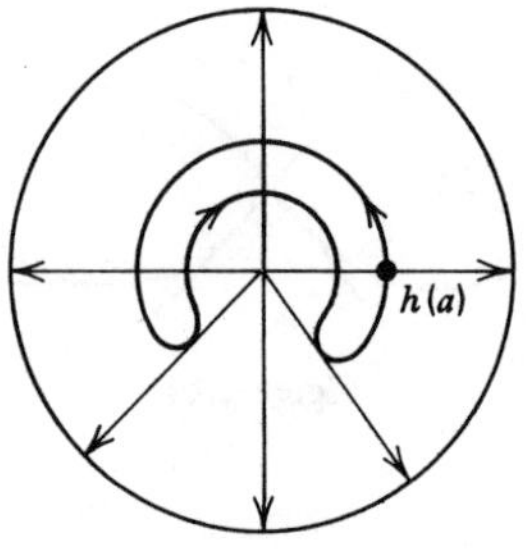

Figure 2.17

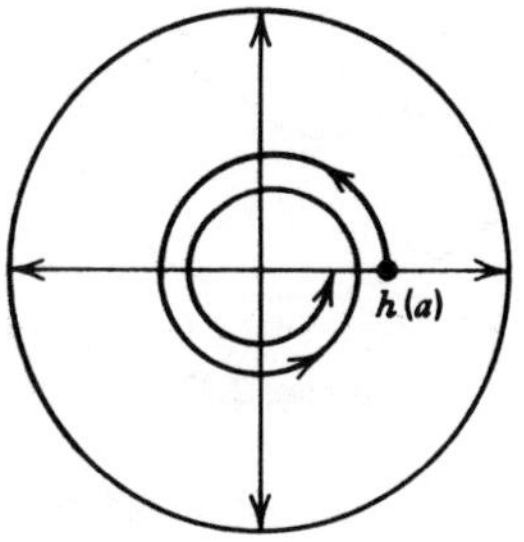

Figure 2.18

in Figs. 2.13 to 2.15 are -1, 0, and 2, respectively, as indicated by the tangential mappings of the curves in Figs. 2.16 to 2.18.

2.3. Envelopes of Curves

2.3.1. Definition. An *envelope* of a one-parameter family of curves is a curve that has the following two properties: (i) at every one of its points it is tangent to at least one curve of the family; (ii) it is tangent to every curve of the family at at least one point.

2.3.2. Theorem. *Let*

$$f(x_1, x_2, t)=0 \tag{2.3}$$

be the equation of a one-parameter family of curves in the $\mathbf{x}_1\mathbf{x}_2$*-plane, where* t *is the parameter, and let an envelope* C *of this family be given by*

$$x_1=\phi(t), \qquad x_2=\psi(t). \tag{2.4}$$

Assume further that f, ϕ, *and* ψ *are continuously differentiable and that for each value of* t *the curve* (2.3) *and* C *have a common tangent at the point* $(\phi(t), \psi(t))$. *Then the coordinates* (2.4) *of each point of* C *satisfy* (2.3) *and*

$$f_t(x_1, x_2, t)=0, \tag{2.5}$$

where f_t *is the partial derivative of* f *with respect to* t.

Proof. Since (2.3) and C have a common tangent at the point $(\phi(t), \psi(t))$, this point certainly lies on (2.3), for each t. Therefore

$$f(\phi(t), \psi(t), t)=0 \tag{2.6}$$

is an identity in t. Differentiating (2.6) with respect to t and using the chain rule (Theorem 2.3.6, Chapter 1), we obtain

$$\begin{aligned} f_1(\phi(t), \psi(t), t)\phi'(t)+f_2(\phi(t), \psi(t), t)\psi'(t)& \\ +f_t(\phi(t), \psi(t), t)=0,& \end{aligned} \tag{2.7}$$

where f_1, f_2 are the partial derivatives of f with respect to x_1, x_2, respectively, and the prime denotes the derivative with respect to t.

On the other hand, the slope of a curve of the family (2.3) is

$$\frac{dx_2}{dx_1} = -\frac{f_1(x_1, x_2, t)}{f_2(x_1, x_2, t)}, \tag{2.8}$$

while the slope of the curve C defined by (2.4) is

$$\frac{dx_2}{dx_1} = \frac{\psi'(t)}{\phi'(t)}. \tag{2.9}$$

At the point where C is tangent to a curve of the family (2.3) we then have

$$\frac{\psi'(t)}{\phi'(t)} = -\frac{f_1(\phi(t), \psi(t), t)}{f_2(\phi(t), \psi(t), t)}$$

or

$$f_1(\phi(t), \psi(t), t)\phi'(t) + f_2(\phi(t), \psi(t), t)\psi'(t) = 0. \tag{2.10}$$

Combination of (2.10) with (2.7) thus gives (2.5).

2.4. Convex Curves. In plane analytic geometry we have defined the oriented distance from a point to a straight line. In the following we shall use such oriented distance to define first a relationship between a straight line and a curve, then an interesting class of curves.

2.4.1. Definition. In a plane E^2, points *lie on one side of a straight line* l if the oriented distances from them to l are of the same sign, and a curve C *lies on one side of* l if all points of C do so.

2.4.2. Definition. A curve is *convex* if it lies on one side of each one of its tangent lines.

Circles and ellipses are convex, and the curves in Figs. 2.13 to 2.15 are not convex.

2.4.3. Definition. The *diameter* of a closed convex curve C is the greatest distance between two points of C. The *width of C in a direction* is the distance between two tangents of C parallel to the given direction. C is of *constant width* if its width is independent of the direction.

A circle is of constant width, but there are curves of constant width that are not circles, for instance, the curves in the following two examples and Exercise 12 at the end of Section 2.

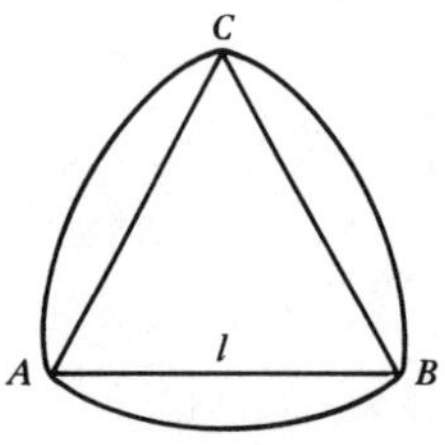

Figure 2.19

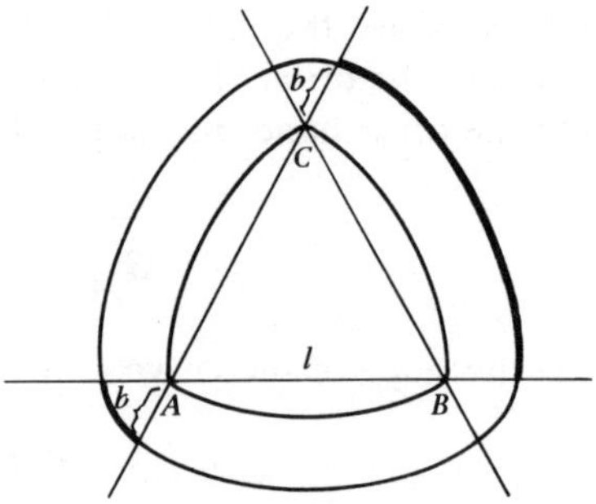

Figure 2.20

Examples. 1. (See Fig. 2.19.) ABC is an equilateral triangle with each of its sides equal to l. The curve consisting of three circular arcs, each of which has center at a vertex of ABC and radius l, is a piecewise differentiable curve of constant width l.

2. (See Fig. 2.20.) From Example 1 we can construct a C^1 curve of arbitrary constant width, say, $l+2b$. It is formed by six circular arcs, two of which have center at each vertex of the triangle ABC and radii $l+b$ and b, respectively, as indicated by the thicker arcs for the vertex A.

2.4.4. Length and Area of a Closed Convex Curve. Now let us consider a positively oriented closed convex curve C, and take any point inside the curve C to be the origin $\mathbf{0}$ of our fixed frame. If p is the oriented perpendicular distance from $\mathbf{0}$ to the positively oriented tangent l at a point $\mathbf{x}(x_1, x_2)$, and θ is the angle between this perpendicular line and the positive half of the $\mathbf{x}_1$-axis (see Fig. 2.21), p is a single-valued function of θ and therefore is periodic with period 2π, and the equation of the tangent line l can be written in the form

$$x_1 \cos\theta + x_2 \sin\theta = p(\theta). \tag{2.11}$$

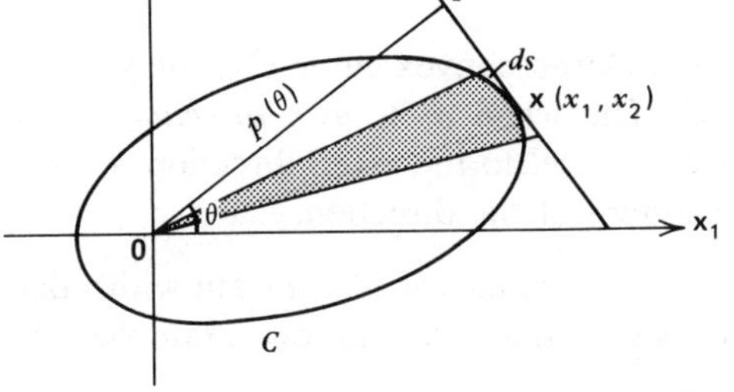

Figure 2.21

The envelope of all the lines (2.11) is the curve C. To determine the point $\mathbf{x}$ of contact of the tangent (2.11) with the curve C, we differentiate (2.11) partially with respect to θ so that we have

$$-x_1\sin\theta + x_2\cos\theta = p'(\theta), \tag{2.12}$$

where the prime denotes the derivative with respect to θ. Solving (2.11) and (2.12) for x_1, x_2 and using Theorem 2.3.2, we have the parametric representation of C:

$$\begin{aligned} x_1 &= p(\theta)\cos\theta - p'(\theta)\cos\theta, \\ x_2 &= p(\theta)\sin\theta + p'(\theta)\cos\theta, \end{aligned} \tag{2.13}$$

where θ is the parameter. Thus if θ and $p(\theta)$ are given, we can determine a unique point $\mathbf{x}(x_1, x_2)$ on the closed convex curve C, and the converse is also true. Hence we call $(\theta, p(\theta))$ the *polar tangential coordinates* of the point $\mathbf{x}$ on C.

Differentiation of (2.13) with respect to θ gives

$$\begin{aligned} x_1' &= -[\,p(\theta) + p''(\theta)]\sin\theta, \\ x_2' &= [\,p(\theta) + p''(\theta)]\cos\theta. \end{aligned} \tag{2.14}$$

Let s be the arc length of the curve C. Then s is a monotone increasing function of θ $(0 \leqslant \theta < 2\pi)$. Thus from Theorem 1.3.3 and equations 2.14 it follows that the radius of curvature of C is

$$\rho = \frac{ds}{d\theta} = p(\theta) + p''(\theta) > 0, \qquad 0 \leqslant \theta < 2\pi, \tag{2.15}$$

which can also be used to define a positively oriented convex curve C.

Let L be the length of C. Then by (2.15) we obtain

$$L = \int_C ds = \int_0^{2\pi} \frac{ds}{d\theta}\,d\theta = \int_0^{2\pi} p(\theta)\,d\theta + [\,p'(\theta)]_0^{2\pi}. \tag{2.16}$$

Since $p'(\theta)$ is also a periodic function of θ with period 2π, we have $p'(0) = p'(2\pi)$ and therefore

$$L = \int_0^{2\pi} p(\theta)\,d\theta, \tag{2.17}$$

a result known as *Cauchy's formula*.

Let A be the area bounded by C. Then by approximating the shaded area in Fig. 2.21 by using the area of the nearest triangle, we obtain, in consequence of (2.15),

$$A = \frac{1}{2}\int_C p(\theta)\,ds = \frac{1}{2}\int_0^{2\pi} p(\theta)[\,p(\theta) + p''(\theta)]\,d\theta. \tag{2.18}$$

But

$$\int_0^{2\pi} p(\theta)p''(\theta)\,d\theta = [\,p(\theta)p'(\theta)]_0^{2\pi} - \int_0^{2\pi} p'^2(\theta)\,d\theta$$
$$= -\int_0^{2\pi} p'^2(\theta)\,d\theta.$$

Hence we have *Blaschke's formula*

$$A = \frac{1}{2}\int_0^{2\pi}\left[\,p^2(\theta) - p'^2(\theta)\right]d\theta. \tag{2.19}$$

2.4.5. *Parallel Curves.* For any positive number n, suppose that a length n is measured off along each normal and in the exterior of a closed convex curve C. The end points of these segments describe a parallel curve C_n of C (for the definition of general parallel space curves, see Exercise 26 of Section 1.3). Now we prove that C_n is also a closed convex curve. Let $(\theta, p_n(\theta))$ be the polar tangential coordinates of the point on the curve C_n corresponding to a point with polar tangential coordinates $(\theta, p(\theta))$ on C. Then

$$p_n(\theta) = p(\theta) + n, \tag{2.20}$$

and by (2.15) we obtain the radius of curvature of C_n:

$$\rho_n = \rho + n. \tag{2.21}$$

From (2.20), $p_n(\theta)$ is a periodic function of θ with period 2π, and from (2.21), $\rho_n > 0$ since both ρ and n are positive. Hence, C_n is also a closed convex curve.

Let L_n and A_n be, respectively, the length and the area of C_n. Then using (2.17), (2.19), and (2.20) we can easily obtain

$$L_n = L + 2n\pi, \tag{2.22}$$

$$A_n = A + nL + n^2\pi. \tag{2.23}$$

Equations 2.22 and 2.23 are called *Steiner's formulas* for parallel curves of a closed convex curve.

2.4.6. *Theorem.* *The length (respectively, area) of a closed convex curve lies between the lengths (respectively, areas) of the osculating circles of the curve with the greatest and the least radii of curvature, respectively.*

Proof. Let ρ_1 and ρ_2 be, respectively, the greatest and least radii of curvature of a closed convex curve C with length L and area A. Then by (2.15), (2.16), and (2.18) we obtain

$$2\pi\rho_1 \geqslant L \geqslant 2\pi\rho_2, \tag{2.24}$$

$$\rho_1\int_0^{2\pi} p(\theta)\,d\theta \geqslant 2A \geqslant \rho_2\int_0^{2\pi} p(\theta)\,d\theta. \tag{2.25}$$

Since (2.16) implies

$$\int_0^{2\pi} p(\theta)\,d\theta = \int_0^{2\pi} \rho(\theta)\,d\theta,$$

we have

$$2\pi\rho_1 \geqslant \int_0^{2\pi} p(\theta)\,d\theta \geqslant 2\pi\rho_2. \tag{2.26}$$

Hence the combination of (2.26) with (2.25) gives

$$\pi\rho_1^2 \geqslant A \geqslant \pi\rho_2^2. \tag{2.27}$$

2.4.7. Theorem [*A. Rosenthal and O. Szasz, 47*]. *Among all closed convex curves of a given diameter D, the curves of constant width D have the greatest length.*

Proof. Let C be a closed convex curve of length L given by the polar tangential coordinates $(\theta, p(\theta))$ defined in Section 2.4.4. Since $p(\theta)$ is a periodic function with period 2π, it can be represented by the Fourier series

$$p(\theta) = \frac{1}{2}a_0 + \sum_{n=1}^{\infty} (a_n \cos n\theta + b_n \sin n\theta), \tag{2.28}$$

where

$$a_0 = \frac{1}{\pi}\int_0^{2\pi} p(\theta)\,d\theta. \tag{2.29}$$

Then using (2.17) we readily see that

$$L = \pi a_0. \tag{2.30}$$

On the other hand, (2.29) can be written as

$$a_0 = \frac{1}{\pi}\int_0^{\pi} [\,p(\theta) + p(\theta+\pi)\,]\,d\theta.$$

Since the width of C in any direction is not greater than D, we have (see Fig. 2.22)

$$p(\theta) + p(\theta + \pi) \leqslant D.$$

Thus $a_0 \leqslant D$ and, in consequence of (2.30),

$$L \leqslant \pi D. \tag{2.31}$$

It is obvious that the equality in (2.31) holds when and only when C is of constant width D. Q.E.D.

Equation 2.31 implies Corollary 2.4.8.

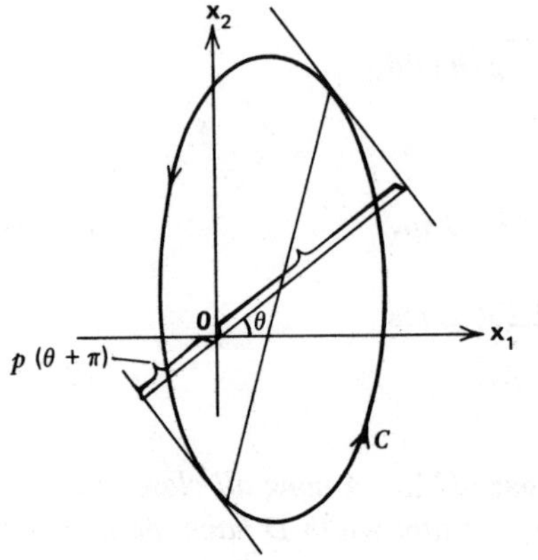

Figure 2.22

***2.4.8. Corollary** [A. **Barber,** 3].* *All closed convex curves of constant width D have the same length πD.*

2.5. The Isoperimetric Inequality

2.5.1. Theorem. *Among all simple closed curves of given length the circle bounds the largest area. In other words, if A is the area bounded by a simple closed curve C of length L, then*

$$L^2 - 4\pi A \geqslant 0, \tag{2.32}$$

where the equality holds when and only when C is a circle.

There are numerous geometric proofs of Theorem 2.5.1, varying in the degree of elegance and in the differentiability or convexity assumption of the curves. The first analytic proof was given by A Hurwitz [35] in 1902, and a simple one by E. Schmidt [50] in 1939.

Theorem 2.5.1 has been extended to curves on surfaces of variable Gaussian curvature by A Huber [34] in 1954, to curves in even-dimensional Euclidean spaces by I. J. Schoenberg [51] in 1954, and to curves on surfaces in Euclidean spaces of any dimension by C. C. Hsiung [33] in 1961.

We shall give the proofs of Schmidt and Hurwitz.

Schmidt's Proof. We enclose C between two parallel lines l_1, l_2 such that C lies between l_1, l_2 and is tangent to them at points P, Q, respectively. Construct a circle $\bar{C}$ that does not meet C but is tangent to l_1, l_2 at points $\bar{P}, \bar{Q}$ respectively. Denote the radius and the center of $\bar{C}$ by r and $\mathbf{0}$, respectively. Take $\mathbf{0}$ to be the origin of a coordinate frame $\mathbf{0x}_1\mathbf{x}_2$ such that

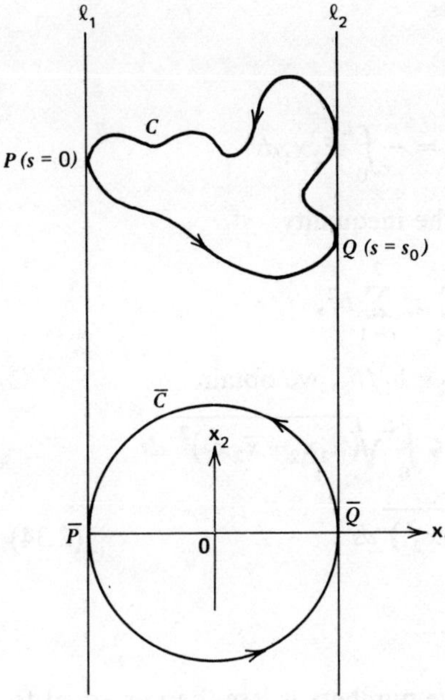

Figure 2.23

the $\mathbf{x}_1$-axis is orthogonal to l_1 and l_2. Let C be positively oriented and given by $\mathbf{x}(s)=(x_1(s), x_2(s))$ with arc length s such that $s=O$ and $s=s_O$ for P and Q, respectively (Fig. 2.23) and

$$(x_1(0), x_2(0))=(x_1(L), x_2(L)).$$

Let the circle $\bar{C}$ be given by $\bar{x}(s)=(\bar{x}_1(s), \bar{x}_2(s))$ such that

$$\begin{aligned} \bar{x}_1(s)&=x_1(s), \\ \bar{x}_2(s)&=-\sqrt{r^2-x_1^2(s)}\,, \qquad 0\leqslant s\leqslant s_0, \\ &=+\sqrt{r^2-x_1^2(s)}\,, \qquad s_0\leqslant s\leqslant L. \end{aligned} \tag{2.33}$$

Denote $\bar{A}$ the area bounded by $\bar{C}$. In general, the area bounded by a closed curve $\mathbf{y}(s)=(y_1(s), y_2(s))$ of length L with arc length s can be expressed by the line integrals

$$\int_0^L y_1 y_2'\,ds=-\int_0^L y_2 y_1'\,ds=\frac{1}{2}\int_0^L (y_1 y_2'-y_2 y_1')\,ds,$$

where the prime denotes the derivative with respect to s. For a proof of this formula see, for instance, R. Courant [16, p. 273]. Application of this formula to our two curves C and $\bar{C}$ gives

$$A=\int_0^L x_1 x_2'\,ds,$$

$$\bar{A}=\pi r^2=-\int_0^l \bar{x}_2\bar{x}_1'\,ds=-\int_0^L \bar{x}_2 x_1'\,ds.$$

Adding these two equations and using the inequality

$$\left(\sum_{i=1}^{2} a_i b_i\right)^2 \leqslant \sum_{i=1}^{2} a_i^2 \sum_{i=1}^{2} b_i^2,$$

the equality holding if and only if $a_1/a_2=b_1/b_2$, we obtain

$$\begin{aligned} A+\pi r^2 &= \int_0^L (x_1x_2'-\bar{x}_2x_1')\,ds \leqslant \int_0^L \sqrt{(x_1x_2'-\bar{x}_2x_1')^2}\,ds \\ &\leqslant \int_0^L \sqrt{(x_1^2+x_2^2)(x_1'^2+x_2'^2)}\,ds \qquad (2.34) \\ &= \int_0^L \sqrt{x_1^2+\bar{x}_2^2}\,ds = Lr. \end{aligned}$$

Since the geometric mean of two positive numbers is less than or equal to their arithmetic mean, and the equality holds if and only if the two numbers are equal, it follows that

$$\sqrt{A}\,\sqrt{\pi r^2}=\frac{1}{2}(A+\pi r^2)\leqslant\frac{1}{2}Lr. \qquad (2.35)$$

By squaring the inequality above and canceling r^2, we thus obtain the inequality (2.32).

Suppose now that the equality sign in (2.32) holds. Then the inequality must hold everywhere in (2.34) and (2.35). The equality in (2.35) implies that A and πr^2 have the same geometric and arithmetic mean, so that

$$A=\pi r^2. \qquad (2.36)$$

Equations 2.36 and 2.32 with equality give

$$L=2\pi r, \qquad (2.37)$$

which shows that r is constant. The third equality in (2.34) implies that

$$(x_1x_2'-\bar{x}_2x_1')^2=(x_1^2+\bar{x}_2^2)(x_1'^2+x_2'^2),$$

which gives $(x_1x_1'+\bar{x}_2x_2')^2=0$ or

$$\frac{x_1}{x_2'}=\frac{-\bar{x}_2}{x_1'}=\frac{\pm\sqrt{x_1^2+\bar{x}_2^2}}{\sqrt{x_1'^2+x_2'^2}}=\pm r. \qquad (2.38)$$

The second equality in (2.34) shows that

$$x_1x_2' - \bar{x}_2x_1' > 0.$$

But from (2.38) we have

$$x_1x_2' - \bar{x}_2x_1' = \pm r(x_2'^2 + x_1'^2) = \pm r.$$

Thus we must take the positive sign on the right-hand side of (2.38) so that

$$x_1 = rx_2', \qquad \bar{x}_2 = -rx_1'. \tag{2.39}$$

Let $\mathbf{e}_1, \bar{\mathbf{e}}_1$ be, respectively, the unit tangent vectors of $C, \bar{C}$ at the points (x_1, x_2), $(x_1, \bar{x}_2)$ related by (2.39). Then

$$\mathbf{e}_1 = \left(\frac{dx_1}{ds}, \frac{dx_2}{ds}\right), \qquad \bar{\mathbf{e}}_1 = \left(\frac{dx_1}{d\bar{s}}, \frac{d\bar{x}_2}{d\bar{s}}\right), \tag{2.40}$$

where $\bar{s}$ is the arc length of $\bar{C}$. Since $\bar{\mathbf{e}}_1$ is a unit vector orthogonal to the position vector of the point $(x_1, \bar{x}_2)$, we have

$$\bar{\mathbf{e}}_1 = \left(-\frac{\bar{x}_2}{r}, \frac{x_1}{r}\right),$$

which, together with (2.39) and (2.40), implies that $\mathbf{e}_1 = \bar{\mathbf{e}}_1$ so that

$$\frac{dx_1}{ds} = \frac{dx_1}{d\bar{s}}, \qquad \frac{dx_2}{ds} = \frac{d\bar{x}_2}{d\bar{s}}. \tag{2.41}$$

From the first equation of (2.41) we have $ds = d\bar{s}$, and therefore from the second we have $x_2 = \bar{x}_2 + a$, where a is constant. Substituting the last equation in the second equation of (2.33) hence shows that the curve C is a circle given by

$$x_1^2 + (x_2 - a)^2 = r^2.$$

Hurwitz's Proof. At first we prove Wirtinger's lemma.

2.5.2. Wirtinger's Lemma. *Let $f(t)$ be a continuous periodic function of period 2π with continuous derivative $f'(t)$. If $\int_0^{2\pi} f(t)\,dt = 0$, then*

$$\int_0^{2\pi} [f'(t)]^2\,dt \geqslant \int_0^{2\pi} [f(t)]^2\,dt, \tag{2.42}$$

where the equality holds if and only if

$$f(t) = a\cos t + b\sin t, \tag{2.43}$$

where a and b are constant.

To prove Lemma 2.5.2 let the Fourier series expansion of $f(t)$ be

$$f(t)=\frac{1}{2}a_0+\sum_{n=1}^{\infty}(a_n\cos nt+b_n\sin nt),$$

where

$$a_0=\frac{1}{\pi}\int_0^{2\pi}f(t)\,dt,$$

which is zero by our hypothesis. Since $f'(t)$ is continuous, by term-by-term differentiation we obtain its Fourier series:

$$f'(t)=\sum_{n=1}^{\infty}n(b_n\cos nt-a_n\sin nt).$$

Thus from Parseval's formulas for $f(t)$ and $f'(t)$:

$$\begin{aligned}\frac{1}{\pi}\int_0^{2\pi}[f(t)]^2\,dt&=\sum_{n=1}^{\infty}(a_n^2+b_n^2),\\ \frac{1}{\pi}\int_0^{2\pi}[f'(t)]^2\,dt&=\sum_{n=1}^{\infty}n^2(a_n^2+b_n^2),\end{aligned}\tag{2.44}$$

it follows that

$$\int_0^{2\pi}[f'(t)]^2\,dt-\int_0^{2\pi}[f(t)]^2\,dt=\pi\sum_{n=1}^{\infty}(n^2-1)(a_n^2+b_n^2),$$

which is greater than or equal to zero, and is equal to zero if and only if $a_n=b_n=0$ for all $n>1$, that is, if and only if $f(t)=a_1\cos t+b_1\sin t$. Hence the lemma is proved.

It should be remarked that the lemma is also true if the function $f(t)$ has a general period L instead of 2π, since from the theory of Fourier series every equation in the proof above still holds for the general period L.

Now we can give Hurwitz's proof of Theorem 2.5.1. For this purpose, let the simple closed curve C be given by $\mathbf{x}(s)=(x_1(s),x_2(s))$ with arc length s with respect to a coordinate frame $\mathbf{0x}_1\mathbf{x}_2$. Since the proof depends entirely on Wirtinger's lemma, from the foregoing remark we can assume without loss of generality that $L=2\pi$. We further assume that

$$\int_0^{2\pi}x_1(s)\,ds=0,$$

which means that the center of gravity of C lies on the $\mathbf{x}_1$-axis, and which can always be achieved by a proper choice of the coordinate frame. The length and the area of C are given by the integrals

$$2\pi=\int_0^{2\pi}(x_1'^2+x_2'^2)\,ds,\qquad A=\int_0^{2\pi}x_1x_2'\,ds,$$

the prime denoting the derivative with respect to s, from which we obtain

$$2(\pi - A) = \int_0^{2\pi} (x_1'^2 - x_1^2)\, ds + \int_0^{2\pi} (x_1 - x_2')^2\, ds.$$

By Wirtinger's lemma the first integral on the right-hand side of the equation above is greater than or equal to zero, and the second integral is clearly greater than or equal to zero. Hence

$$\pi \geqslant A, \tag{2.45}$$

which is our isoperimetric inequality (2.32) under the assumption that $L = 2\pi$.

From (2.43) it follows that the equality in (2.45) holds if and only if

$$x_1 = a\cos s + b\sin s, \qquad x_2' = x_1,$$

that is, if and only if

$$x_1 = a\cos s + b\sin s, \qquad x_2 = a\sin s - b\cos s + c,$$

where c is constant. Hence C is a circle given by $x_1^2 + (x_2 - c)^2 = a^2 + b^2$.

Q.E.D.

By squaring (2.31) and using (2.32) we obtain Corollary 2.5.3.

2.5.3. Corollary [**L. Bieberbach, 4**]. *For a closed convex curve C of area A and diameter D,*

$$A \leqslant \frac{1}{4}\pi D^2, \tag{2.46}$$

where the equality holds when and only when C is a circle.

2.6. The Four-Vertex Theorem. *A closed convex curve has at least four vertices, each of which is defined to be a point where the curvature of the curve has a relative extremum.*

This theorem was first given by S. Mukhopadhyaya [41] in 1909. It is also true for some nonconvex curves, but the proof is more difficult; for simple closed nonconvex curves see S. B. Jackson [36] and L. Vietoris [58]. We also remark that the theorem cannot be improved, since an ellipse with unequal axes has exactly four vertices, which are its points of intersection with the axes (see Exercise 14 of Section 2).

Here we give two proofs, which were the work of G. Herglotz.

Proofs. From Theorem 2.3.5 of Chapter 1 we know that a closed curve has at least two vertices corresponding, respectively, to a relative maximum and a relative minimum of the curvature of the curve, and the number of vertices of the curve is even. In our two proofs we shall show that the assumption of a closed convex curve with only two vertices will produce a contradiction.

1. Let C be a closed curve given by $\mathbf{x}(s)=(x_1(s), x_2(s))$ with arc length s and curvature κ. Then substituting (2.2) in the second equation of (2.1) we obtain

$$x_1''=-\kappa x_2', \qquad x_2''=\kappa x_1', \tag{2.47}$$

where the primes denote the derivatives with respect to s. From (2.47) it follows that

$$\int_C \kappa'\,ds=\int_C d\kappa=0,$$

$$\int_C x_1\kappa'\,ds=\int_C(-x_2'+\kappa x_1)'\,ds=0,$$

$$\int_C x_2\kappa'\,ds=\int_C(x_1'+\kappa x_2)'\,ds=0.$$

Hence for any constants a_0, a_1, a_2 we have

$$\int_C(a_0+a_1x_1+a_2x_2)\kappa'\,ds=0. \tag{2.48}$$

Now suppose that C is convex with nonconstant curvature κ and that κ attains a relative maximum and a relative minimum at points M and N, respectively. Let l be the straight line joining M and N, and suppose that l divides C into two parts C_1 and C_2. Then each of these two parts lies on a definite side of l. Otherwise, l will meet C at at least another point P distinct from M and N (Fig. 2.24). By convexity, among the three distinct points M, N, P the tangent of C at the intermediate point, say N , must coincide with l; therefore by convexity again l is tangent to C at the three points M, N, P. But then the tangent of C at a point near N will have M and P on distinct sides unless the whole segment MP of l belongs to C (Fig. 2.25). This implies that $\kappa=0$ at M and N, hence that $\kappa\equiv 0$ on C, a contradiction.

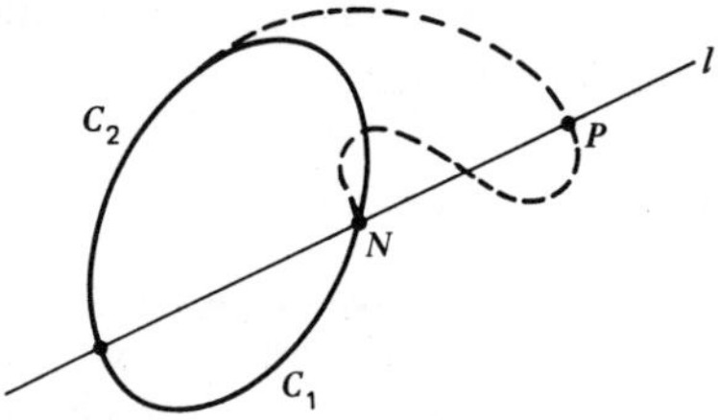

Figure 2.24

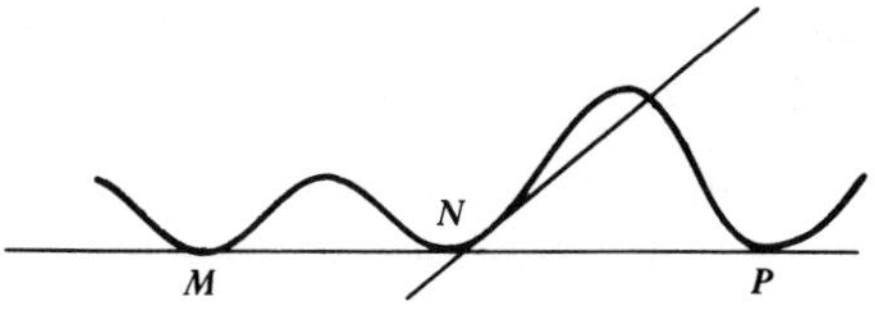

Figure 2.25

Now let the equation of l be

$$a_0 + a_1 x_1 + a_2 x_2 = 0.$$

If C has no other vertices, then

$$\kappa' \begin{cases} <0 & \text{along} \quad C_1, \\ >0 & \text{along} \quad C_2. \end{cases}$$

Since the distances to l from all points of C_1 are of one sign, and those from all points of C_2 are of the opposite sign, we may assume

$$a_0 + a_1 x_1 + a_2 x_2 \begin{cases} <0 & \text{for all points of } C_1, \\ >0 & \text{for all points of } C_2. \end{cases}$$

Thus the integrand in (2.48) is always positive, contradicting the fact that its integral is zero. Hence the theorem is proved.

2. Suppose that a closed convex curve C of length L and curvature κ has only two vertices M and N, corresponding to a relative maximum and a relative minimum of κ, respectively. Then by the argument in proof 1, the line MN does not meet C at any other point. Let C be given by $\mathbf{x}(x_1(s), x_2(s))$ with arc length s, and take MN to be on the $\mathbf{x}_1$-axis with $s=0$ and $s=s_0$ for M and N, respectively. Then we can suppose

$$x_2(s) \begin{cases} <0, & 0<s<s_0, \\ >0, & s_0<s<L. \end{cases}$$

From (2.47) it follows that

$$\int_0^L \kappa x_2' \, ds = -x_1' \big|_0^L = 0. \tag{2.49}$$

The integral on the left-hand side of (2.49) can be written as

$$\int_0^L \kappa x_2' \, ds = \int_0^{s_0} \kappa x_2' \, ds + \int_{s_0}^L \kappa x_2' \, ds. \tag{2.50}$$

Since $\kappa(s)$ is monotone in each of the intervals $0 \leqslant s \leqslant s_0$ and $s_0 \leqslant s \leqslant L$, an application of the second mean value theorem (Theorem 2.1.4, Chapter 1) to each term on the right-hand side of (2.50) gives

$$\begin{aligned} \int_0^{s_0} \kappa x_2' \, ds &= \kappa(0) \int_0^{\xi_1} x_2' \, ds + \kappa(s_0) \int_{\xi_1}^{s_0} x_2' \, ds \\ &= x_2(\xi_1)[\kappa(0) - \kappa(s_0)], \qquad 0<\xi_1<s_0, \\ \int_{s_0}^L \kappa x_2' \, ds &= \kappa(s_0) \int_{s_0}^{\xi_2} x_2' \, ds + \kappa(L) \int_{\xi_2}^L x_2' \, ds \\ &= x_2(\xi_2)[\kappa(s_0) - \kappa(0)], \qquad s_0<\xi_2<L. \end{aligned}$$

Adding the two equations above and using (2.49) and (2.50), we obtain

$$[x_2(\xi_1)-x_2(\xi_2)][\kappa(0)-\kappa(s_0)]=0,$$

which is a contradiction, since

$$x_2(\xi_1)-x_2(\xi_2)<0, \qquad \kappa(0)-\kappa(s_0)>0.$$

Hence the theorem is proved.

2.7. The Measure of a Set of Lines. The measure $M(S)$ of a set S of points in a plane E^2 can naturally be defined to be the area $A(S)$ of S, which is invariant under all orientation-preserving isometries of E^2 (for the definition, see Definition 5.4.6, Chapter 1). Let the coordinates of a point in E^2 with respect to a fixed frame $\mathbf{0x}_1\mathbf{x}_2$ be (x_1, x_2). Then we define

$$M(S)=\int\int_S dx_1\wedge dx_2, \tag{2.51}$$

where $\wedge$ denotes the exterior product (Section 6.2, Chapter 1).

Similarly, we can define a measure for a set of lines in E^2. As was defined in Section 2.6.4, the equation of a line l (Fig. 2.26) with respect to the frame $\mathbf{0x}_1\mathbf{x}_2$ can be written as

$$x_1\cos\theta+x_2\sin\theta=p. \tag{2.52}$$

Every line in E^2 is determined completely by (p,θ), so that (p,θ) may be considered to be the coordinates of the line. Thus we can replace the set of all lines in E^2 by the set

$$\mathcal{L}=\{(p,\theta)\in E^2 \mid p\geqslant 0,\ 0\leqslant\theta<2\pi\}, \tag{2.53}$$

and it is natural to define a measure of a subset $\mathbb{S}\subset\mathcal{L}$ to be

$$\mathfrak{M}(\mathbb{S})=\int\int_{\mathbb{S}} dp\wedge d\theta. \tag{2.54}$$

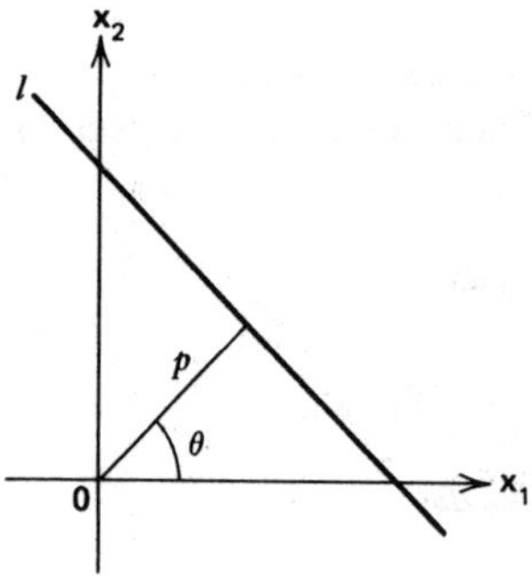

Figure 2.26

Actually, up to a constant factor this is the only measure on $\mathfrak{L}$, which is invariant under all orientation-preserving isometries of E^2. In more detail we have Theorem 2.7.1.

2.7.1. Theorem. *Let $f(p,\theta)$ be a continuous function on $\mathfrak{L}$. For any set $\mathfrak{S}\subset\mathfrak{L}$ if*

$$\mathfrak{M}(\mathfrak{S})=\int\int_{\mathfrak{S}} f(p,\theta)\,dp\wedge d\theta \tag{2.55}$$

is invariant under all orientation-preserving isometries of E^2, then $f(p,\theta)$ is constant.

Proof. Consider a general differentiable mapping of $\mathfrak{L}$ onto $\mathfrak{L}$:

$$p=p(\bar{p},\bar{\theta}),\qquad \theta=\theta(\bar{p},\bar{\theta}),$$

where $p(\bar{p},\bar{\theta})$ and $\theta(\bar{p},\bar{\theta})$ are differentiable functions of variables $\bar{p},\bar{\theta}$. Then by (6.1.13) of Chapter 1 we have

$$dp=\frac{\partial}{\partial\bar{p}}d\bar{p}+\frac{\partial p}{\partial\bar{\theta}}d\bar{\theta},\qquad d\theta=\frac{\partial\theta}{\partial\bar{p}}d\bar{p}+\frac{\partial\theta}{\partial\bar{\theta}}d\bar{\theta},$$

and therefore

$$dp\wedge d\theta=\frac{\partial(p,\theta)}{\partial(\bar{p},\bar{\theta})}d\bar{p}\wedge d\bar{\theta},$$

which implies that

$$f(p,\theta)\,dp\wedge d\theta=f\big(p(\bar{p},\bar{\theta}),\theta(\bar{p},\bar{\theta})\big)\frac{\partial(p,\theta)}{\partial(\bar{p},\bar{\theta})}d\bar{p}\wedge d\bar{\theta}, \tag{2.56}$$

where $\partial(p,\theta)/\partial(\bar{p},\bar{\theta})$ is the Jacobian of the mapping defined by

$$\frac{\partial(p,\theta)}{\partial(\bar{p},\bar{\theta})}=\begin{vmatrix}\dfrac{\partial p}{\partial\bar{p}} & \dfrac{\partial p}{\partial\bar{\theta}}\\[2mm] \dfrac{\partial\theta}{\partial\bar{p}} & \dfrac{\partial\theta}{\partial\bar{\theta}}\end{vmatrix}.$$

On the other hand, a general orientation-preserving isometry of E^2 is given by

$$\begin{aligned} x_1&=a+\bar{x}_1\cos\phi-\bar{x}_2\sin\phi,\\ x_2&=b+\bar{x}_1\sin\phi+\bar{x}_2\cos\phi, \end{aligned} \tag{2.57}$$

where a, b and ϕ are constants. Under this isometry, (2.5.2) becomes

$$\bar{x}_1\cos\bar{\theta}+\bar{x}_2\sin\bar{\theta}=\bar{p},$$

where

$$\bar{\theta}=\theta-\phi, \qquad \bar{p}=p-a\cos\theta-b\sin\theta.$$

From the two equations above it follows easily that $\partial(p,\theta)/\partial(\bar{p},\bar{\theta})=1$, which reduces (2.56) to

$$f(p,\theta)\,dp\wedge d\theta=f\big(p(\bar{p},\bar{\theta}),\theta(\bar{p},\bar{\theta})\big)\,d\bar{p}\wedge d\bar{\theta}.$$

Thus

$$\int\int_{\mathbb{S}} f(p,\theta)\,dp\wedge d\theta=\int\int_{\mathbb{S}} f\big(p(\bar{p},\bar{\theta}),\theta(\bar{p},\bar{\theta})\big)\,d\bar{p}\wedge d\bar{\theta}. \qquad (2.58)$$

But the assumption that (2.55) is invariant under the isometry (2.57) means that

$$\int\int_{\mathbb{S}} f(p,\theta)\,dp\wedge d\theta=\int\int_{\mathbb{S}} f(\bar{p},\bar{\theta})\,d\bar{p}\wedge d\bar{\theta}. \qquad (2.59)$$

From (2.58) and (2.59) it follows that

$$\int\int_{\mathbb{S}} f\big(p(\bar{p},\bar{\theta}),\theta(\bar{p},\bar{\theta})\big)\,d\bar{p}\wedge d\bar{\theta}=\int\int_{\mathbb{S}} f(\bar{p},\bar{\theta})\,d\bar{p}\wedge d\bar{\theta}, \qquad (2.60)$$

and for (2.60) to be true for all $\mathbb{S}$, we must have

$$f\big(p(\bar{p},\bar{\theta}),\theta(\bar{p},\bar{\theta})\big)=f(\bar{p},\bar{\theta}).$$

Since for any pair of lines (p,θ) and $(\bar{p},\bar{\theta})$ in E^2 there is an orientation-preserving isometry F such that $F(\bar{p},\bar{\theta})=(p,\theta)$, we have

$$f(p,\theta)=(f\circ F)(\bar{p},\bar{\theta})=f(\bar{p},\bar{\theta}),$$

and $f(p,\theta)=\text{const}$. Hence the theorem is proved.

2.7.2. *Buffon's Needle Problem.* Let the plane E^2 be equipped with a system of parallel lines such that every pair of consecutive lines has a constant distance a. The well-known Buffon's needle problem is as follows. If a needle of length L (here we suppose $L<a$) is thrown onto the plane, what is the probability that the needle will intersect one of the parallel lines? A solution to this problem can be deduced from the following more general theorem.

2.7.3. *Theorem (The Cauchy-Crofton formula).* *Let C be a regular curve of length L. Then the measure of the set of lines (counted with the multiplicities, each of which is the number of the common points of a line and C) which intersect C is equal to $2L$.*

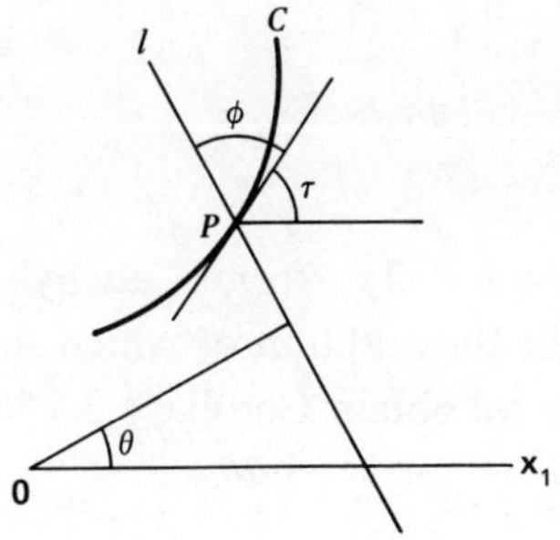

Figure 2.27

Proof. Let C be given by $\mathbf{x}(s)=(x_1(s), x_2(s))$ with arc length s with respect to a fixed frame $\mathbf{0x}_1\mathbf{x}_2$. Suppose that a line l with coordinates (p, θ) intersects C at a point $P(x_1, x_2)$, and let ϕ, τ be the angles between the tangent of C at P with l and the $\mathbf{x}_1$-axis, respectively (Fig. 2.27). Then

$$x_1' = \cos\tau, \qquad x_2' = \sin\tau, \tag{2.61}$$

$$\theta = \tau + \phi \pm \frac{1}{2}\pi, \tag{2.62}$$

where the prime denotes the derivative with respect to s. Using (2.52), (2.61), and (2.62), and noting that $d\theta = \tau' ds + d\phi$, we can easily obtain

$$dp \wedge d\theta = \cos\left(\phi \pm \tfrac{1}{2}\pi\right) ds \wedge d\phi.$$

Let $\mathbb{S}$ be the set of lines that intersect C. Since all measures are supposed to be positive, we have the measure of $\mathbb{S}$:

$$\mathfrak{M}(\mathbb{S}) = \int\int_{\mathbb{S}} dp \wedge d\theta = \int\int_{\mathbb{S}} |\sin\phi| \, ds \wedge d\phi$$
$$= \int_0^L ds \int_0^{\pi} |\sin\phi| \, d\phi = 2L.$$

2.7.4. Corollary. *The measure of the set of lines that intersect a convex curve C of length L is equal to L.*

Proof. For each tangent of C the number of the points of intersection is 1, and ϕ in Fig. 2.17 is zero, so that $ds \wedge d\phi = 0$. Thus each tangent is of measure zero. Q.E.D.

Corollary 2.7.4 means that for a closed convex curve C

$$\int\int_{\mathbb{S}} dp \wedge d\theta = L,$$

where the left-hand side is (see Fig. 2.21)

$$\int_0^{2\pi} d\theta \int_0^{P} dp = \int_0^{2\pi} d\theta\,[\,p\,]_0^P = \int_0^{2\pi} p(\theta)\,d\theta.$$

Thus we obtain Cauchy's formula (2.17).

Furthermore, since $p(\theta)$ is periodic with period 2π, from Cauchy's formula (2.17) it follows that $\int_0^{2\pi} p(\theta+\pi)\,d\theta = L$, the addition of which to (2.17) gives $\int_0^{2\pi}[p(\theta)+p(\theta+\pi)]\,d\theta = 2L$. Hence we obtain Corollary 2.4.8.

2.7.5. Corollary. *The answer to Buffon's needle problem* 2.7.2 *is* $2L/\pi a$.

Proof. Let us put the needle on a circular disk of diameter a. Then we throw the disk instead of the needle alone onto the plane. Except for a set of measure zero, the disk always intersects one and only one line. Thus our problem is entirely equivalent to that of comparing the measure of the lines intersecting the disk with that intersecting the needle in the disk. By Corollary 2.7.4 the first measure is equal to the length πa of the boundary of the disk, whereas by Theorem 2.7.3 the second is $2L$. This means that in the set of lines of measure πa, only those lines of measure $2L$ intersect the needle. Hence the probability is $2L/\pi a$.

2.8. More on Rotation Index. We have presented an intuitive definition for the rotation index of an oriented closed curve in Section 2.2, and shall give a rigorous one in this section and prove an important theorem on rotation index (or turning tangents) in the next section.

Let an oriented closed curve C of length L be given by $\mathbf{x}(s)$ with arc length s, and let $\mathbf{0}$ be a fixed point. Choose a fixed direction through $\mathbf{0}$, say $\mathbf{0x}_1$, and denote by $\tau(s)$ the angle from $\mathbf{0x}_1$ to the unit tangent vector $\mathbf{e}_1(s)$ of C at $\mathbf{x}(s)$, measured counterclockwise. We assume that $0 \leqslant \tau(s) < 2\pi$, so that $\tau(s)$ is uniquely determined. It should be remarked that $\tau(s)$ is not continuous, since in every neighborhood of s_0 at which $\tau(s_0)=0$ there may be values of $\tau(s)$ differing from 2π by an arbitrarily small quantity (see Fig. 2.28). There exists, however, a continuous function $\tilde{\tau}(s)$ closely related to $\tau(s)$, as given by the following lemma.

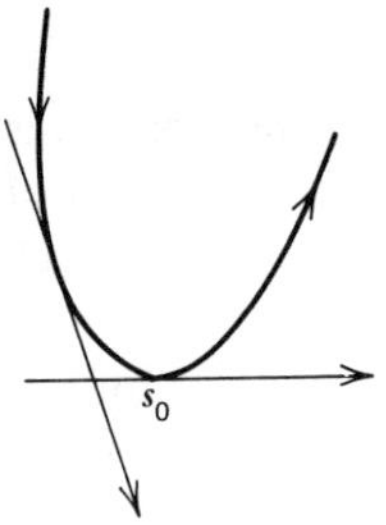

Figure 2.28

2.8.1. Lemma. *There exists a continuous function $\tilde{\tau}(s)$ such that $\tilde{\tau}(s)-\tau(s)$ is an integral multiple of 2π or, in notation, $\tilde{\tau}(s)\equiv\tau(s)$, mod 2π.*

Proof. Since a function continuous in a closed bounded set is uniformly continuous, the tangential mapping h (Definition 2.2.3) of the curve C is uniformly continuous. Therefore there exists a number $\delta>0$ such that for $|s_1-s_2|<\delta$, $h(s_1)$ and $h(s_2)$ lie in the same half-plane. From our conditions on $\tilde{\tau}(s)$, it follows that if $\tilde{\tau}(s_1)$ is known, $\tilde{\tau}(s_2)$ is completely determined. We then divide the interval $0\leqslant s\leqslant L$ by the points $s_0(=0)<s_1<\cdots<s_m(=L)$ such that $|s_i-s_{i-1}|<\delta, i=1,\cdots,m$. To define $\tilde{\tau}(s)$, we assign to $\tilde{\tau}(s_0)$ the value $\tau(s_0)$. Then $\tilde{\tau}(s)$ is determined in the subinterval $s_0\leqslant s\leqslant s_1$, in particular, at s_1. Using this $\tilde{\tau}(s_1)$ we can determine $\tilde{\tau}(s)$ in the second subinterval, and so on. The function $\tilde{\tau}(s)$ so defined clearly satisfies the conditions of the lemma.

Since $\tau(L)=\tau(0)$, the difference $\tilde{\tau}(L)-\tilde{\tau}(0)$ is an integral multiple of 2π, say, $\gamma\cdot 2\pi$. We assert that the integer γ is independent of the choice of the function $\tilde{\tau}(s)$. In fact, let $\tilde{\tau}^*(s)$ be a function satisfying the same conditions. Then we have

$$\tilde{\tau}^*(s)-\tilde{\tau}(s)=n(s)\cdot 2\pi,$$

where $n(s)$ is an integer. Since $n(s)$ is continuous in s, it must be a constant. It follows that

$$\tilde{\tau}^*(L)-\tilde{\tau}^*(0)=\tilde{\tau}(L)-\tilde{\tau}(0),$$

which shows that γ is independent of the choice of $\tilde{\tau}(s)$. We define γ to be the *rotation index* of C.

2.8.2. Theorem (Theorem on turning tangents). *The rotation index of a simple closed curve C is ± 1.*

Proof. Consider the mapping Σ, which maps an ordered pair of points $\mathbf{x}(s_1), \mathbf{x}(s_2), 0\leqslant s_1\leqslant s_2\leqslant L$, of C to the end point of the unit vector through $\mathbf{0}$ parallel to the line segment joining $\mathbf{x}(s_1)$ to $\mathbf{x}(s_2)$. These ordered pairs of points can be represented as a triangle Δ in the (s_1,s_2)-plane defined by $0\leqslant s_1\leqslant s_2\leqslant L$, (Fig. 2.29). The mapping Σ of Δ into the unit circle Γ is continuous, and its restriction to the side $s_1=s_2$ is the tangential mapping h.

For a point $p\in\Delta$ let $\tau(p)$ be the angle between $\mathbf{0x}_1$ and $\mathbf{0}\Sigma(p)$ such that $0\leqslant\tau(p)<2\pi$. Again this function need not be continuous. We shall, however, show that there exists a continuous function $\tilde{\tau}(p), p\in\Delta$, such that $\tilde{\tau}(p)\equiv\tau(p)$, mod 2π.

In fact, let m be an interior point of Δ. We cover Δ by the radii through m. By the arguments used in the proof of Lemma 2.8.1 (here we use each

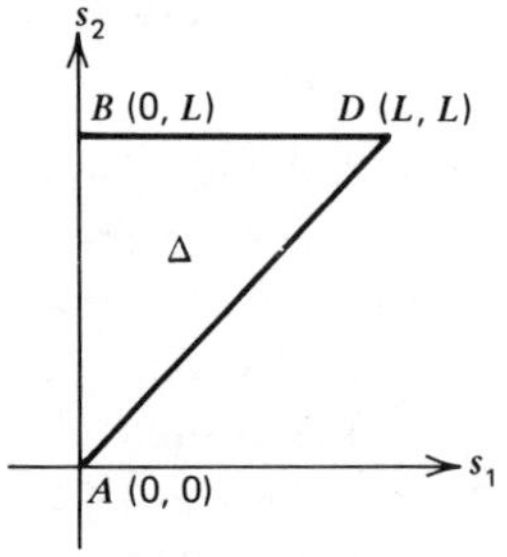

Figure 2.29

radius for the curve and divide it into m parts, etc.), we can define a function $\tilde{\tau}(p), p\in\Delta$, such that $\tilde{\tau}(p)\equiv\tau(p)$, mod 2π, and such that it is continuous along every radius through m. It remains to prove that it is continuous in Δ. For this purpose let p_0 be a point of Δ. Since Σ is continuous and the segment mp_0 is a closed bounded (compact) set, by use of the Heine-Borel covering theorem (Theorem 1.5.5, Chapter 1: every set of open intervals covering a closed bounded set has a finite subset), we see that there exists a number $\eta=\eta(p_0)>0$ such that for $q_0\in mp_0$ and any point $q\in\Delta$ for which the distance $d(q,q_0)<\eta$, the points $\Sigma(q)$ and $\Sigma(q_0)$ are never antipodal. The latter condition is equivalent to the relation

$$\tilde{\tau}(q)-\tilde{\tau}(q_0)\not\equiv 0, \qquad \text{mod}\,\pi. \tag{2.63}$$

Suppose that $\pi/2>\varepsilon>0$ be given. We choose a neighborhood U of p_0 such that U is contained in the η-neighborhood of p_0 and such that for $p\in U$ the angle between $\mathbf{0}\Sigma(p_0)$ and $\mathbf{0}\Sigma(p)$ is less than ε. This is possible because the mapping Σ is continuous. The last condition can be expressed in the form

$$\tilde{\tau}(p)-\tilde{\tau}(p_0)=\varepsilon'+2k(p)\pi, \qquad |\varepsilon'|<\varepsilon, \tag{2.64}$$

where $k(p)$ is an integer. Let q_0 be any point on the segment mp_0. Draw the segment q_0q parallel to p_0p with q on mp. The function $\tilde{\tau}(q)-\tilde{\tau}(q_0)$ is continuous in q along mp and is zero when q coincides with q_0. Since $d(q,q_0)<\eta$, it follows from (2.63) that $|\tilde{\tau}(q)-\tilde{\tau}(q_0)|<\pi$. In particular, for $q_0=p_0$, $|\tilde{\tau}(p)-\tilde{\tau}(p_0)|<\pi$. Combining this with (2.64) we obtain $k(p)=0$, which shows that $\tilde{\tau}(p)$ is differentiable.

Now let $A(0,0)$, $B(0,L)$, $D(L,L)$ be the vertices of Δ (Fig. 2.29). The rotation index γ of C is defined by the line integral

$$2\pi\gamma=\int_{AD} d\tilde{\tau}.$$

Since $\tilde{\tau}(p)$ is defined in Δ,

$$\int_{AD} d\tau=\tilde{\tau}(p)|_A^D,$$

which means that the line integral $\int_{AD} d\tilde{\tau}$ is independent of the path so that

$$\int_{AD} d\tilde{\tau} = \int_{AB} d\tilde{\tau} + \int_{BD} d\tilde{\tau}.$$

To evaluate the line integrals on the right-hand side we make use of a suitable coordinate system. We can suppose $\mathbf{x}(0)$ to be the *lowest* point of the curve C, that is, the point whose $\mathbf{x}_2$-coordinate is a minimum, and we choose $\mathbf{x}(0)$ to be the origin $\mathbf{0}$. Then the tangent vector of C at $\mathbf{x}(0)$ is horizontal and can be taken to be in the direction of $\mathbf{0x}_1$, as otherwise we reverse the orientation of C. Thus C lies in the upper half-plane bounded by $\mathbf{0x}_1$, and the line integral along AB is equal to the angle rotated by $\mathbf{0}P$ as $P \varepsilon C$ goes once along C. Since C lies in the upper plane, the vector $\mathbf{0}P$ never points downward. It follows that the integral along AB is equal to π. On the other hand, the line integral along BD is the angle rotated by $P\mathbf{0}$ as P goes once along C. Since the vector $P\mathbf{0}$ never points upward, this integral is also equal to π. Thus their sum is 2π and the rotation index is $+1$. Since we may have reversed the orientation of C, the rotation index of C is ± 1. Hence our proof of the theorem is complete.

Theorem 2.8.2 was essentially known to Riemann, but the proof above was given by H. Hopf [25].

The rotation index can also be defined by an integral formula. In fact, using the function $\tilde{\tau}(s)$ in our lemma, we can express the components of the unit tangent and normal vectors as follows:

$$\mathbf{e}_1 = (\cos\tilde{\tau}(s), \sin\tilde{\tau}(s)), \qquad \mathbf{e}_2 = (-\sin\tilde{\tau}(s), \cos\tilde{\tau}(s)).$$

It follows from Frenet formulas (1.3.11) that

$$d\tilde{\tau}(s) = d\mathbf{e}_1 \cdot \mathbf{e}_2 = \kappa\, ds. \tag{2.65}$$

From this we derive the following formula for the rotation index:

$$2\pi\gamma = \int_C \kappa\, ds. \tag{2.66}$$

This formula holds for closed curves that are not necessarily simple.

We have given the rotation indices of some closed curves in Section 2.2.

2.8.3. Piecewise Smooth Curves (see Definition 1.2.12). The notion of rotation index and the theorem on turning tangents can be extended to closed piecewise smooth curves. We summarize the results without proof as follows. Let s_i, $i = 1, \cdots, m$, be the arc length of a closed oriented piecewise smooth curve C, measured from A_0 to A_i, so that $s_m = L$ is the length of C. Since C is oriented, its tangential mapping h is defined at all points different from A_i. At a vertex A_i there are two unit vectors, tangent respectively to $A_{i-1}A_i$ and A_iA_{i+1} (we define $A_{m+1}=A_1$), and their corresponding points on Γ under h are denoted by $h(A_i)^-$ and $h(A_i)^+$. Let ϕ_i be the angle from $h(A_i)^-$ to $h(A_i)^+$, with $0 \leqslant \phi_i < \pi$, briefly the exterior angle from the tangent of $A_{i-1}A_i$ to the tangent of A_iA_{i+1}. For each arc $A_{i-1}A_i$ we can define a continuous function $\tilde{\tau}(s)$, which is one of the determinations of the angle from $\mathbf{0x}_1$ to the tangent of C at $\mathbf{x}(s)$. Again the difference $\tilde{\tau}(s_i)-\tilde{\tau}(s_{i-1})$ is independent of the choice of this function. The number γ defined by the equation

$$2\pi\gamma = \sum_{i=1}^{m} \left[\tilde{\tau}(s_i) - \tilde{\tau}(s_{i-1}) \right] + \sum_{i=1}^{m} \phi_i \tag{2.67}$$

is an integer, which will be called the *rotation index* of the curve C. The theorem on turning tangents (Theorem 2.8.2) is again valid:

2.8.4. Theorem. *The rotation index of a simple closed piecewise smooth curve is equal to* ± 1.

It is natural to expect that for a curve there is a relationship between the convexity and the sign of the curvature, which is well defined for oriented plane curves in an oriented plane. That this is so is confirmed by the following theorem as an application of the theorem on turning tangents.

2.8.5. Theorem. *A simple closed curve C is convex, if and only if it can be so oriented that its curvature* $\kappa > 0$.

Let us first remark that the theorem is not true without the assumption that the curve is simple. In fact, Fig. 2.30 gives a nonconvex curve with $\kappa > 0$.

Proof of Theorem 2.8.5. Let $\tilde{\tau}(s)$ be the function constructed in Lemma 2.8.1, so that from (2.65) we have $\kappa = d\tilde{\tau}/ds$. The condition $\kappa \geqslant 0$ is therefore equivalent to the assertion that $\tilde{\tau}(s)$ is a monotone nondecreasing function. We can assume that $\tilde{\tau}(0)=0$. From the theorem on turning tangents it follows that the curve C can be so oriented that $\tilde{\tau}(L)=2\pi$.

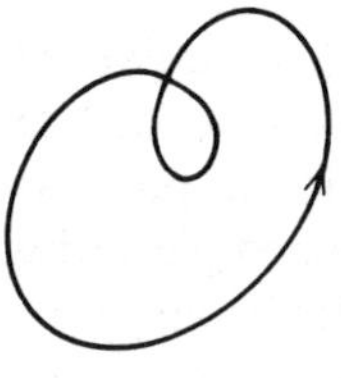

Figure 2.30

Suppose that $\tilde{\tau}(s)$, $0 \leqslant s \leqslant L$, is monotone nondecreasing and that C is not convex. Then there is a point $A = \mathbf{x}(s_0)$ on C such that there are points of C on both sides of the tangent t of C at A. Choose a positive side of t and consider the oriented orthogonal distance p from a point $\mathbf{x}(s)$ of C to t. Since the curve C is closed and p is a continuous function in s, by Weierstrass' theorem on continuous functions (Corollary 1.5.10, Chapter 1) this distance p attains a maximum and a minimum at points, say, M and N of C, respectively. Clearly, M and N are not on t. Since $p = (\mathbf{x}(s) - \mathbf{x}(s_0)) \cdot \mathbf{e}_2(s_0)$, at M and N we have $dp/ds = \mathbf{e}_1(s) \cdot \mathbf{e}_2(s_0) = 0$, which means that the tangents t_1, t_2 of C at M, N respectively are parallel to t. Furthermore, among t and t_1, t_2 there are two tangents parallel in the same sense; let the values of the parameter at the corresponding points of contact of these two tangents be s_1 and s_2 with $s_1 < s_2$. This happens only when $\tilde{\tau}(s) = \tilde{\tau}(s_1)$ for all $s_1 \leqslant s \leqslant s_2$, since $\tilde{\tau}(s)$ is monotone nondecreasing. It follows that the arc $s_1 \leqslant s \leqslant s_2$ of C is a line segment parallel to t. But this is obviously impossible.

Next let C be convex. Then C can be in a half-plane as was shown in the proof of Theorem 2.8.2. To prove that $\tilde{\tau}(s)$ is monotone let us suppose $\tilde{\tau}(s_1) = \tilde{\tau}(s_2)$, $s_1 < s_2$. Then the tangents of C at $\mathbf{x}(s_1)$ and $\mathbf{x}(s_2)$ are parallel in the same sense; therefore there must exist a tangent of C parallel to them in the opposite sense. From the convexity of C it follows that two of the three tangents coincide.

We are thus led to the consideration of a line t tangent to C at two distinct points A and B. We claim that the line segment AB must be a part of C. In fact, suppose this is not the case and let D be a point of AB not on C. Through D draw a line u orthogonal to t in the half-plane that contains C. Then u intersects C at at least two points. Among these points of intersection let F be farthest from t and G nearest, so that $F \neq G$. Then G is an interior point of the triangle ABF. The tangent to C at G must have points of C on both sides, which contradicts the convexity of C.

It follows that under the hypothesis of the last paragraph, the line segment AB is a part of C, and the tangents of C at A and B are parallel in the same sense. This proves that the segment that joins $\mathbf{x}(s_1)$ to $\mathbf{x}(s_2)$ belongs to C, implying that $\tilde{\tau}(s)$ remains constant in the interval $s_1 \leqslant s \leqslant s_2$. Hence the function $\tilde{\tau}(s)$ is monotone, and our theorem is proved.

The first half of Theorem 2.8.5 can also be stated as follows.

2.8.6. Theorem. *A closed curve with curvature $\kappa(s) \geqslant 0$ and rotation index equal to 1 is convex.*

Exercises

1. Prove that a necessary and sufficient condition that a curve $\mathbf{x}(t)$ be plane is

$$|\mathbf{x}', \mathbf{x}'', \mathbf{x}'''| = 0,$$

where the left-hand side is a determinant, and the primes denote the derivatives with respect to t.

2. Determine $f(t)$ so that the curve

$$\mathbf{x}(t) = (a\cos t, a\sin t, f(t)), \qquad a = \text{const},$$

shall be plane, and name the curve.

3. Find the winding numbers A, B, C, D, E, F, and G, relative to various components and the rotation index I of each of the following curves.

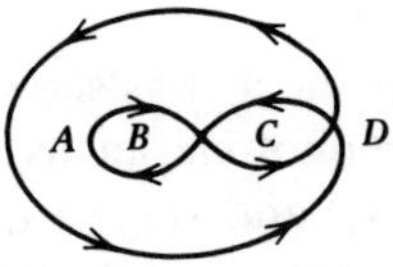

(a)

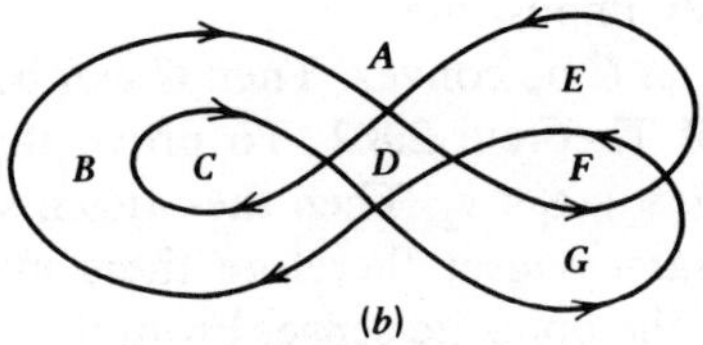

(b)

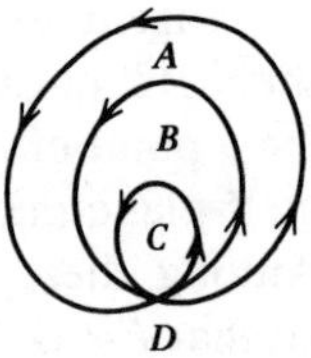

(c)

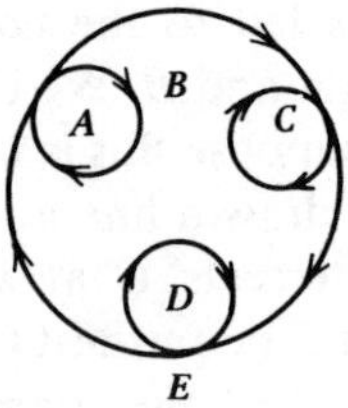

(d)

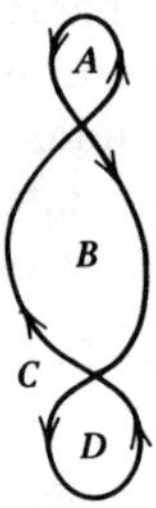

(e)

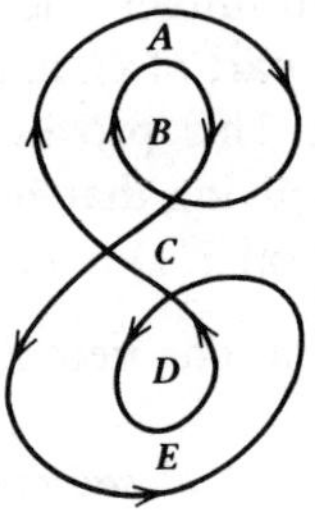

(f)

4. Prove geometrically that a convex curve has no self-intersections.

5. *(a) A set $S \subset E^2$ is *convex* if given any two points $p, q \in S$ the line segment joining p to q is contained in S. Prove that a closed convex curve bounds a convex set.

(b) Converse of part a. Let C be a regular closed curve. Prove that if C is the boundary of a convex region, then C is convex.

6. If a line l intersects a closed convex curve C, show that l either is tangent to C or intersects C at exactly two points.

7. Prove (2.40).

8. Is there any simple closed curve 5 inches long that bounds an area of 2 square inches?

***9.** The *convex* hull H of a nonconvex simple closed curve C, defined to be the boundary of the smallest convex set containing the interior of C, is formed by the arcs of C and the line segments of the tangents to C, which bridge the *convex gaps* (Fig. 2.31). H can be proved to be a C^1 closed convex curve. Use this to show that the isoperimetric inequality (2.28) can be restricted to convex curves.

10. Let C be a regular closed convex curve, and S the convex set bounded by C. Let $p_0 \in S$, and $p_0 \notin C$.

(a) Show that the line joining p_0 to an arbitrary point $q \in C$ is not tangent to C at q.

(b) From part a deduce that the rotation index of C is equal to the winding number of C relative to p_0.

(c) From part b obtain a simple proof for the fact that the rotation index of a closed convex curve is ± 1.

11. If the tangents of a closed convex curve C at two points $\mathbf{x}$ and $\mathbf{y}$ are parallel, then $\mathbf{x}$ and $\mathbf{y}$ are called *opposite points*. A convenient set of formulas for C can be obtained by using

$$(*) \qquad \mathbf{y} = \mathbf{x} + \lambda \mathbf{e}_1 + \mu \mathbf{e}_2, \qquad \mathbf{x} = \mathbf{x}(s),$$

where s is the arc length, $\mathbf{x}\mathbf{e}_1\mathbf{e}_2$ is a Frenet frame, and μ is the width of

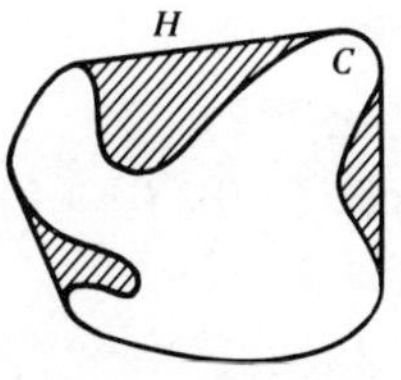

Figure 2.31

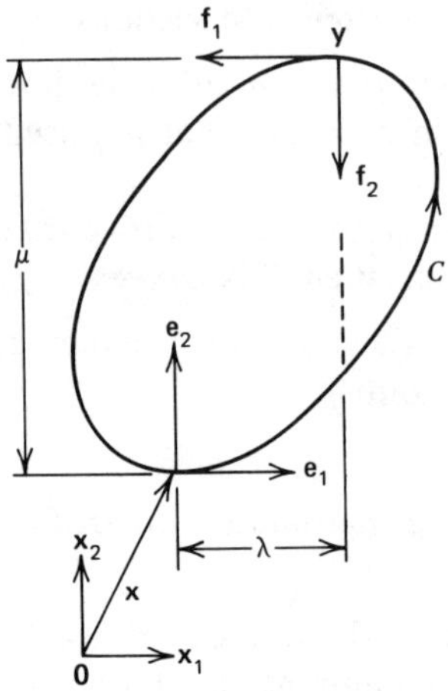

Figure 2.32

C at $\mathbf{x}$, that is, the width of C in the direction of the tangent at $\mathbf{x}$ (Fig. 2.32). Denote the Frenet frame of C at $\mathbf{y}$ by $\mathbf{y}\mathbf{f}_1\mathbf{f}_2$. Then

$$\mathbf{f}_1 = -\mathbf{e}_1, \qquad \mathbf{f}_2 = -\mathbf{e}_2.$$

If C is of constant width μ, prove each of the following:

*(a) The chord $\mathbf{xy}$ is a double normal to C, that is, a normal to C at both $\mathbf{x}$ and $\mathbf{y}$ (L. Euler).

*(b) The length of C is $\mu\pi$ (Corollary 2.4.8, Barber), or

$$\int_{\theta=0}^{\theta=\pi} (ds + ds_1) = \mu\pi,$$

where s_1 is the arc length of $\mathbf{y}$, and θ is the angle between the tangent of C at a point with the $\mathbf{x}_1$-axis.

(c) The sum of the radii of curvature of C at $\mathbf{x}$ and $\mathbf{y}$ is equal to the constant μ.

***12.** Any involute of a closed piecewise C^2 curve with n cusps and no parallel tangents is a curve of constant width with $2n$ vertices. Prove this statement for $n=3$ (see Fig. 2.33).

***13.** Show that there is at least one circumscribed square tangent to a closed convex curve (A. Emch, 1926).

14. Find the curvature of the ellipse with unequal axes:

$$\mathbf{x}(\theta) = (a\cos\theta, b\sin\theta), \qquad \theta \in [0, 2\pi), \qquad a \neq b,$$

and show that the ellipse has exactly four vertices $(a,0)$, $(-a,0)$, $(0,b)$, $(0,-b)$.

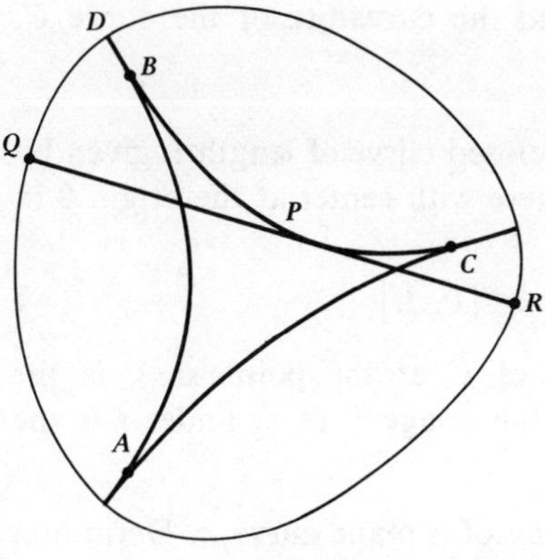

Figure 2.33

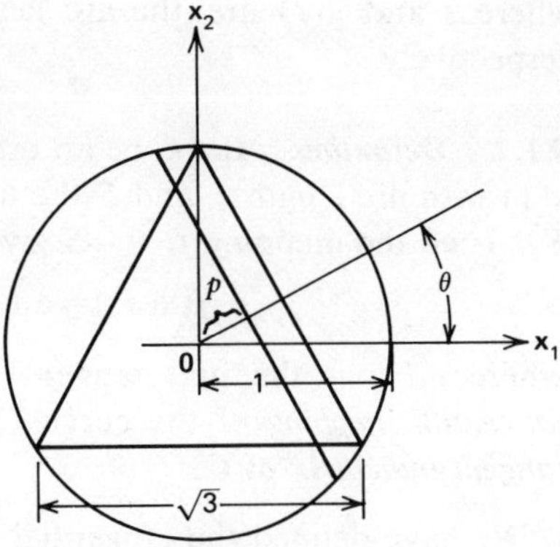

Figure 2.34

15. Show that the curve given by the equation

$$x_1^4 + x_2^4 = 1$$

has eight vertices situated on the lines $x_1 = 0$, $x_2 = 0$, $x_1 \pm x_2 = 0$.

16. Let C be a unit circle in a plane, let $\mathfrak{M}_2$ be the measure of the set of lines in the plane, which intersect C, and let $\mathfrak{M}_1$ be the measure of the set of those lines, each of which determines in C a chord of length $> \sqrt{3}$. Find the ratio $\mathfrak{M}_1 / \mathfrak{M}_2$, which intuitively is the probability that a line that intersects C determines in C a chord longer than the side of an equilateral triangle inscribed in C (Fig. 2.34).

3. GLOBAL THEOREMS FOR SPACE CURVES

This section gives several global theorems for general curves in space E^3. Throughout this section all curves are assumed to be general space curves unless stated otherwise.

3.1. Total Curvature

3.1.1. Definition. The *total curvature* of a closed curve C of length L is defined by the integral

$$K = \int_0^L |\kappa(s)|\, ds, \tag{3.1}$$

where s and $\kappa(s)$ are the arc length and the curvature of the curve C, respectively.

3.1.2. Definition. Let C be an oriented closed curve of length L given by $\mathbf{x}(s)$ with arc length s, and S the unit sphere with center at the origin $\mathbf{0}$ in E^3. Then the mapping f: $C \to S$ given by

$$f(\mathbf{x}(s)) = \mathbf{e}_1(s), \qquad s \in [0, L],$$

where $\mathbf{e}_1(s)$ is the unit tangent vector of C at the point $\mathbf{x}(s)$, is the *tangential mapping* of the curve C, and the image Γ of C under f is the *tangent indicatrix* of C.

We have defined the tangential mapping of a plane curve in Definition 2.2.2.

By the Frenet formulas (1.3.11) we can easily see that a point of Γ is singular (i.e., with either no tangent or a tangent of higher contact) if it is the image under f of a point of zero curvature of C. Moreover, we have

3.1.3. Lemma. *The total curvature of a closed curve C is equal to the length of the tangent indicatrix Γ of C.*

Concerning the total curvature we have the following theorem.

3.1.4. Theorem **[*W. Fenchel*, 20].** *Let K be the total curvature of a closed curve C. Then $K \geq 2\pi$, where the equality holds if and only if C is a plane convex curve.*

The following proof was given independently by B. Segre [52] and by H. Rutishauser and H. Samelson [48]. See also W. Fenchel [21]. The proof depends on the following lemma.

3.1.5. Lemma. *Let Γ be a closed rectifiable curve of length L on the unit sphere S. If $L < 2\pi$, there exists a point $m \in S$ such that the sperical distance $\overline{mx} \leq L/4$ for all points $x \in \Gamma$. If $L = 2\pi$, but Γ is not the union of two great semicircles, there exists a point $m \in S$ such that $\overline{mx} < \pi/2$ for all $x \in \Gamma$.*

We use the notation $\overline{ab}$ to denote the spherical distance of two points a, b on the unit sphere, that is, the shorter length of the great circular arc between a and b. If $\overline{ab} < \pi$, the midpoint m of a and b is the point on the shortest great circular arc between a and b defined by the conditions $\overline{am} = \overline{bm} = \frac{1}{2}\overline{ab}$. Let x be a point on S such that $\overline{mx} \leq \frac{1}{2}\pi$. Then $2\overline{mx} \leq \overline{ax} + \overline{bx}$. In fact, let x' be the symmetry of x relative to m, that is, let x' be the point on the great circle through m and x such that $\overline{mx} = \overline{x'm}$. Since the theorems on the congruence of two triangles in a plane are also true for two spherical triangles, the two spherical triangles Δbxm and $\Delta ax'm$ are congruent, so that

$$\overline{x'a} = \overline{xb}, \qquad \overline{x'x} = \overline{x'm} + \overline{mx} = 2\,\overline{mx},$$

from which it follows by the triangle inequality that

$$2\,\overline{mx} = \overline{x'x} \leqslant \overline{x'a} + \overline{ax} = \overline{bx} + \overline{ax}, \tag{3.2}$$

as will be proved.

Proof of Lemma 3.1.5. To prove the first part of the lemma we take two points a, b on Γ, which divide Γ into two equal parts. Then $\overline{ab} \leqslant \widehat{ab}$ (length along Γ)$=L/2<\pi$. Denote the midpoint of a, b by m, and let x be a point of Γ such that $2\overline{mx}<\pi$. Such points x do exist, for instance, the point a. Since the spherical distance is the shortest distance between two points on a sphere (see Exercise 6, Section 10, Chapter 3), we have

$$\overline{ax} \leqslant \widehat{ax}, \qquad \overline{bx} \leqslant \widehat{bx}.$$

From (3.2) it follows that

$$2\,\overline{mx} \leqslant ax + bx = ab = \frac{L}{2}.$$

Hence the function $f(x)=\overline{mx}$, $x\in\Gamma$, does not take values in the open interval $(L/4, \pi/2)$ for all x. Since Γ is connected and $f(x)$ is a continuous function on Γ, it is known (Theorem 1.3.4, Chapter 1) that the range of the function $f(x)$ is connected in the interval $(0, \pi)$. But $f(x) \leqslant L/4$ for at least one x, namely, $x=a$, hence we have $f(x)=\overline{mx}\leqslant L/4$ for all x on Γ.

Suppose next that Γ is of length 2π but is not the union of two great semicircles. Again let a, b be two points on Γ, which bisect Γ. Then a, b can not be antipodal, and for any point x on Γ we have the strict inequality

$$\overline{ax} + \overline{bx} < \widehat{ax} + \widehat{bx} = \pi.$$

Again let m denote the midpoint of a, b. If $f(x)=\overline{mx}<\pi/2$ (such x exists; e.g., $x=a$), then from (3.2) we have

$$2\,\overline{mx} \leqslant \overline{ax} + \overline{bx} < \pi,$$

which means that $f(x)$ cannot take the value $\pi/2$. Since the range of $f(x)$ is connected and $f(a)<\pi/2$, we have $f(x)<\pi/2$ for all $x\in\Gamma$. Hence the second statement of the lemma is proved.

Proof of Theorem 3.1.4. Let the curve C be given by $\mathbf{x}(s)$ with arc length s, take a fixed unit vector $\mathbf{a}$, and put

$$g(s) = \mathbf{a}\cdot\mathbf{x}(s).$$

$g(s)$ is a continuous function on C, hence must have a maximum and a minimum. Since $g'(s)$ exists, at such an extremum s_0 we have

$$g'(s_0)=\mathbf{a}\cdot\mathbf{e}_1(s_0)=0, \tag{3.3}$$

where $\mathbf{e}_1(s)$ is the unit tangent vector of C at the point $\mathbf{x}(s)$. Equation 3.3 means that $\mathbf{a}$, as a point on the unit sphere, has a spherical distance $\pi/2$ from at least two points of the tangent indicatrix Γ of C. Since $\mathbf{a}$ is arbitrary, the tangent indicatrix Γ is met by every great circle. By Lemma 3.1.5, the length of the tangent indicatrix Γ is greater than or equal to 2π, and is equal to 2π when and only when Γ is the union of two great semicircles. In the former case, from Lemma 3.1.3 it follows that the total curvature of C is greater than or equal to 2π. In the latter case, C itself is the union of two plane arcs, therefore must be a plane curve, since it has a tangent everywhere. Moreover, suppose C be so oriented that its rotation index

$$\frac{1}{2\pi}\int_0^L \kappa\, ds \geqslant 0.$$

Then we have

$$\int_0^L (|\kappa|-\kappa)\, ds = 2\pi - \int_0^L \kappa\, ds \geqslant 0,$$

since the integrand on the left-hand side is continuous and nonnegative. Thus the rotation index of C is either 0 or 1. For every vector in the plane, parallel to it there is a tangent t of C such that C lies to the left of t. Then t is parallel to the vector in the same sense, and at the point of contact of t we have $\kappa(s)\geqslant 0$, implying that $\int_{\kappa>0}\kappa\, ds \geqslant 2\pi$. Since $\int_C|\kappa|\, ds = 2\pi$, there is no point of C with $\kappa<0$, and therefore $\int_C \kappa\, ds = 2\pi$. By Theorem 2.23 we conclude that C is convex. Hence Theorem 3.1.4 is completely proved.

3.1.6. Corollary. *For a closed curve C if $|\kappa(s)|\leqslant 1/R$, where s and $\kappa(s)$ are the arc length and the curvature of C, respectively, and R is a positive constant, C has a length $L\geqslant 2\pi R$.*

Proof. $L=\int_0^L ds \geqslant \int_0^L R|\kappa|\, ds = R\int_0^L|\kappa|\, ds \geqslant 2\pi R$.

3.1.7. Remark. Theorem 3.14 holds also for piecewise smooth curves. As the total curvature of such a curve C we define

$$K=\int_0^L|\kappa|\, ds + \sum_i a_i,$$

where a_i are the angles at the vertices of C. In other words, in this case the tangent indicatrix of C consists of a number of arcs, each corresponding to a smooth arc of C, and we join each pair of successive vertices by the shortest great circular arc on the unit sphere. The length of the curve so

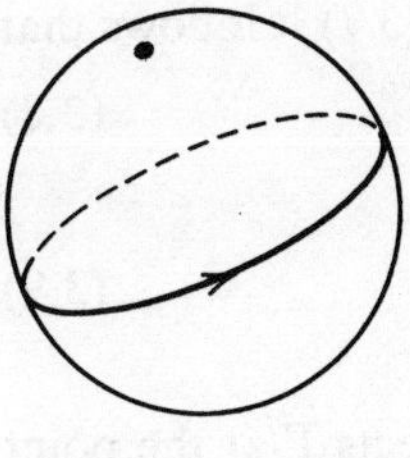

Figure 2.35

obtained is the total curvature of C. It can be proved that for a closed piecewise smooth curve we have also $K \geqslant 2\pi$.

In the following we give another proof of Fenchel's theorem (Theorem 3.1.4) and a related theorem of Fary and Milnor on the total curvature of a nontrivial knot. These are based on the following Crofton's theorem on the measure of great circles that intersect an arc on the unit sphere.

3.1.8. Definition. Every oriented great circle on the unit sphere S determines uniquely a *pole*, the end point of the unit vector normal to the plane of the circle such that the pole remains to one's left side as one goes along the circle (see Fig. 2.35). By the *measure* of a set of great circles on S we mean the area of the domain of their poles.

3.1.9. Theorem (M. W. Crofton). *Let Γ be a smooth arc of length L on the unit sphere S, and let M be the measure of the oriented great circles of S intersecting Γ, each counted with a multiplicity which is the number of its points of intersection with Γ. Then*

$$M = 4L. \tag{3.4}$$

Proof. Suppose that Γ is given by a unit vector $\mathbf{e}_1(s)$ with arc length s. Locally (i.e., in a certain neighborhood of s), let $\mathbf{e}_2(s)$ and $\mathbf{e}_3(s)$ be unit vectors depending smoothly on s such that

$$\mathbf{e}_i \cdot \mathbf{e}_j = \delta_{ij}, \qquad 1 \leqslant i, j \leqslant 3, \tag{3.5}$$

$$|\mathbf{e}_1, \mathbf{e}_2, \mathbf{e}_3| = +1. \tag{3.6}$$

Then by (1.3.3) and (1.3.5) we have

$$\begin{aligned} \frac{d\mathbf{e}_1}{ds} &= a_2\mathbf{e}_2 + a_3\mathbf{e}_3, \\ \frac{d\mathbf{e}_2}{ds} &= -a_2\mathbf{e}_1 + a_1\mathbf{e}_3, \\ \frac{d\mathbf{e}_3}{ds} &= -a_3\mathbf{e}_1 - a_1\mathbf{e}_2. \end{aligned} \tag{3.7}$$

Since s is the arc length of Γ, from the first equation of (3.7) it follows that

$$a_2^2 + a_3^2 = 1, \tag{3.8}$$

so that we can put

$$a_2 = \cos\tau(s), \qquad a_3 = \sin\tau(s). \tag{3.9}$$

If an oriented great circle on the unit sphere S intersects Γ at the point $\mathbf{e}_1(s)$, the pole of the great circle is of the form

$$\mathbf{y}(s) = \mathbf{e}_2(s)\cos\theta + \mathbf{e}_3(s)\sin\theta, \tag{3.10}$$

and vice versa. Thus (s,θ) can serve as local coordinates in the domain D of these poles. To find an expression for the element of area of this domain D, we differentiate (3.10) and, using (3.7), obtain

$$\begin{aligned} d\mathbf{y} &= (-\mathbf{e}_2\sin\theta + \mathbf{e}_3\cos\theta)(d\theta + a_1\,ds) \\ &\quad - \mathbf{e}_1(a_2\cos\theta + a_3\sin\theta)\,ds. \end{aligned} \tag{3.11}$$

Since

$$d\mathbf{y} = \mathbf{y}_\theta\,d\theta + \mathbf{y}_s\,ds,$$

where each subscript denotes the partial derivative with respect to it, we have

$$\begin{aligned} \mathbf{y}_\theta &= -\mathbf{e}_2\sin\theta + \mathbf{e}_3\cos\theta, \\ \mathbf{y}_s &= a_1(-\mathbf{e}_2\sin\theta + \mathbf{e}_3\cos\theta) - \mathbf{e}_1(a_2\cos\theta + a_3\sin\theta). \end{aligned} \tag{3.12}$$

On the other hand, the element of area dA of the domain D is [see (8.13), Chapter 3]

$$dA = \sqrt{(\mathbf{y}_\theta \times \mathbf{y}_s)^2}\;d\theta\,ds. \tag{3.13}$$

Making use of (5.4.2) of Chapter 1 and (3.12), from (3.13) we can easily obtain

$$|dA| = |a_2\cos\theta + a_3\sin\theta|\,d\theta\,ds = |\cos(\tau-\theta)|\,d\theta\,ds, \tag{3.14}$$

where the absolute value on the left-hand side means that the area is calculated in the measure-theoretic sense regardless of the orientation.

Now let $\mathbf{y}^\perp$ be the oriented great circle with $\mathbf{y}$ as its pole, and let $n(\mathbf{y}^\perp)$ be the (arithmetic) number of common points of $\mathbf{y}^\perp$ and Γ. Then

$$M = \int_D\!\!\int n(\mathbf{y}^\perp)|dA| = \int_0^L ds \int_0^{2\pi} |\cos(\tau-\theta)|\,d\theta. \tag{3.15}$$

But

$$\int_0^{2\pi} |\cos(\tau-\theta)|\, d\theta = \int_{-\tau}^{2\pi-\tau} |\cos\phi|\, d\phi = 4 \quad \text{for each fixed } s, \tag{3.16}$$

since the absolute value of the area under the cosine curve over each quadrant is 1. Substitution of (3.16) in (3.15) hence gives (3.4).

3.1.10. Remark. By applying Theorem 3.1.9 to each subarc and adding the results, we see that the theorem also holds for a piecewise smooth curve Γ on the unit sphere. Actually, the theorem is true for any rectifiable arc on the unit sphere, but the proof is much longer.

3.1.11. Another Proof of Theorem 3.1.4. Since the tangent indicatrix of a closed curve satisfies the conditions of Theorem 3.1.9, Theorem 3.1.4 follows from Theorem 3.1.9 as an easy consequence. In fact, in the former proof of Theorem 3.1.4 we have shown that the tangent indicatrix Γ of a closed curve C of length L intersects every great circle of the unit sphere S in at least two points, so that in this case every point of S belongs to the domain D and $n(\mathbf{y}^{\perp}) \geqslant 2$. Since the area of the unit sphere S is 4π, from Theorem 3.1.9 we hence obtain the length of Γ:

$$K = \int_0^L |\kappa(s)|\, ds = \frac{1}{4}\int_D\int n(\mathbf{y}^{\perp})|dA| \geqslant 2\pi, \tag{3.17}$$

where s and $\kappa(s)$ are respectively the arc length and the curvature of the curve C. Q.E.D.

Theorem 3.1.9 also leads to the following theorem, which gives a necessary condition on the total curvature of a nontrivial knot.

3.1.12. Theorem [I. Fary and J. W. Milnor, 19, 39]. *The total curvature of a nontrivial knot is greater than or equal to 4π.*

Proof. Let a nontrivial knot C be given by $\mathbf{x}(s)$ with arc length s, let $\mathbf{y}$ be the pole of a great circle, denoted by $\mathbf{y}^{\perp}$, on the unit sphere S intersecting the tangent indicatrix Γ of C on S, and let $n(\mathbf{y}^{\perp})$ be the number of common points of $\mathbf{y}^{\perp}$ and Γ. From (3.3) it follows that $n(\mathbf{y}^{\perp})$ is the number of the relative maxima or minima of the *height function* $\mathbf{y}\cdot\mathbf{x}(s)$, and is even by Theorem 2.3.5 of Chapter 1.

Suppose that the total curvature of C is less than 4π. Then there exists a point $\mathbf{y} \in S$ such that $n(\mathbf{y}^{\perp}) = 2$, as otherwise $n(\mathbf{y}^{\perp}) \geqslant 4$ for all poles and the right-hand side of the inequality (3.17) becomes 4π. By a rotation if necessary, we can assume $\mathbf{y}$ to be the point $(0,0,1)$, so that the coordinate function $x_3(s)$ of a general point $\mathbf{x}(s) = (x_1(s), x_2(s), x_3(s))$ of C has only

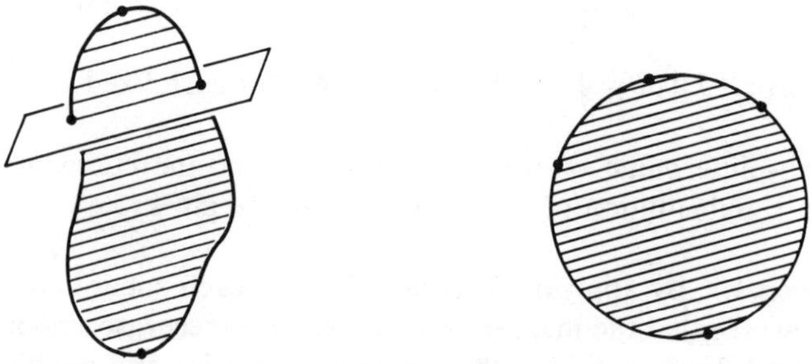

Figure 2.36

one maximum M and one minimum m. These two points divide C into two arcs, such that x_3 increases on the one and decreases on the other. Since every horizontal plane intersects each arc in at most one point, every horizontal plane between the two extremal horizontal planes $x_3 = M$ and $x_3 = m$ intersects C in exactly two points. We join each pair of those points by a line segment, and by all those line segments we form a surface bounded by C, which is homeomorphic to a circular disk; see Fig. 2.36. Thus C is a trivial knot, and this contradiction completes the proof.

K. Borsuk [6] has generalized Theorem 3.1.4 to curves in an n-dimensional ($n \geqslant 3$) Euclidean space E^n. F. Brickell and C. C. Hsiung [7] have jointly further generalized Borsuk's result to curves in an n-dimensional nonelliptic space H^n, and have also generalized Theorem 3.1.12 to nontrivial knots in a space H^3.

3.1.13. Remark. After having studied the total curvature of a closed curve C, we naturally would like to investigate the *total torsion* of C, which is defined to be the integral $\int_C \tau\, ds$, where s and τ are the arc length and the torsion of C, respectively. However *for any real number r there is a closed*

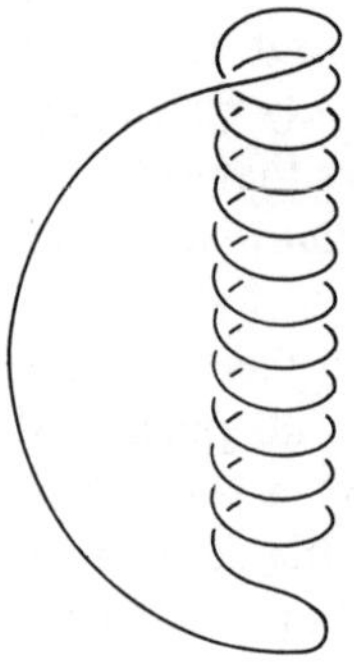

Figure 2.37

curve C such that its total torsion is equal to r. To prove this, we need only to consider curves made up of a circular helix with the ends joined by a plane curve as in Fig. 2.37, and to vary the pitch, the number of coils, and the right- or left-handedness of the helix.

On the other hand, we can have Geppert's theorem.

3.1.14. Theorem [*H. Geppert, 23*]. *The total torsion of a closed curve on a unit sphere is zero.*

This theorem actually gives a sufficient condition for a nonplanar curve to lie on a unit sphere; its proof requires some material in Chapter 3 and is therefore omitted here. B. Segre [53] has studied the total absolute torsion $\int_C |\tau|\, ds$ of a closed curve C and has obtained some results on lower bounds on the integral.

3.2. Deformations. We proved in Section 1.5 that a bijection between two curves, under which the arc length s, the curvature κ, and the torsion τ are preserved, can be established only by an orientation-preserving isometry. It is natural to study the mappings under which only s, κ are preserved. We shall call such a mapping a *deformation* of the curve. The most notable result along this direction is the following theorem of A. Schur. It formulates the geometric fact that if an arc is *stretched*, the distance between its end points becomes longer. By the curvature of a curve in this section, we shall always mean its absolute value.

3.2.1. Theorem (A. Schur). *Let C be a plane arc with curvature $\kappa(s)$ and arc length s, which forms a convex curve with its chord AB. Let C^* be an arc of the same length referred to the same parameter s such that its curvature*

$$\kappa^*(s) \leqslant \kappa(s). \tag{3.18}$$

If d, d^* denote the lengths of the chords joining the end points of C, C^*, respectively, then $d \leqslant d^*$, where the equality holds when and only when C and C^* are congruent.

Proof. Let Γ and Γ^* be the tangent indicatrices of C and C^*, respectively. Let P_1, P_2 be two points on Γ, and P_1^*, P_2^* their corresponding points on Γ^*. Denote the arc lengths of the arcs P_1P_2 and $P_1^*P_2^*$ on Γ and Γ^* by $\widehat{P_1P_2}$, $\widehat{P_1^*P_2^*}$ and their spherical distances by $\overline{P_1P_2}, \overline{P_1^*P_2^*}$, respectively. Then we have

$$\overline{P_1P_2} \leqslant \widehat{P_1P_2}, \qquad \overline{P_1^*P_2^*} \leqslant \widehat{P_1^*P_2^*}.$$

The inequality (3.18) implies (see Exercise 1)

$$P_1^*P_2^* \leqslant \widehat{P_1P_2}. \tag{3.19}$$

Since C is plane, Γ lies on a great circle, and we have

$$\overline{P_1P_2} = \widehat{P_1P_2},$$

provided $P_1P_2 \leqslant \pi$. Now let Q be the point on C at which the tangent of C is parallel to the chord AB. Denote the image of Q on Γ by P_0. Then the condition $\widehat{P_0P} \leqslant \pi$ is satisfied for any point P on Γ, since $\widehat{P_0P}$ is equal to the angle between the tangents of C at Q and the point corresponding to P. Let P_0^*, P^* be the points on Γ^* corresponding to P_0, P, respectively. Then we have

$$\overline{P_0^*P^*} \leqslant \overline{P_0P}, \tag{3.20}$$

from which it follows that

$$\cos \overline{P_0^*P^*} \geqslant \cos \overline{P_0P}, \tag{3.21}$$

since the cosine function is a monotone decreasing function of its argument in the range between 0 and π.

Since C is convex, d is equal to the projection of C on its chord AB:

$$d = \int_0^L \cos \overline{P_0P}\, ds. \tag{3.22}$$

On the other hand, we have

$$d^* \geqslant \int_0^L \cos \overline{P_0^*P^*}\, ds, \tag{3.23}$$

since the integral on the right-hand side is equal to the projection of C^*, hence of the chord joining its end points, on the tangent of C^* at the point Q^* corresponding to Q. Combining (3.21), (3.22), and (3.23) we obtain $d^* \geqslant d$.

Suppose that $d = d^*$. Then the inequalities (3.20), (3.21), and (3.23) all become equalities. Thus the chord joining the end points A^*, B^* of C^* must be parallel to the tangent of C^* at Q^*. Since the equality must hold in (3.20), we have

$$\overline{P_0^*P^*} = \overline{P_0P}, \tag{3.24}$$

which implies that the images of A^*Q^* and B^*Q^* are on the unit circle, or the arcs A^*Q^* and B^*Q^* are plane arcs. On the other hand, by using (3.19) we have

$$\overline{P_0^*P^*} \leqslant \widehat{P_0^*P^*} \leqslant \widehat{P_0P} = \overline{P_0P},$$

and therefore, in consequence of (3.24),

$$\widehat{P_0^*P^*} = \widehat{P_0P}.$$

Hence the arcs A^*Q^*, B^*Q^* have the same curvature as AQ, BQ at corresponding points; therefore A^*Q^*, B^*Q^* are congruent to AQ, BQ, respectively.

It remains to prove that the arcs A^*Q^*, B^*Q^* lie in the same plane. Suppose the contrary. Then A^*Q^* and B^*Q^* must be tangent at Q^* to the line of intersection of the two distinct planes on which they lie. Since this line of intersection is parallel to A^*B^*, the only possibility is that it contains A^*, B^*; then the tangent to C at Q must also contain the end points A, B, which contradicts the hypothesis that C together with the chord AB is a convex curve. Hence C^* is a plane arc and is congruent to C.

Schur's theorem has many applications. One application is the derivation of the following theorem of Schwarz, which discusses the lengths of arcs joining two given points with curvature bounded from above by a fixed constant.

3.2.2. Theorem (H. A. Schwarz). *Let C be an arc of length L joining two given points A, B with curvature $\kappa \leqslant 1/R$ such that $R \geqslant \frac{1}{2}d$, where R is a positive constant and d is the length of the line segment AB. Let $\mathbb{S}$ be a circle of radius R through A, B. Then either $L \leqslant$ the shorter arc $\widehat{AB}$ or $L \geqslant$ the longer arc $\widehat{AB}$ on $\mathbb{S}$.*

Proof. We first remark that the assumption $R \geqslant \frac{1}{2}d$ is necessary for the existence of the circle $\mathbb{S}$. To prove the theorem we can assume that $L < 2\pi R$, since otherwise there is nothing to prove. Then we compare C with an arc C' of the same length of $\mathbb{S}$. Let d' be the length of the chord of C'. The conditions of Schur's theorem are satisfied, and we obtain $d' \leqslant d$. Hence either $L \geqslant$ the longer arc of $\mathbb{S}$ with the chord $\widehat{AB}$ or $L \leqslant$ the shorter arc of $\mathbb{S}$ with the chord $\widehat{AB}$. Q.E.D.

In particular, we consider arcs joining A, B with curvature $\kappa = 1/R$, $R \geqslant d/2$. Then the lengths of such arcs have no upper bound, as shown by the example of a helix. They have d as a lower bound, but can be as close to d as possible. This is therefore an example of a minimum problem with no solution.

The discussion above leads to the minimum problem: determine the shortest closed curve with curvature $\kappa \leqslant 1/R$, R being a given positive constant. As a second application of Schur's theorem we have the following solution of this problem.

3.2.3. Theorem. *The shortest closed curve with curvature $\kappa \leqslant 1/R$, R being a positive constant, is a circle of radius R.*

This theorem is motivated by the physical problem of finding the shortest piece of wire whose end points can be put together without breaking the wire, that is, without increasing its curvature at any point beyond $1/R$.

Proof. By Corollary 3.1.6 the required curve has length $L=2\pi R$. Moreover, by Fenchel's theorem we have

$$2\pi \leqslant \int_0^L \kappa\, ds \leqslant \frac{1}{R}\int_0^L ds = \frac{1}{R}L = 2\pi,$$

where s is the arc length of the curve. Therefore $\kappa = 1/R$; hence the curve is a circle. We can also show that this curve is a circle of radius R by comparing it with a circle of radius R and using Schur's theorem with $d^* = d = 0$. Q.E.D.

Finally, we remark that Schur's theorem can be generalized to piecewise smooth curves. We give here a statement of this generalization without proof.

3.2.4. Theorem. *Let C and C^* be two piecewise smooth curves of the same length such that C with its chord forms a convex plane curve. Referred to the arc length s from one end point as parameter let $\kappa(s)$ be the curvature of C at a regular point, and $\theta_i(s)$ the angle between the oriented tangents at a vertex, and denote the corresponding quantities for C^* by the same notation with an asterisk. Let d and d^* be the distances between the end points of C and C^*, respectively. If $\kappa^*(s) \leqslant \kappa(s)$, $\theta_i^*(s) \leqslant \theta_i(s)$, then $d^* \geqslant d$, where the equality holds if and only if $\kappa^*(s) = \kappa(s)$, $\theta_i^*(s) = \theta_i(s)$.*

We remark that the last set of conditions in Theorem 3.2.4 does not necessarily imply that C and C^* are congruent. In fact, there are simple rectilinear polygons in space that have equal sides and equal angles but are not congruent.

Exercises

1. Show that a space (not plane) curve is a cylindrical helix if and only if its tangent indicatrix or binormal indicatrix (which can be defined in the same way as the tangent indicatrix) is a plane curve, hence a circle.
2. Prove Lemma 3.1.3.
3. Verify Fenchel's theorem for the ellispse given by

$$\mathbf{x}(s) = (2\cos t, \sin t, 0), \qquad 0 \leqslant t \leqslant 2\pi.$$

4. Let C be a curve on the unit sphere given by

$$\mathbf{x}(s) = (\cos s, \sin s, 0), \qquad 0 \leqslant s \leqslant 2\pi.$$

Verify that Crofton's formula is true for C.

3

Local Theory of Surfaces

This chapter begins with the definition of a surface in a space E^3; then we establish a general local theory of those surfaces, which is used throughout the remainder of the book.

1. PARAMETRIZATIONS

There are various notions of a surface in E^3. However in this book unless stated otherwise we deal only with the surfaces given by the following definition: roughly speaking, our surfaces have no sharp points, edges, or self-intersections.

1.1. Definition. A surface in E^3 is a subset $S \subset E^3$ such that for each point $\mathbf{p} \in S$ there are a neighborhood V of $\mathbf{p}$ in E^3 and a mapping $\mathbf{x}$: $U \to V \cap S$ of an open set $U \subset E^2$ onto $V \cap S \subset E^3$ subject to the following three conditions (Fig. 3.1):

$$\mathbf{x} \text{ is of class } C^3 \text{ (see Definition 4.1.1 of Chapter 1).} \tag{1.1}$$

$$\mathbf{x} \text{ is a homeomorphism.} \tag{1.2}$$

$$\mathbf{x} \text{ is regular at each point } \mathbf{q} \in U. \tag{1.3}$$

The mapping $\mathbf{x}$ is called a *parametrization* or a *(local) coordinate system* at $\mathbf{p}$ or in a neighborhood of $\mathbf{p}$, and the neighborhood $V \cap S$ of $\mathbf{p}$ in S a *coordinate neighborhood*.

1.2. Remarks. 1. $\mathbf{x}$ of class C^2 instead of class C^3 in condition 1.1 above would suffice for many considerations of this book.

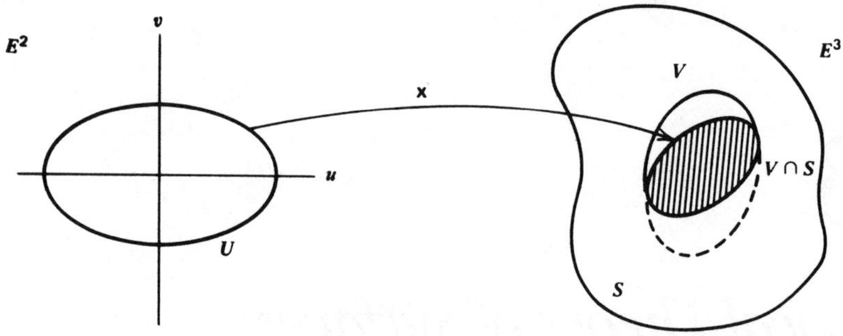

Figure 3.1

2. By Definition 1.2.8 of Chapter 1, condition 1.2 means that $\mathbf{x}$ has a continuous inverse $\mathbf{x}^{-1}$.

3. By Definition 4.2.6 of Chapter 1 and the paragraph just below it, condition 1.3 means that for each $\mathbf{q} \in U$, the differential mapping $d\mathbf{x}$: $E^2 \to E^3$ is injective or equivalently that the Jacobian matrix $J_x(\mathbf{q})$ of the mapping $\mathbf{x}$ at each $\mathbf{q} \in U$ has rank 2. This implies that at each $\mathbf{q} \in U$ the vector product

$$\frac{\partial \mathbf{x}}{\partial u} \times \frac{\partial \mathbf{x}}{\partial v} \neq 0, \tag{1.3$'$}$$

where $(u,v) \in U$. Thus $\mathbf{x}$ is neither constant nor a function of u or v alone, so that the surface S is neither a point nor a curve. Furthermore, the vectors $\partial \mathbf{x}/\partial u$ and $\partial \mathbf{x}/\partial v$ are linearly independent at each $\mathbf{q}$.

1.3. Example. Show that the sphere

$$S^2 = \{(x_1, x_2, x_3) \in E^3 | x_1^2 + x_2^2 + x_3^2 = a^2\},$$

which consists of all points at distance a from the origin $\mathbf{0}(0,0,0)$, is a surface.

First we find a parametrization in a neighborhood of the north pole $(0,0,a)$. For each point (q_1, q_2, q_3) of the northern hemisphere S_+^2 of S^2 given by $\{(x_1, x_2, x_3) \in S^2 | x_3 > 0\}$ we take its orthogonal projection $(q_1, q_2, 0)$ onto the $\mathbf{x}_1\mathbf{x}_2$-plane (see Fig. 3.2). By identifying this plane with E^2 we thus obtain an injection

$$\mathbf{x}^1 \colon U = \{(x_1, x_2) \in E^2 | x_1^2 + x_2^2 < a^2\} \to S_+^2$$

given by

$$\mathbf{x}^1(x_1, x_2) = \left(x_1, x_2, +\sqrt{a^2 - (x_1^2 + x_2^2)}\,\right), \qquad (x_1, x_2) \in U. \tag{1.4}$$

Since $x_1^2 + x_2^2 < a^2$, the function $f = +\sqrt{a^2 - (x_1^2 + x_2^2)}$ has continuous partial derivatives of all orders. Thus $\mathbf{x}^1$ satisfies condition 1.1.

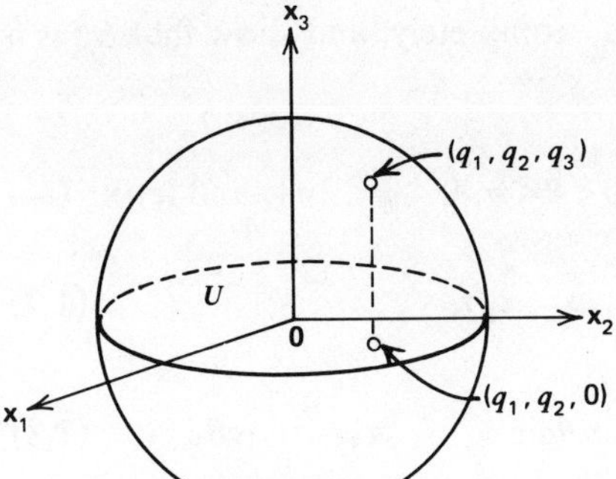

Figure 3.2

Since $\mathbf{x}^1$ is an injection, and $(\mathbf{x}^1)^{-1}$ is the restriction to the set $\mathbf{x}^1(U)$ of the projection: $(q_1, q_2, q_3) \to (q_1, q_2, 0)$, $(\mathbf{x}^1)^{-1}$ is continuous and satisfies condition 1.2.

To show that $\mathbf{x}^1$ is regular, we compute its Jacobian matrix

$$\begin{bmatrix} \dfrac{\partial x_1}{\partial x_1} & \dfrac{\partial x_2}{\partial x_1} & \dfrac{\partial f}{\partial x_1} \\ \dfrac{\partial x_1}{\partial x_2} & \dfrac{\partial x_2}{\partial x_2} & \dfrac{\partial f}{\partial x_2} \end{bmatrix} = \begin{bmatrix} 1 & 0 & \dfrac{\partial f}{\partial x_1} \\ 0 & 1 & \dfrac{\partial f}{\partial x_2} \end{bmatrix},$$

whose rank is obviously 2 at each point $(x_1, x_2, x_3) \in S^2_+$. Hence $\mathbf{x}^1$ is a parametrization of S^2_+.

Next we cover the whole sphere S^2 with similar parametrizations as follows. Define $\mathbf{x}^2$: $U \subset E^2 \to E^3$ by

$$\mathbf{x}^2(x_1, x_2) = \left(x_1, x_2, -\sqrt{a^2 - (x_1^2 + x_2^2)}\right), \tag{1.5}$$

check that $\mathbf{x}^2$ is a parametrization, and observe that $\mathbf{x}^1(U) \cup \mathbf{x}^2(U)$ covers S^2 minus the equator

$$\left\{(x_1, x_2, x_3) \in E^3 \,|\, x_1^2 + x_2^2 = a^2, x_3 = 0\right\}.$$

Similarly, by using $\mathbf{x}_1\mathbf{x}_3$- and $\mathbf{x}_2\mathbf{x}_3$-planes we have the parametrizations:

$$\begin{aligned} \mathbf{x}^3(x_1, x_3) &= \left(x_1, +\sqrt{a^2 - (x_1^2 + x_3^2)}, x_3\right), \\ \mathbf{x}^4(x_1, x_3) &= \left(x_1, -\sqrt{a^2 - (x_1^2 + x_3^2)}, x_3\right), \\ \mathbf{x}^5(x_2, x_3) &= \left(+\sqrt{a^2 - (x_2^2 + x_3^2)}, x_2, x_3\right), \\ \mathbf{x}^6(x_2, x_3) &= \left(-\sqrt{a^2 - (x_2^2 + x_3^2)}, x_2, x_3\right), \end{aligned} \tag{1.6}$$

which, together with $\mathbf{x}^1$ and $\mathbf{x}^2$, cover S^2 completely, and show that S^2 is a surface.

1.4. Example. Let $U=\{(\theta,\phi)\in E^2 | 0<\theta<\pi,\ 0<\phi<2\pi\}$, and let $\mathbf{x}: U\to E^3$ be given by

$$\mathbf{x}(\theta,\phi)=(x_1,x_2,x_3), \tag{1.7}$$

where

$$x_1=a\sin\theta\cos\phi, \qquad x_2=a\sin\theta\sin\phi, \qquad x_3=a\cos\theta, \tag{1.8}$$

a being constant; θ is called the *colatitude*, $\phi=$*longitude*, the curves $\phi=$constant the *meridians*, and the curves $\theta=$constant the *parallels of colatitude*; the parallel $\theta=0$ is also called the *equator* (Fig. 3.3). In the following we shall prove that $\mathbf{x}$ is a parametrization of the sphere S^2 with center at the origin $\mathbf{0}(0,0,0)$ and radius a.

Clearly, $\mathbf{x}(U)\subset S^2$. It is easy to see that condition 1.1 is satisfied. We can also easily verify condition 1.3 by noticing the three Jacobian determinants of the mapping $\mathbf{x}$:

$$\frac{\partial(x_1,x_2)}{\partial(\theta,\phi)}=\begin{vmatrix}\dfrac{\partial x_1}{\partial\theta} & \dfrac{\partial x_1}{\partial\phi}\\[2mm] \dfrac{\partial x_2}{\partial\theta} & \dfrac{\partial x_2}{\partial\phi}\end{vmatrix}=a^2\sin\theta\cos\theta,$$

$$\frac{\partial(x_2,x_3)}{\partial(\theta,\phi)}=a^2\sin^2\theta\cos\phi, \qquad \frac{\partial(x_1,x_3)}{\partial(\theta,\phi)}=a^2\sin^2\theta\sin\phi,$$

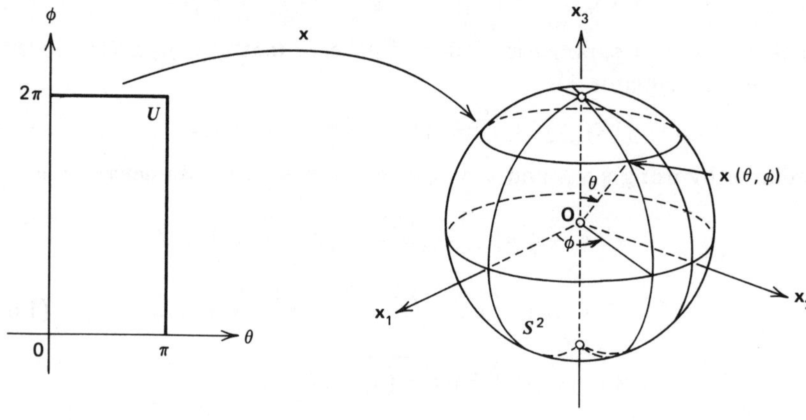

Figure 3.3

whose simultaneous vanishing gives

$$a^4(\sin^2\theta\cos^2\theta+\sin^4\theta\cos^2\phi+\sin^4\theta\sin^2\phi)=a^4\sin^2\theta=0,$$

which does not hold in U.

Next, let $(x_1,x_2,x_3)\in S^2-C$, where C is the semicircle

$$C=\{(x_1,x_2,x_3)\in S^2|x_2=0,\ x_1\geqslant 0\}.$$

Then θ is uniquely determined by $\theta=\cos^{-1}(x_3/a)$, since $0<\theta<\pi$. Using this known θ we find $\sin\phi$ and $\cos\phi$ from the first two equations of (1.8) and obtain $\phi(0<\phi<2\pi)$ uniquely. Thus $\mathbf{x}$ has an inverse $\mathbf{x}^{-1}$. Since S^2 is a surface, by Lemma 1.13 we see that $\mathbf{x}^{-1}$ is continuous, and hence condition 1.2 is verified. Q.E.D.

It should be remarked that $\mathbf{x}(U)=S^2-$semicircle through the two poles $(0,0,a)$ and $(0,0,-a)$ and that we can cover S^2 by the coordinate neighborhoods of two parametrizations of this type (see Exercise 10).

As is seen from Example 1.3 it is not a simple matter to show that a given subset of E^3 is a surface. Next we give a lemma and a theorem that can be used to simplify such work in some cases. At first we extend Example 1.3 to surfaces with only one parametrization.

1.5. Lemma. *If f: $U\to E^1$ is a C^3 function in an open set U of E^2, then the graph of f, that is, the subset of E^3 given by $(x_1,x_2,f(x_1,x_2))$ for $(x_1,x_2)\in U$ is a surface.*

Proof. Using the argument of Example 1.3 with q_3 replaced by $f(q_1,q_2)$ we can readily see that the mapping $\mathbf{x}$: $U\to E^3$ given by

$$\mathbf{x}(x_1,x_2)=(x_1,x_2,f(x_1,x_2))$$

is a parametrization of the graph of f, whose coordinate neighborhood covers every point of the graph.

This parametrization $\mathbf{x}$ is called a *Monge parametrization*, and the corresponding surface a *simple surface*, so that a general surface in E^3 can be constructed by gluing together simple surfaces.

1.6. Theorem. *Let f: $U\subset E^3\to E^1$ be a C^3 function in an open subset U of E^3, and let $c\in f(U)$, $S=f^{-1}(c)$. If every point of S is not a critical point of f (such c is called a regular value of f), then S is a surface.*

Proof. Let $\mathbf{p}=(\bar{x}_1,\bar{x}_2,\bar{x}_3)$ be any point of S, and let $(x_1,x_2,x_3)\in U$. Then from the definition of a critical point (Definition 2.3.3 of Chapter 1), at least one of the three partial derivatives f_{x_1},f_{x_2},f_{x_3} is not zero at $\mathbf{p}$, say $f_{x_3}(\mathbf{p})\neq 0$.

Define a mapping F: $U \to E^3$ by

$$F(x_1, x_2, x_3) = (x_1, x_2, f(x_1, x_2, x_3)),$$

and let $(u_1, u_2, u_3) \in F(U)$. Then the Jacobian matrix of F is given by

$$J_F = \begin{bmatrix} 1 & 0 & 0 \\ 0 & 1 & 0 \\ f_{x_1} & f_{x_2} & f_{x_3} \end{bmatrix},$$

whose rank is 3 at $\mathbf{p}$ since $f_{x_3}(\mathbf{p}) \neq 0$, so that the differential mapping dF is injective at $\mathbf{p}$ (see the paragraph just below Definition 4.2.6 of Chapter 1). Thus by the inverse function theorem (Theorem 4.2.7 of Chapter 1) there exist neighborhoods V of $\mathbf{p}$ in U and W of $F(\mathbf{p})$ in E^3 such that F: $V \to W$ is invertible and the inverse F^{-1}: $W \to V$ is of class C^3 (Fig. 3.4). It follows that the coordinate functions of F^{-1}

$$x_1 = u_1, \qquad x_2 = u_2, \qquad x_3 = g(u_1, u_2, u_3), \qquad (u_1, u_2, u_3) \in W,$$

are of class C^3. In particular, $x_3 = g(u_1, u_2, c) = h(x_1, x_2)$ is a C^3 function defined in the projection of V onto the $\mathbf{x}_1\mathbf{x}_2$-plane. Since

$$F(S \cap V) = \{(u_1, u_2, u_3) | u_3 = c\} \cap W,$$

the graph of f is $S \cap V$. By Lemma 1.5, $S \cap V$ is a coordinate neighborhood of $\mathbf{p}$, so that every point $\mathbf{p} \in S$ can be covered by a coordinate neighborhood. Hence S is a surface. Q.E.D.

It should be remarked that in the proof above we use the inverse function theorem to solve for x_3 from the equation $f(x_1, x_2, x_3) = c$ in a neighborhood of $\mathbf{p}$, since $f_{x_3}(\mathbf{p}) \neq 0$. However, this fact is a special case of the general implicit function theorem, which is equivalent to the inverse function theorem.

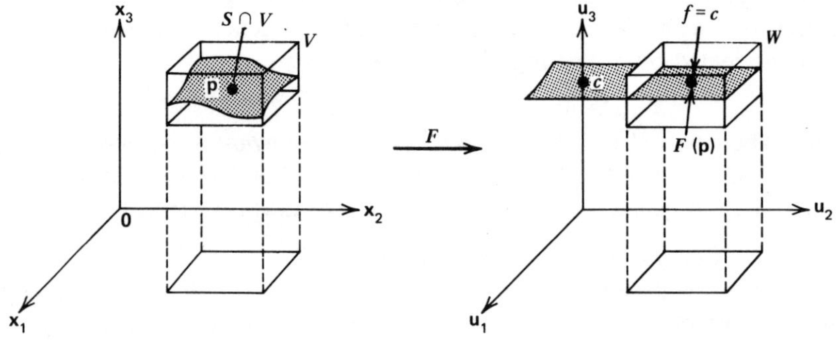

Figure 3.4

1.7. Example. The ellipsoid

$$S: \frac{x_1^2}{a^2}+\frac{x_2^2}{b^2}+\frac{x_3^2}{c^2}=1$$

is a surface. In fact, let

$$f(x_1,x_2,x_3)=\frac{x_1^2}{a^2}+\frac{x_2^2}{b^2}+\frac{x_3^2}{c^2}-1.$$

Then S is given by $f(x_1,x_2,x_3)=0$, and the partial derivatives $f_{x_1}, f_{x_2}, f_{x_3}$ vanish simultaneously only at the point $(0,0,0)$, which does not belong to $f^{-1}(0)$. This example includes spheres as a special case of $a=b=c$.

The surfaces discussed so far in our examples are connected subsets of E^3 (for the definition of connectedness, see Definition 1.3.1 of Chapter 1). However, in Definition 1.1 of a surface there is no connectedness restriction, and the following example shows that the surfaces given by Theorem 1.6 may not be connected.

1.8. Example. The hyperboloid of two sheets

$$S: \frac{x_1^2}{a^2}-\frac{x_2^2}{b^2}-\frac{x_3^2}{c^2}=1$$

is not connected and is a surface, since $S=f^{-1}(0)$, where 0 is a regular value of the function $f(x_1,x_2,x_3)=x_1^2/a^2-x_2^2/b^2-x_3^2/c^2-1$.

1.9. Example: Cylinders. Let C be a curve in the $\mathbf{x}_1\mathbf{x}_2$-plane defined by

$$f(x_1,x_2)=c, \qquad x_3=0.$$

Then the cylinder S (Fig. 3.5) generated by the line L perpendicular to the

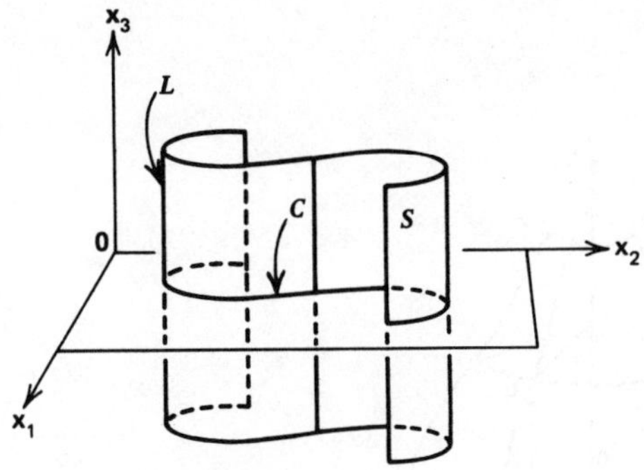

Figure 3.5

$\mathbf{x}_1\mathbf{x}_2$-plane along the curve C is given by

$$f(x_1, x_2) = c. \tag{1.9}$$

Since the slope of the tangent to the curve C is $dx_2/dx_1 = -f_{x_1}/f_{x_2}$, f_{x_1} and f_{x_2} can never be zero simultaneously along C. Thus by Theorem 1.6, the cylinder S is a surface.

We can easily see that if C is not a closed curve, and $\mathbf{y}(u) = (y_1(u), y_2(u), 0)$ is a parametrization of C, then

$$\mathbf{x}(u,v) = (y_1(u), y_2(u), v) \tag{1.10}$$

is a parametrization of the cylinder S.

1.10. Example: Surfaces of Revolution. The surface of revolution S obtained by revolving the plane curve C:

$$x_2 = 0, \qquad x_3 = f(x_1) \tag{1.11}$$

about the $\mathbf{x}_3$-axis is given by (Fig. 3.6)

$$x_1 = u\cos v, \qquad x_2 = u\sin v, \qquad x_3 = f(u). \tag{1.12}$$

The $\mathbf{x}_3$-axis is called the axis of *revolution*, the curves u = constant are called the *parallels*, and the curves v = constant the *meridians* (equivalent to the generating curve C) of S.

Let $\mathbf{p} = (p_1, 0, p_3)$ be a general point on the curve C, and $\mathbf{q} = (q_1, q_2, q_3)$ a corresponding point of $\mathbf{p}$ on S. Then

$$p_3 = q_3, \quad q_1^2 + q_2^2 = p_1^2. \tag{1.13}$$

Since (1.11) can be written as

$$x_2 = 0, \quad x_3 - f(x_1) = 0,$$

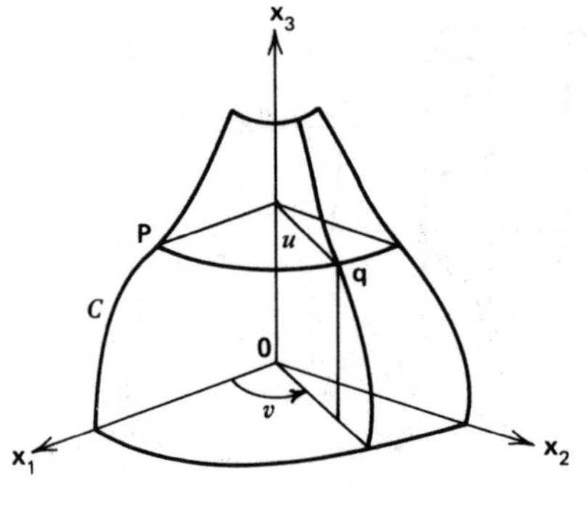

Figure 3.6

we define a function g in E^3 by

$$g(x_1,x_2,x_3)=x_3-f\left(\sqrt{x_1^2+x_2^2}\right). \tag{1.14}$$

Thus from (1.13) it follows that S is given by

$$g(x_1,x_2,x_3)=0.$$

Let $y=\sqrt{x_1^2+x_2^2}$. Then using the chain rule (Theorem 2.3.6 of Chapter 1), from (1.14) we obtain the partial derivatives

$$g_{x_1}=\frac{-x_1 f_y}{y}, \qquad g_{x_2}=\frac{-x_2 f_y}{y}, \qquad g_{x_3}=1,$$

which show that g_{x_1}, g_{x_2}, and g_{x_3} cannot vanish simultaneously on S. Hence S is a surface by Theorem 1.6.

Let

$$\mathbf{y}(w)=(g(w),0,h(w)), \qquad a<w<b, g(w)>0, \tag{1.15}$$

be a parametrization of the curve C. Then the mapping

$$\mathbf{x}(v,w)=(g(w)\cos v, g(w)\sin v, h(w)) \tag{1.16}$$

of the open set $U=\{(u,w)\in E^2|0<v<2\pi,\ a<w<b\}$ into the surface of revolution S is a parametrization of S. To show this we need to verify conditions 1.1, 1.2, and 1.3. Conditions 1.1 and 1.3 are straightforward. To verify condition 1.2 we first see that $\mathbf{x}$ is injective, since (1.15) is a parametrization of the curve C, and (1.16) gives

$$x_3=h(w), \qquad x_1^2+x_2^2=(g(w))^2,$$

where $(x_1,x_2,x_3)\in S$. Moreover, w is a continuous function of x_3 and of $\sqrt{x_1^2+x_2^2}$ and is therefore a continuous function of (x_1,x_2,x_3). To prove that $\mathbf{x}^{-1}$ is continuous, it remains to prove that v is a continuous function of (x_1,x_2,x_3). For $v\neq\pi$, since $g(w)>0$ by considering

$$\tan\frac{v}{2}=\frac{\sin v}{1+\cos v}=\frac{x_2/g(w)}{1+x_1/g(w)}=\frac{x_2}{x_1+\sqrt{x_1^2+x_2^2}}$$

we obtain

$$v=2\tan^{-1}\frac{x_2}{x_1+\sqrt{x_1^2+x_2^2}},$$

showing that if $v\neq\pi$, v is a continuous function of (x_1,x_2,x_3). Similarly, if v is in a small interval about π, then

$$v=2\cot^{-1}\frac{x_2}{-x_1+\sqrt{x_1^2+x_2^2}}.$$

Thus v is a continuous function of (x_1,x_2,x_3).

The following lemma is a local converse of Theorem 1.6.

1.11. Lemma. *Let $S \subset E^3$ be a surface and $\mathbf{p} \in S$. Then there exists a neighborhood V of $\mathbf{p}$ in S such that V is the graph of a C^3 function that takes one of the following three forms: $x_3 = f(x_1, x_2)$, $x_2 = g(x_1, x_3)$, $x_1 = h(x_2, x_3)$.*

Proof. Let $\mathbf{x}\colon U \subset E^2 \to S$ be a parametrization of S at $\mathbf{p}$, and write $\mathbf{x}(u, v) = (x_1(u, v), x_2(u, v), x_3(u, v))$, where $(u, v) \in U$. By condition 1.3, one of the Jacobian determinants

$$\frac{\partial(x_1, x_2)}{\partial(u, v)}, \qquad \frac{\partial(x_2, x_3)}{\partial(u, v)}, \qquad \frac{\partial(x_1, x_3)}{\partial(u, v)}$$

is not zero at $\mathbf{x}^{-1}(\mathbf{p}) = q$. Suppose that

$$\frac{\partial(x_1, x_2)}{\partial(u, v)}(\mathbf{q}) \neq 0, \tag{1.17}$$

and consider the mapping $\pi \circ \mathbf{x}\colon U \to E^2$, where π is the projection $\pi(x_1, x_2, x_3) = (x_1, x_2)$. Then $\pi \circ \mathbf{x}\colon (u, v) = (x_1(u, v), x_2(u, v))$, and by the inverse function theorem and (1.17) there exist neighborhoods V_1 of $\mathbf{q}$ and V_2 of $\pi \circ \mathbf{x}(\mathbf{q})$ such that $\pi \circ \mathbf{x}\colon V_1 \to V_2$ is a diffeomorphism of class C^3, that is, $\pi \circ \mathbf{x}$ is of class C^3 and has a C^3 inverse (Fig. 3.7). Thus $\pi|(\mathbf{x}(V_1) = V)$ is injective and there is a C^3 inverse $(\pi \circ \mathbf{x})^{-1}\colon V_2 \to V_1$. Since $\mathbf{x}$ is a homeomorphism, V is a neighborhood of $\mathbf{p}$ in S, and by composing the mapping $(\pi \circ \mathbf{x})^{-1}\colon (x_1, x_2) \to (u(x_1, x_2), v(x_1, x_2))$ with the function $(u, v) \to x_3(u, v)$, we find that V is the graph of the C^3 function $x_3 = x_3(u(x_1, x_2), v(x_1, x_2)) = f(x_1, x_2)$. Hence the first case is proved.

The other two cases, in which $x_2 = g(x_1, x_3)$ and $x_1 = h(x_2, x_3)$, can be treated similarly.

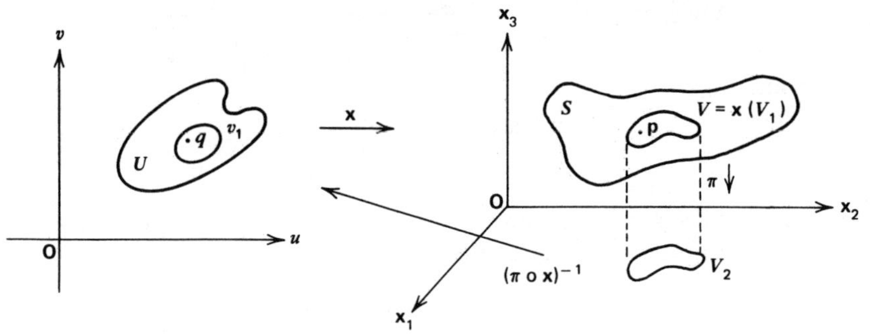

Figure 3.7

1.12. Example. The one-sheeted cone C, given by

$$x_3 = +\sqrt{x_1^2 + x_2^2}\,, \qquad (x_1, x_2) \in E^2,$$

is not a surface. We cannot conclude this just from the fact that the natural parametrization

$$(x_1, x_2) \to \left(x_1, x_2, +\sqrt{x_1^2 + x_2^2}\,\right)$$

is not of class C^3; there could be other parametrizations satisfying Definition 1.1. However, this is not the case. In fact, if C were a surface then, by Lemma 1.11, in a neighborhood of $(0,0,0) \in C$ the surface would be the graph of a C^3 function taking one of the three forms: $x_2 = h(x_1, x_3)$, $x_1 = g(x_2, x_3)$, $x_3 = f(x_1, x_2)$. The first two forms can be discarded, since the projections of C onto the $\mathbf{x}_1\mathbf{x}_3$- and $\mathbf{x}_2\mathbf{x}_3$-planes are not injective. In a neighborhood of $(0,0,0)$ the last form would have to agree with $x_3 = +\sqrt{x_1^2 + x_2^2}$, which is not differentiable at $(0,0,0)$.

The following lemma reduces the conditions for a mapping to be a parametrization of a given surface.

1.13. Lemma. *Let* $\mathbf{p}$ *be a point of a surface* $S \subset E^3$, *and* $\mathbf{x}$: $U \to E^3$ *a mapping of an open set* $U \subset E^2$ *satisfying conditions* 1.1 *and* 1.3. *If* $\mathbf{x}$ *is injective, then* $\mathbf{x}^{-1}$ *is continuous.*

Proof. The first part of the proof is similar to the proof of Lemma 1.12. Let $\mathbf{x}(u, v) = (x_1(u, v), x_2(u, v), x_3(u, v))$, where $(u, v) \in U$, and let $\mathbf{q} \in U$. By conditions 1.1 and 1.3 we may assume, interchanging the coordinate axes of E^3 if necessary, that $(\partial(x_1, x_2)/\partial(u, v))(\mathbf{q}) \neq 0$. Let π: $E^3 \to E^2$ be the projection $\pi(x_1, x_2, x_3) = (x_1, x_2)$. Using the inverse function theorem, we obtain neighborhoods V_1 of $\mathbf{q}$ in U and V_1 of $\pi \circ \mathbf{x}(\mathbf{q})$ in E^2 such that $\pi \circ \mathbf{x}$: $V_1 \to V_2$ is a diffeomorphism of class C^3.

Now assume that $\mathbf{x}$ is injective. Then, restricted to $\mathbf{x}(V_1)$,

$$\mathbf{x}^{-1} = (\pi \circ \mathbf{x})^{-1} \circ \pi,$$

(see Fig. 3.7). Thus $\mathbf{x}^{-1}$, as a composition of continuous mappings, is continuous. Since $\mathbf{q}$ is arbitrary, $\mathbf{x}^{-1}$ is continuous in $\mathbf{x}(U)$.

1.14. Example: Tori. The surface T obtained by revolving a circle about an axis in the plane of the circle but not intersecting the circle is a torus. As a special case of the surfaces of revolution discussed in Example 1.10, a torus is a surface.

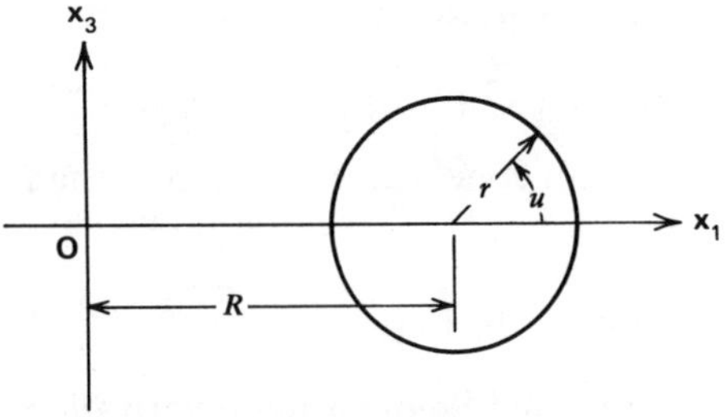

Figure 3.8

Let C be the circle in the $\mathbf{x}_1\mathbf{x}_3$-plane with radius $r>0$ and center $(R,0,0)$. To revolve C about the $\mathbf{x}_3$-axis, we must assume $R>r$ to keep C from meeting the $\mathbf{x}_3$-axis. Let

$$\mathbf{y}(u)=(R+r\cos u, r\sin u)$$

be the natural parametrization (Fig. 3.8) for C. Then we can easily obtain the following parametrization (Fig. 3.9) for the torus T:

$$\mathbf{x}(u,v)=((R+r\cos u)\cos v, (R+r\cos u)\sin v, r\sin u), \qquad (1.18)$$

where $0<u<2\pi$, $0<v<2\pi$. The verification of condition 1.1 for this mapping $\mathbf{x}$ is easy, and that of condition 1.3 is only a straightforward computation. Since T is a surface, to verify condition 1.2 we need only, by Lemma 1.13, show that $\mathbf{x}$ is injective. For this purpose we observe that

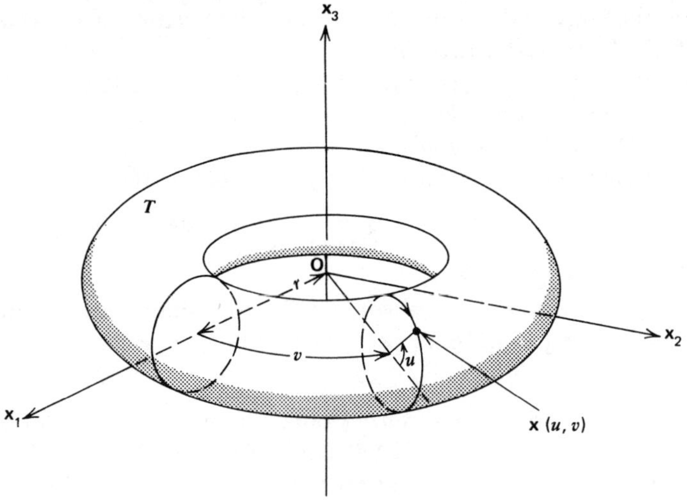

Figure 3.9

$\sin u = x_3/r$, and also that if $\sqrt{x_1^2+x_2^2} \leqslant R$, then $\pi/2 \leqslant u \leqslant 3\pi/2$, and if $\sqrt{x_1^2+x_2^2} \geqslant R$, then either $0<u\leqslant \pi/2$ or $3\pi/2 \leqslant u < 2\pi$. Thus given (x_1, x_2, x_3), we can determine u, $0<u<2\pi$, uniquely. By knowing u, x_1, x_2 we find $\cos v$ and $\sin v$, and therefore determine v, $0<v<2\pi$, uniquely. Hence $\mathbf{x}$ is injective.

It should be remarked that the torus T can be covered by the coordinate neighborhoods of three parametrizations of the type (1.18). (See Exercise 10 of Section 1.)

1.15. Example: Möbius Bands. Given a rectangle $ABB'A'$ with $AA'=2a\pi$ and $AB=2b$, where a and b are positive numbers, we obtain a cylinder (for definition, see also Example 1.9) by gluing the opposite edges AB and $A'B'$, that is, by identifying A, B with A', B', respectively. On the other hand, we obtain a Möbius band by gluing, after one twist, the opposite edges AB and $A'B'$, that is, by identifying A, B with B', A', respectively (Fig. 3.10).

The construction of a Möbius band may be given geometrically as follows (see Fig. 3.11). Let S^1 be a circle in the $\mathbf{x}_1\mathbf{x}_2$-plane given by $x_1^2+x_2^2=a^2$, and let AB be an open segment in the $\mathbf{x}_1\mathbf{x}_3$-plane given by $x_1=a$, $-b<x_3<b$. Revolve the $\mathbf{x}_1\mathbf{x}_3$-plane about the $\mathbf{x}_3$-axis so that the midpoint C of the segment AB moves along S^1, and at the same time rotate the segment AB about C in the $\mathbf{x}_1\mathbf{x}_3$-plane in such a way that when the segment $\mathbf{0}C$ is rotated through an angle u, the segment AB is rotated through an angle $u/2$. Thus when C moves along the circle S^1 once, AB returns to the initial position inverted, and describes a Möbius band M. Geometrically it is obvious that M is a surface. We shall also show this fact analytically as follows.

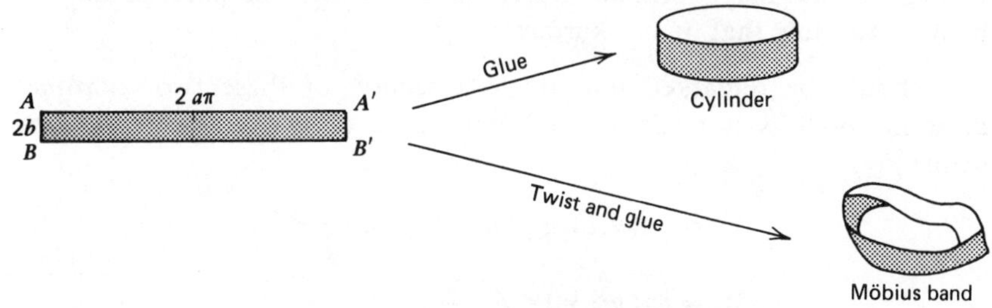

Figure 3.10

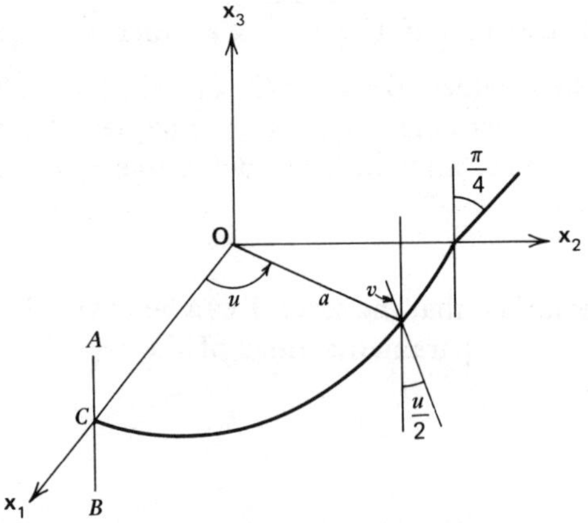

Figure 3.11

Using Fig. 3.11 we can easily obtain a parametrization $\mathbf{x}$: $U \subset E^2 \to M$ of the Möbius band M given by

$$\mathbf{x}(u,v) = \left(\left(a - v\sin\frac{u}{2}\right)\cos u, \left(a - v\sin\frac{u}{2}\right)\sin u, v\cos\frac{u}{2}\right), \quad (1.19)$$

where $0 < u < 2\pi$ and $-b < v < b$ with $M - AB|_{u=0}$ as coordinate neighborhood. It should also be noted that for this parametrization u is measured counterclockwise from the positive $\mathbf{x}_1$-axis.

Similarly, we can obtain another parametrization $\mathbf{x}(\bar{u}, \bar{v})$ of M given by

$$\bar{\mathbf{x}}(\bar{u},\bar{v}) = \left(\left(a - \bar{v}\sin\left(\frac{\pi}{4} + \frac{\bar{u}}{2}\right)\sin\bar{u},\right.\right.$$
$$\left.\left(a - \bar{v}\sin\left(\frac{\pi}{4} + \frac{\bar{u}}{2}\right)\cos\bar{u}, \quad \bar{v}\cos\left(\frac{\pi}{4} + \frac{\bar{u}}{2}\right)\right), \quad (1.20)$$

where $0 < \bar{u} < 2\pi$, $-b < \bar{v} < b$, and $\bar{u}$ is measured counterclockwise from the positive $\mathbf{x}_2$-axis, with $M - AB|_{u=\pi/2}$ as coordinate neighborhood. These two coordinate neighborhoods cover the whole Möbius band M and can be used to show that M is a surface.

It should be remarked that the intersection of these two coordinate neighborhoods is not connected but consists of two connected components:

$$W_1 = \left\{\mathbf{x}(u,v) \mid \frac{\pi}{2} < u < 2\pi, -b < v < b\right\},$$
$$W_2 = \left\{\mathbf{x}(u,v) \mid 0 < u < \frac{\pi}{2}, -b < v < b\right\}. \quad (1.21)$$

Furthermore, the change between the coordinates (u,v) and $(\bar{u},\bar{v})$ is given by

$$\bar{u}=u-\frac{\pi}{2},\qquad \bar{v}=v\qquad \text{in}\quad W_1, \tag{1.22}$$

$$\bar{u}=u+\frac{3\pi}{2},\qquad \bar{v}=-v\qquad \text{in}\quad W_2, \tag{1.23}$$

from which it follows that the Jacobian determinant

$$\frac{\partial(\bar{u},\bar{v})}{\partial(u,v)}=\begin{cases} 1>0 & \text{in}\quad W_1,\\ -1<0 & \text{in}\quad W_2.\end{cases} \tag{1.24}$$

In general, for changes of coordinates we have Lemma 1.16.

1.16. Lemma (Change of parameters). *Let* $\mathbf{p}$ *be a point of a surface S, and let* $\mathbf{x}$: $U\subset E^2\to S$, y: $V\subset E^2\to S$ *be two parametrizations of S such that* $\mathbf{p}\in\mathbf{x}(U)\cap\mathbf{y}(V)=W$. *Then the change of coordinates* $\mathbf{h}=\mathbf{x}^{-1}\circ\mathbf{y}$: $\mathbf{y}^{-1}(W)\to\mathbf{x}^{-1}(W)$ *is a diffeomorphism of class* C^3 *(Fig. 3.12).*

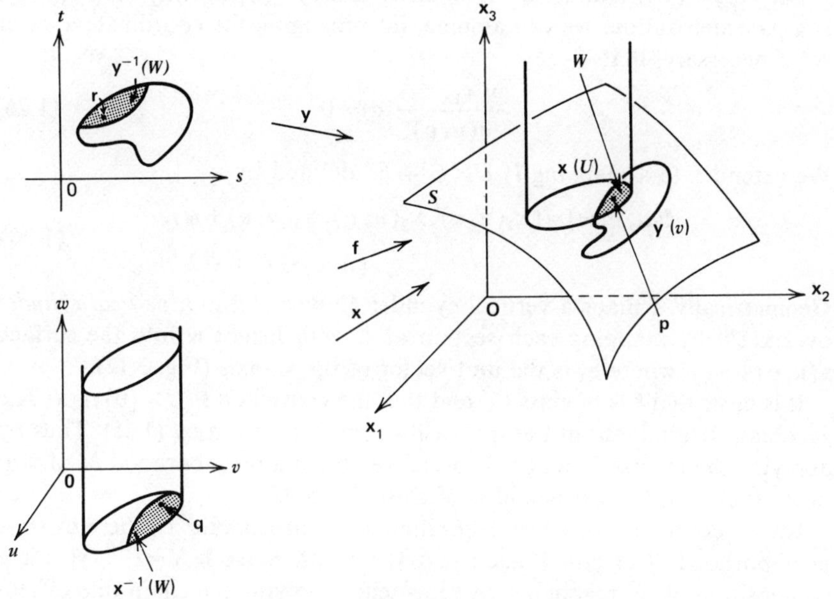

Figure 3.12

In other words, if $\mathbf{x}$ and $\mathbf{y}$ are given by

$$\mathbf{x}(u,v)=(x_1(u,v),x_2(u,v),x_3(u,v)), \qquad (u,v)\in U,$$
$$\mathbf{y}(s,t)=(x_1(s,t),x_2(s,t),x_3(s,t)), \qquad (s,t)\in V,$$

then the change of coordinates $\mathbf{h}$ can be given by

$$u=u(s,t), \qquad v=v(s,t), \qquad (s,t)\in \mathbf{y}^{-1}(W),$$

u and v being of class C^3, and has an inverse $\mathbf{h}^{-1}$ given by

$$s=s(u,v), \qquad t=t(u,v), \qquad (u,v)\in \mathbf{x}^{-1}(W),$$

s and t being of class C^3. Since

$$\frac{\partial(u,v)}{\partial(s,t)}\frac{\partial(s,t)}{\partial(u,v)}=1,$$

the Jacobian determinants of both $\mathbf{h}$ and $\mathbf{h}^{-1}$ are nonzero everywhere.

Proof of Lemma 1.16. Since the composition of continuous mappings is continuous (Exercise 6 of Section 1.2, Chapter 1), it follows that $\mathbf{h}=\mathbf{x}^{-1}$ $\mathbf{y}$ is a homeomorphism. Since we have not yet defined C^3 functions on S, in the following we shall show, but not by the same argument as above, that $\mathbf{h}$ is of class C^3.

Let $\mathbf{r}\in\mathbf{y}^{-1}(W)$ and set $\mathbf{q}=\mathbf{h}(r)$. Since $\mathbf{x}(u,v)=(x_1(u,v),x_2(u,v),x_3(u,v))$ is a parametrization, we can assume, interchanging the coordinate axes of E^3 if necessary, that

$$\frac{\partial(x_1,x_2)}{\partial(u,v)}(\mathbf{q})\neq 0. \tag{1.25}$$

We extend $\mathbf{x}$ to a mapping $\mathbf{f}$: $U\times E^1\to E^3$ defined by

$$\mathbf{f}(u,v,w)=(x_1(u,v),x_2(u,v),x_3(u,v)+w), \qquad (u,v)\in U, w\in E^1. \tag{1.26}$$

Geometrically, $\mathbf{f}$ maps a vertical cylinder C over U into a *vertical cylinder* over $\mathbf{x}(U)$ by mapping each section of C with height w into the surface $\mathbf{x}(u,v)+w\mathbf{e}_3$, where $\mathbf{e}_3$ is the unit vector of the $\mathbf{x}_3$-axis (Fig. 3.12).

It is clear that $\mathbf{f}$ is of class C^3 and that the restriction $\mathbf{f}|(U\times\{0\})=\mathbf{x}$. The Jacobian determinant of $\mathbf{f}$ at $\mathbf{q}$ is easily seen to be given by (1.25). Thus by applying the inverse function theorem we obtain a neighborhood M of $\mathbf{x}(\mathbf{q})$ in E^3 such that $\mathbf{f}^{-1}$ exists and is of class C^3 in M.

By the continuity of $\mathbf{y}$ (see Definition 1..2.7 of Chapter 1), there exists a neighborhood N of $\mathbf{r}$ in V such that $\mathbf{y}(N)\subset M$. Since $\mathbf{h}|N=\mathbf{f}^{-1}\circ\mathbf{y}|N$ is a composition of C^3 mappings, by repeatedly applying the chain rule (Theorem 2.3.6 of Chapter 1) we see that $\mathbf{h}$ is of class C^3 at $\mathbf{r}$. Because $\mathbf{r}$ is arbitrary, $\mathbf{h}$ is of class C^3 on $\mathbf{y}^{-1}(W)$.

Similarly, we can show that $\mathbf{h}^{-1}$ is of class C^3. Hence $\mathbf{h}$ is a diffeomorphism of class C^3. Q.E.D.

As in the case of curves, locally a surface is also often defined to be a mapping into E^3, instead of a subset of E^3, as follows.

1.17. Definition. A *parametrized surface* $\mathbf{x}$: $U \subset E^2 \to E^3$ is a C^3 mapping $\mathbf{x}$ of an open set $U \subset E^2$ into E^3. $\mathbf{x}$ is *regular* if the differential mapping $d\mathbf{x}_q$: $E^2 \to E^3$ is injective for all $\mathbf{q} \in U$ [i.e., if the vectors $\partial \mathbf{x}/\partial u, \partial \mathbf{x}/\partial v$ are linearly independent for all $\mathbf{q} = (u, v) \in U$]. A point $\mathbf{p} \in U$ where $d\mathbf{x}_p$ is not injective is a *singular point* of $\mathbf{x}$.

It should be noted that a parametrized surface, even when regular, may have self-intersections in the image set of the surface.

The following lemma gives a relationship between parametrized surfaces and our general surfaces.

1.18. Lemma. *Let* $\mathbf{x}$: $U \subset E^2 \to E^3$ *be a regular parametrized surface and let* $\mathbf{q} \in U$. *Then there exists a neighborhood* V *of* $\mathbf{q}$ *in* E^2 *such that* $\mathbf{x}(V) \subset E^3$ *is a surface.*

Proof. Let $\mathbf{x}$ be given by

$$\mathbf{x}(u, v) = (x_1(u, v), x_2(u, v), x_3(u, v)).$$

Then by regularity we can suppose (1.25) to be true. Define a mapping $\mathbf{f}$: $U \times E^1 \to E^3$ by (1.26). Since the Jacobian determinant of $\mathbf{f}$ at $\mathbf{q}$ is given by (1.25), applying the inverse function theorem we obtain neighborhoods W_1 of $\mathbf{q}$ and W_2 of $\mathbf{f}(\mathbf{q})$ such that $\mathbf{f}$: $W_1 \to W_2$ is a diffeomorphism. Set $V = W_1 \cap U$. Then $\mathbf{f}|V = \mathbf{x}|V$, from which it follows that $\mathbf{x}(V)$ is diffeomorphic to V and is therefore a surface.

Exercises

1. Prove the equivalence of conditions 1.3 and 1.3′.
2. Show that the cylinder $\{x_1, x_2, x_3) \in E^3 | x_1^2 + x_2^2 = 1\}$ is a surface, and find parametrizations whose coordinate neighborhoods cover it.
3. None of the following subset S of E^3 are surfaces. Which points do not satisfy at least one of the three conditions 1.1, 1.2, 1.3?
   (a) Cone S: $x_3^2 = x_1^2 + x_2^2$.
   (b) Closed disk S: $x_1^2 + x_2^2 \leqslant 1$, $x_3 = 0$.
   (c) Folded plane S: $x_1 x_2 = 0, x_1 \geqslant 0, x_2 \geqslant 0$.

4. (a) Show that the set $\{(x_1,x_2,x_3)\in E^3 | x_1^3+3(x_2^2+x_3^2)^2=2\}$ is a surface.

(b) For which values of c is the set $\{(x_1,x_2,x_3)\in E^3 | x_3(x_3+4)-3x_1x_2=c\}$ a surface?

5. Let $U=\{(u,v)\in E^2 | u>0,\ 0<v<2\pi\}$. Show that the subset $\mathbf{x}(U)\subset E^3$ given by

$$\mathbf{x}(u,v)=(u\cos v, u\sin v, u+v)$$

is a simple surface.

6. In which of the following cases is the mapping $\mathbf{x}$: $E^2\to E^3$ a parametrization? (a) $\mathbf{x}(u,v)=(u,v,uv)$; (b) $\mathbf{x}(u,v)=(u,v^2,v^3)$; (c) $\mathbf{x}(u,v)=(u+u^2,v,v^2)$; (d) $\mathbf{x}(u,v)=(u,\cos v,\sin v)$.

7. Prove that (1.10) is a parametrization of the cylinder S for a non-closed curve C.

8. Find a parametrization of the surface obtained by revolving: (a) C: $x_1=\cosh x_3$, $x_2=0$ about the $\mathbf{x}_3$-axis (catenoid Fig. 3.13); (b) C: $(x_1-3)^2+x_2^2=4$, $x_3=0$ about the $\mathbf{x}_2$-axis (torus); (c) $C: x_3=x_1^2$, $x_2=0$ about the $\mathbf{x}_3$-axis (paraboloid of revolution).

9. In each of the following show that $\mathbf{x}$: $U\subset E^2\to E^3$ is a parametrization of a surface S in E^3, and that S is the surface as indicated:

(a) $\mathbf{x}(u,v)=(a\sin u\cos v, b\sin u\sin v, c\cos u)$, where $0<u<\pi, 0<v<2\pi$, the ellipsoid in Example 1.7.

(b) $\mathbf{x}(u,v)=(a\cosh u, b\sinh u\sin v, c\sinh u\cos v)$, $u>0, 0<v<2\pi$, the hyperboloid of two sheets in Example 1.8.

(c) $\mathbf{x}(u,v)=(au\cos v, bu\sin v, u^2)$, $u>0$, $0<v<2\pi$, the elliptic paraboloid

$$x_3=\frac{x_1^2}{a^2}+\frac{x_2^2}{b^2}.$$

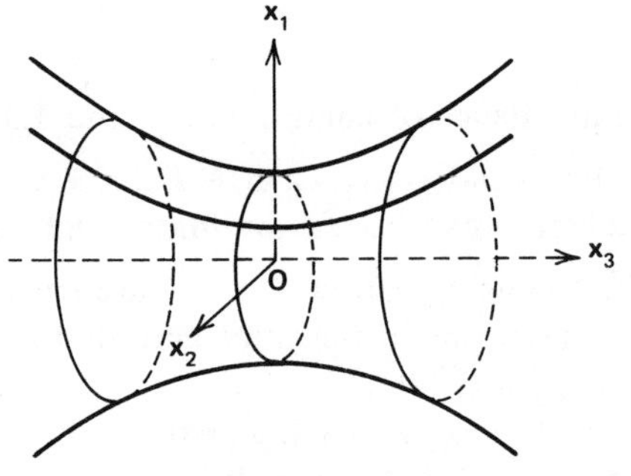

Figure 3.13

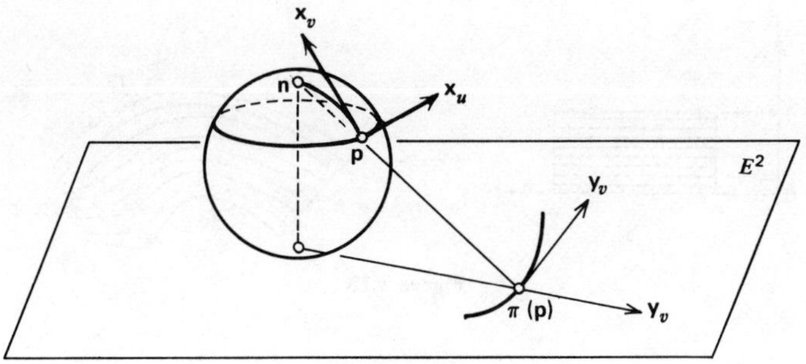

Figure 3.14

(d) $\mathbf{x}(u,v)=a(u+v),b(u-v),4uv),\ U=E^2-(0,0)$, the hyperbolic paraboloid

$$x_3=\frac{x_1^2}{a^2}-\frac{x_2^2}{b^2}.$$

10. Show that a torus can be covered by the coordinate neighborhoods of three parametrizations of the type (1.18).

11. Stereographic projection is a mapping π of the punctured sphere (i.e., the unit sphere S^2 minus a point $\mathbf{n}$) onto the plane E^2, which maps a point $\mathbf{p}\in S^2-\{\mathbf{n}\}$ to the point at which the line through $\mathbf{n}$ and $\mathbf{p}$ intersects the plane E^2 (Fig. 3.14). If we take E^2 to be the $\mathbf{x}_1\mathbf{x}_2$-plane, and $\mathbf{n}$ the north pole with coordinates (0,0,2), then the sphere is given by the equation

$$x_1^2+x_2^2+(x_3-2)^2=1.$$

(a) Show that π: $S^2\to E^2$ and $\pi^{-1}: E^2\to S^2$ are given, respectively, by

$$u=\frac{2x_1}{2-x_3},\qquad v=\frac{2x_2}{2-x_3};$$

$$x_1=\frac{4u}{D},\qquad x_2=\frac{4v}{D},\qquad x_3=\frac{2(u^2+v^2)}{D},$$

where $\mathbf{p}=(x_1,x_2,x_3)$, $\pi(\mathbf{p})=(u,v)$, and $D=u^2+v^2+4$.

*(b) Show that the stereographic projection π is a diffeomorphism.

(c) Show that using stereographic projection, we can cover S^2 by two parametrizations.

***12.** Let f_0, f_1: $[a,b]\to S$ be two curves on a surface S. Then a homotopy (Fig. 3.15) of f_0 into f_1 is a continuous mapping F: $[a,b]\times I\to S$,

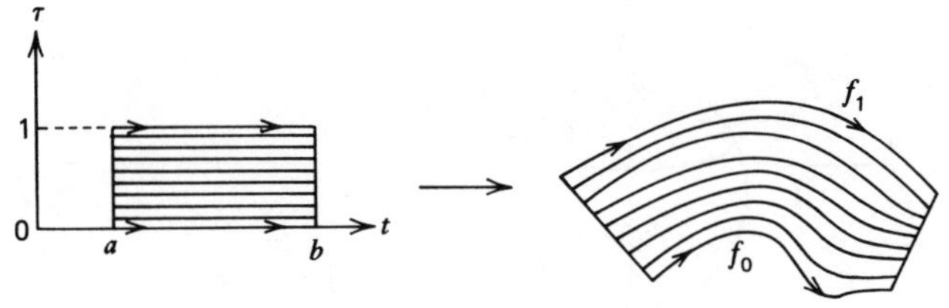

Figure 3.15

$I=[0,1]$, such that

$$F(t,0)=f_0(t),\ F(t,1)=f_1(t) \qquad \text{for all} \quad t\in[a,b].$$

In case f_0, f_1 are closed curves, the homotopy F satisfies the additional condition

$$F(a,\tau)=F(b,\tau) \qquad \text{for all} \quad \tau\in[0,1].$$

Show that:

(a) Any closed curve in the plane E^2 is homotopic to a constant closed curve.

(b) The curve

$$f_0\colon [0,2\pi]\to S=E^2-\{(0,0)\}$$

given by $f_0(t)=(\cos t, \sin t), t\in[0,2\pi]$, is not homotopic to a constant curve

$$f_1\colon [0,2\pi]\to \mathbf{p}\in S.$$

2. FUNCTIONS AND FUNDAMENTAL FORMS

For studying the properties of a surface in a neighborhood of a point, we first define C^k functions of a surface at a point.

2.1. Definition. Let $f\colon V\subset S\to E^1$ be a function defined in an open subset V of a surface S. Then f is *of class* C^k, $k\geqslant 1$, *at* $p\in V$ if, for some parametrization $\mathbf{x}\colon U\subset E^2\to S$ with $\mathbf{p}\in\mathbf{x}(U)\subset V$, the composition $f\circ\mathbf{x}\colon U\subset E^2\to E^1$ is of class C^k at $\mathbf{x}^{-1}(\mathbf{p})$. f is *of class* C^k *in* V if it is so at all points of V.

From Lemma 1.16 it follows immediately that Definition 2.1 for $k\leqslant 3$ does not depend on the choice of the parametrization $\mathbf{x}$. In fact, if $\mathbf{y}$:

$V \subset E^2 \to S$ is another parametrization with $\mathbf{p} \in \mathbf{x}(V)$, and if $h = \mathbf{x}^{-1} \circ \mathbf{y}$, then $f \circ \mathbf{y} = f \circ \mathbf{x} \circ h$ is also of class C^3.

2.2. Remarks. 1. Let $\mathbf{x}$: $U \subset E^2 \to S$ be a parametrization of a surface S, and f a function on S. Since $U \to \mathbf{x}(U)$ is diffeomorphic, we can identify $\mathbf{x}(U)$ with U so that a point (u, v) of U can also be considered as a point of $\mathbf{x}(U)$ with coordinates (u, v). Using this identification we shall often denote f and $f \circ \mathbf{x}$ by the same notation $f(u, v)$, which is called the *expression of f in the coordinate system* $\mathbf{x}$.

2. Definition 2.1 can be easily extended to mappings between surfaces. A continuous mapping f: $V_1 \subset S_1 \to S_2$ of an open set V_1 of a surface S_1 to a surface S_2 is said to be of class $C^k, k \geqslant 1$, at $\mathbf{p} \in V_1$ if, for given parametrizations

$$\mathbf{x}_1\colon U_1 \subset E^2 \to S_1, \qquad \mathbf{x}_2\colon U_2 \subset E^2 \to S_2$$

with $\mathbf{p} \in \mathbf{x}_1(U_1)$ and $f(\mathbf{x}_1(U_1)) \subset \mathbf{x}_2(U_2)$, the mapping

$$\mathbf{x}_2^{-1} \circ f \circ \mathbf{x}_1\colon U_1 \to U_2$$

is of class C^k at $\mathbf{q} = \mathbf{x}_1^{-1}(\mathbf{p})$.

2.3. Definition. A curve (respectively, $C^k, k \geqslant 1$, curve) α: $I \to S$ on a surface S is a differentiable (respectively, C^k) function from a generalized interval I (for the definition see Section 1.1 of Chapter 2) into S.

The following lemma will be useful in applications of this definition.

2.4. Lemma. *Let* $\mathbf{x}$: $U \subset E^2 \to S$ *be a parametrization of a surface S. If* α: $I \to S$ *is a* C^3 *curve whose image set lies in* $\mathbf{x}(U)$, *there exist unique* C^3 *functions* u, v *on I such that*

$$\alpha(t) = \mathbf{x}(u(t), v(t)) \qquad \textit{for all} \quad t, \tag{2.1}$$

or, in functional notation, $\alpha = \mathbf{x}(u, v)$ (*see Fig. 3.16*).

Proof. By definition, the coordinate expression $\mathbf{x}^{-1}\alpha$: $I \to U$ is of class C^3; it is just a curve in E^2 whose image set lies in the domain U of $\mathbf{x}$. If u, v are the Euclidean coordinate functions (for the definition see the definition following (1.2.5) of Chapter 2) of $\mathbf{x}^{-1}\alpha$, then

$$\alpha = \mathbf{x}\mathbf{x}^{-1}\alpha = \mathbf{x}(u, v).$$

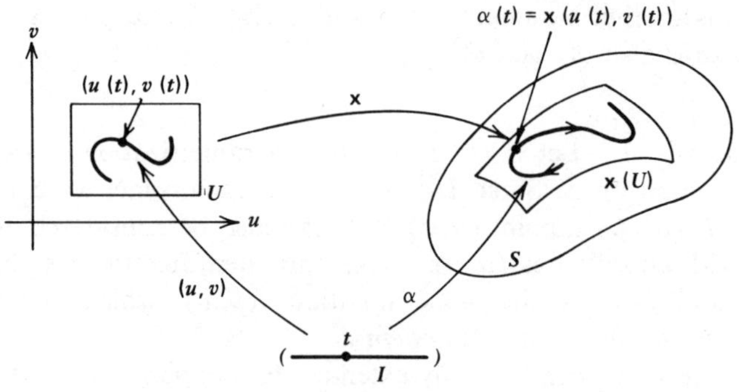

Figure 3.16

These are the only such functions, for if $\alpha = \mathbf{x}(u^*, v^*)$, then

$$(u,v) = \mathbf{x}^{-1}\alpha = \mathbf{x}^{-1}\mathbf{x}(u^*, v^*) = (u^*, v^*). \qquad \text{Q.E.D.}$$

These functions

$$u = u(t), \qquad v = v(t) \qquad t \in I \tag{2.2}$$

are called the *coordinate functions* of the curve α with respect to the parametrization $\mathbf{x}$. For a general curve α we shall always assume that $u'^2 + v'^2 \neq 0$, where the prime denotes the derivative with respect to t, since otherwise the image set of α is just one point. To present an example of coordinate functions, we consider the curve α given by

$$\alpha(t) = (2\cos^2 t, \sin 2t, 2\sin t) \qquad \text{for} \quad 0 < t < \frac{\pi}{2}, \tag{2.3}$$

whose image set lies in the intersection of a cylinder C and the sphere S^2 with radius 2 and center at the origin where C is constructed on the circle in the $\mathbf{x}_1\mathbf{x}_2$-plane with radius 1 and center at $(1,0,0)$, (Fig. 3.17). This curve α lies in the part of the sphere S^2 that is covered by the coordinate

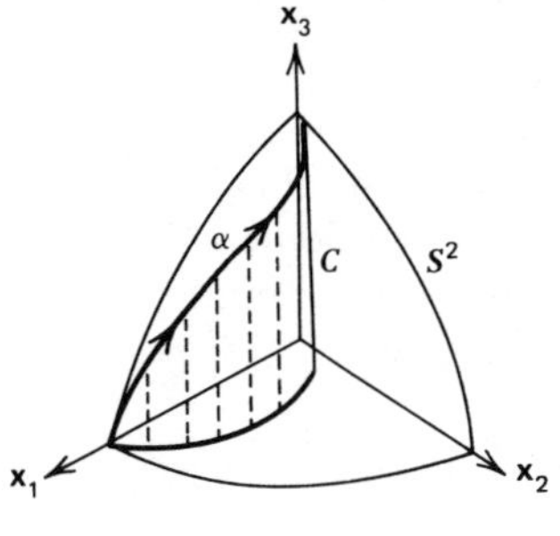

Figure 3.17

neighborhood of the parametrization **x** given by (1.7) and (1.8). In fact, from (2.1) and (2.3) and (1.7) and (1.8) it follows easily that the coordinate functions of the curve α with respect to **x** are $u(t)=\pi/2-t, v(t)=t$.

2.5. Definition. Let **p** be a point of a surface S in E^3. A vector in E^3 is a *tangent vector* of S at **p** if it is tangent to some curve on S at **p**. A *tangent vector field* on a surface S or on some region R of S is a function that assigns to each point **p** of S or R a tangent vector of S at **p**.

2.6. Definition. Let **v** be a tangent vector of a surface S at a point **p**, and let f be a differentiable real-valued function on S. The *derivative* $\mathbf{v}[f]$ of f with respect to **v** is the common value of $(d/dt)(f\mathbf{x})(0)$ for all curves **x** on S, for which the initial tangent vector $\mathbf{x}'(0)=(d\mathbf{x}/dt)(0)$ is **v**.

Directional derivatives on a surface have exactly the same properties as those in the Euclidean case (Theorem 3.6.3 of Chapter 1).

2.7. Definition. Just as for E^3, a *0-form* f on a surface S is simply a (differentiable) real-valued function on S, a *1-form* ϕ on S is a real-valued function on tangent vectors of S, which is linear at each point (Definition 6.1.1 of Chapter 1), and a *2-form* ψ on S is a real-valued function on all ordered pairs of tangent vectors $\mathbf{v},\mathbf{w}$ of S such that $\psi(\mathbf{v},\mathbf{w})$ is bilinear in **v**, **w**, and $\psi(\mathbf{v},\mathbf{w})=-\psi(\mathbf{w},\mathbf{v})$.

2.8. Lemma. *All tangent vectors of a surface S at a point* **p** *form a plane that is called the tangent plane of S at* **p** *and is denoted by $T_p(S)$.*

Proof. Let $\mathbf{x}$: $U\subset E^2\to S$ be a parametrization of the surface S, and let $(u,v)\in U$ and $\mathbf{p}=\mathbf{x}(u_0,v_0)$. If the coordinate functions of any curve C on S through **p** are given by (2.2), then

$$u_0=u(t_0),\qquad v_0=v(t_0),\qquad t_0\in I. \tag{2.4}$$

By the chain rule, the tangent vector of the curve C at **p** is

$$\left(\frac{d\mathbf{x}}{dt}\right)_0=\mathbf{x}_u(u_0,v_0)u'(t_0)+\mathbf{x}_v(u_0,v_0)v'(t_0), \tag{2.5}$$

where the subscripts u, v denote the partial derivatives. Since by condition 1.3′, the vectors $\mathbf{x}_u(u_0,v_0)$ and $\mathbf{x}_v(u_0,v_0)$ span a plane, from (2.5) it follows that all tangent vectors of S at **p** are in this plane, and the lemma is proved. Q.E.D.

Let $\mathbf{x}\colon U \subset E^2 \to S$ be a parametrization of a surface S. On U, let u, v be the natural coordinate functions, and $\mathbf{u}_1, \mathbf{u}_2$ the natural unit vector fields so that

$$du(\mathbf{u}_1)=1,\ du(\mathbf{u}_2)=0; \qquad dv(\mathbf{u}_1)=0,\ dv(\mathbf{u}_2)=1. \tag{2.6}$$

Then using Lemma 6.1.4 of Chapter 1 and part 1 of Remarks 2.2 we readily see that if ϕ is a 1-form on S, then

$$\phi = f_1\,du + f_2\,dv, \tag{2.7}$$

where $f_1 = \phi(\mathbf{u}_1)$, $f_2 = \phi(\mathbf{u}_2)$. Moreover if, on S, τ is a 2-form, f a function, and ψ a 1-form given by

$$\psi = g_1\,du + g_2\,dv, \tag{2.8}$$

then by Section 6.2 of Chapter 1 we have, for the 1-form ϕ given by (2.7),

$$\tau = g\,du\wedge dv, \qquad g = \tau(\mathbf{u}_1, \mathbf{u}_2), \tag{2.9}$$

$$\phi\wedge\psi = (f_1 g_2 - f_2 g_1)\,du\wedge dv = -\tau\wedge\phi, \tag{2.10}$$

$$df = \frac{\partial f_1}{\partial u}du + \frac{\partial f_2}{\partial v}dv, \tag{2.11}$$

$$d\phi = \left(\frac{\partial f_2}{\partial u} - \frac{\partial f_1}{\partial v}\right)du\wedge dv. \tag{2.12}$$

2.9. Definition. The line orthogonal to the tangent plane $T_p(S)$ of a surface S at a point $\mathbf{p}$ is the *normal* to S at $\mathbf{p}$. A *normal vector field* on S or on a region R of S is a function that assigns to each point $\mathbf{p}$ of S or R a normal vector of S at $\mathbf{p}$.

The unit normal vector $\mathbf{e}_3$ of S at $\mathbf{p}$ is defined up to a sign, and is given, in consequence of the Lagrange identity [(3.3.8) of Chapter 1], by

$$\mathbf{e}_3 = \pm\frac{\mathbf{x}_u\times\mathbf{x}_v}{\sqrt{(\mathbf{x}_u\times\mathbf{x}_v)^2}} = \pm\frac{\mathbf{x}_u\times\mathbf{x}_v}{\sqrt{\mathbf{x}_u^2\mathbf{x}_v^2 - (\mathbf{x}_u\cdot\mathbf{x}_v)^2}}. \tag{2.13}$$

Let $\mathbf{x}\colon U \subset E^2 \to S$ be a parametrization of a surface S, and let $(u,v)\in U$. At a point $\mathbf{x} = \mathbf{x}(u,v)$ on S, define two tangent vectors $\mathbf{e}_1, \mathbf{e}_2$ such that: (i) $\mathbf{e}_i(u,v)$, $i=1,2$, are of class C^2; and (ii) the determinant $|\mathbf{e}_1, \mathbf{e}_2, \mathbf{e}_3| > 0$. Here we do not assume $\mathbf{e}_1, \mathbf{e}_2$ to be orthogonal unit vectors, so that the inner products

$$g_{ij} = \mathbf{e}_i\cdot\mathbf{e}_j, \qquad i,j=1,2, \tag{2.14}$$

are not necessarily the Kronecker deltas. Since $\mathbf{e}_1, \mathbf{e}_2$ are linearly independent, any tangent vector of S at the point $\mathbf{x}$ can be expressed as their linear combination, and we can write

$$\mathbf{x}_u = p_1\mathbf{e}_1 + p_2\mathbf{e}_2, \qquad \mathbf{x}_v = q_1\mathbf{e}_1 + q_2\mathbf{e}_2, \tag{2.15}$$

where the p's and q's are functions of u, v. For simplicity, we use the differential notation

$$d\mathbf{x} = \mathbf{x}_u\,du + \mathbf{x}_v\,dv, \tag{2.16}$$

so that (2.15) can be written in the condensed form

$$d\mathbf{x} = \theta^1\mathbf{e}_1 + \theta^2\mathbf{e}_2, \tag{2.17}$$

where

$$\theta^1 = p_1\,du + q_1\,dv, \qquad \theta^2 = p_2\,du + q_2\,dv \tag{2.18}$$

are, by (2.7), linear differential forms or 1-forms on S. Similarly, since $\mathbf{e}_3 \cdot d\mathbf{e}_3 = 0$ for the unit normal vector $\mathbf{e}_3$, we can write

$$d\mathbf{e}_3 = \phi^1\mathbf{e}_1 + \phi^2\mathbf{e}_2, \tag{2.19}$$

where ϕ^1, ϕ^2 are 1-forms.

From (2.16) and (2.19) it is natural to construct the following three inner products:

$$\mathrm{I} = d\mathbf{x}^2 = \sum_{i,j=1}^{2} g_{ij}\theta^i\theta^j,$$

$$\mathrm{II} = -d\mathbf{x}\cdot d\mathbf{e}_3 = -\sum_{i,j=1}^{2} g_{ij}\theta^i\phi^j, \tag{2.20}$$

$$\mathrm{III} = d\mathbf{e}_3^2 = \sum g_{ij}\phi^i\phi^j.$$

Each of these products is a quadratic form in the sense of linear algebra, that is, is not a 2-form defined by (2.9), so that the multiplication between the θ's and the ϕ's is ordinary but not exterior. It is obvious that the first form I is positive definite.

2.10. Definition. The quadratic forms I, II, and III given by (2.20) are respectively the *first*, *second*, and *third fundamental forms* of the surface S at the point $\mathbf{x}$.

The vectors $\mathbf{e}_1, \mathbf{e}_2$ in the discussion above are arbitrary tangent vectors of the surface S at the point $\mathbf{x}$ and can be specified in the following two natural ways:

1. We choose

$$\mathbf{e}_1 = \mathbf{x}_u, \qquad \mathbf{e}_2 = \mathbf{x}_v. \tag{2.21}$$

This choice depends on the parameters u, v. The corresponding trihedrons $\mathbf{x}\mathbf{e}_1\mathbf{e}_2\mathbf{e}_3$ are called the *Gauss trihedrons*. (Since $\mathbf{e}_1, \mathbf{e}_2$ are not assumed to be orthogonal unit vectors here, we use the term "trihedrons" instead of "frames.") From (2.16) and (2.17) it follows that $\theta^1 = du$ and $\theta^2 = dv$, and I becomes

$$\mathrm{I} = E\,du^2 + 2F\,du\,dv + G\,dv^2, \tag{2.22}$$

where we have used the customary notation E, F, G for g_{11}, g_{12}, g_{22}, respectively.

2. The vectors $\mathbf{e}_1, \mathbf{e}_2$ are chosen to be orthogonal unit vectors, so that

$$\mathbf{e}_i \cdot \mathbf{e}_j = \delta_{ij}, \qquad i, j = 1, 2, \tag{2.23}$$

where δ_{ij} are the Kronecker deltas. The corresponding frames $\mathbf{x}\mathbf{e}_1\mathbf{e}_2\mathbf{e}_3$ are called the *Darboux frames*; they are not completely determined by (2.23).

Thus we see that in studying the differential geometry of a surface we are naturally led to a number of quadratic forms, which have also simple geometric interpretations.

2.11. Definition. The first form I is called the *element of arc*, for the arc length of the curve (2.2) on S is given by the formula

$$s = \int_{t_1}^{t_2} \sqrt{(d\mathbf{x}/dt)^2}\; dt = \int \sqrt{\mathrm{I}}\,, \tag{2.24}$$

which can also be written as

$$ds^2 = \mathrm{I}. \tag{2.25}$$

2.12. Definition. Suppose that a surface S admits a unit normal vector field $\mathbf{e}_3$ (a necessary and sufficient condition for this is given in Theorem 1.2 of Chapter 4). The *Gauss mapping* g of the surface S is defined to be the mapping $g: S \to S^2$, where S^2 is the unit sphere with center at a point $\mathbf{0}$ in E^3, which sends each point $\mathbf{p}$ of S to the end point of the unit vector through $\mathbf{0}$ in the direction of the normal vector $\mathbf{e}_3$ of S at $\mathbf{p}$.

By (2.20) and Definition 2.12 we obtain immediately the following geometric interpretation of the form III.

2.13. Lemma. *The third fundamental form* III *of a surface S is the first fundamental form of the image of S under the Gauss mapping g of S.*

To get a geometric interpretation for the second fundamental form II of a surface S, we study the relationship among the curvatures of curves on S at a point $\mathbf{p}$. To do this we shall make use of a family of Gauss trihedrons.

Let $\mathbf{x}: U \subset E^2 \to S$ be a parametrization of the surface S, and let $(u,v) \in U$. Let C be a curve on S through $\mathbf{p} = \mathbf{x}(u,v)$ with arc length s, and let $\mathbf{t}, \mathbf{n}$ be the unit tangent and principal normal vectors of C at $\mathbf{p}$. The curvature κ of C at $\mathbf{p}$ is then given by $d\mathbf{t}/ds = \kappa\mathbf{n}$. If θ denotes the angle between the principal normal vector $\mathbf{n}$ and the surface normal $\mathbf{e}_3$, we have the geometric interpretation of the second fundamental form II:

$$\kappa \cos\theta = \kappa(\mathbf{n} \cdot \mathbf{e}_3) = \frac{d\mathbf{t}}{ds} \cdot \mathbf{e}_3 = -\mathbf{t} \cdot \frac{d\mathbf{e}_3}{ds} = -\frac{d\mathbf{x}}{ds} \cdot \frac{d\mathbf{e}_3}{ds} = \frac{\mathrm{II}}{\mathrm{I}},$$

which can be written as

$$\kappa \cos\theta = \frac{\mathrm{II}}{\mathrm{I}} = \frac{L\,du + 2M\,du\,dv + N\,dv^2}{E\,du^2 + 2F\,du\,dv + G\,dv^2}. \tag{2.26}$$

The right-hand side of (2.26) depends only on the point $\mathbf{p}$ and the vector $\mathbf{t}$.

Suppose the curve C to be such that $\cos\theta \neq 0$; that is, its osculating plane does not coincide with the tangent plane of the surface. For such curves, (2.26) implies Lemma 2.14.

2.14. Lemma. *Curves on a surface S with the same tangent and osculating plane at a point $\mathbf{p}$ have the same curvature at $\mathbf{p}$.*

Let κ_N be the curvature of the normal section, which is defined to be the curve of intersection of S by the plane through $\mathbf{t}$ and $\mathbf{e}_3$. κ_N is called the *normal curvature* of S in the direction of $\mathbf{t}$. From (2.26) we have

$$\kappa \cos\theta = \kappa_N, \tag{2.27}$$

since $\theta = 0$ for the normal section. Thus we have the following Meusnier's theorem, which gives the law of change of the curvature as the osculating plane rotates about $\mathbf{t}$ and can be stated geometrically as follows.

2.15. Theorem (J. B. Meusnier). *The osculating circles of plane sections through the same tangent lie on a sphere.*

Proof. The proof follows readily from (2.27) and Fig. 3.18.

For applications of the first and second fundamental forms to practical problems, it is useful to have formulas for their coefficients. By (2.16), (2.20), and (2.22) we can easily have

$$E = \mathbf{x}_u^2, \qquad F = \mathbf{x}_u \cdot \mathbf{x}_v, \qquad G = \mathbf{x}_v^2. \tag{2.28}$$

On the other hand, we have

$$\mathbf{e}_3 \cdot \mathbf{x}_u = \mathbf{e}_3 \cdot \mathbf{x}_v = 0,$$

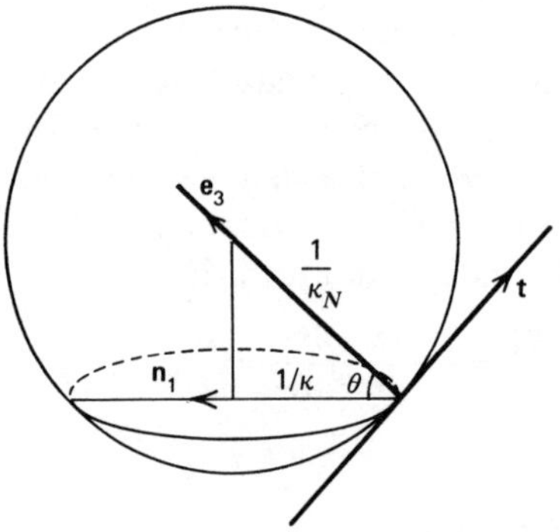

Figure 3.18

which give, by differentiation with respect to u and v,

$$\begin{aligned} \mathbf{e}_3 \cdot \mathbf{x}_{uu} &= -\mathbf{e}_{3u} \cdot \mathbf{x}_u, \\ \mathbf{e}_3 \cdot \mathbf{x}_{uv} &= -\mathbf{e}_{3v} \cdot \mathbf{x}_u = -\mathbf{e}_{3u} \cdot \mathbf{x}_v, \\ \mathbf{e}_3 \cdot \mathbf{x}_{vv} &= -\mathbf{e}_{3v} \cdot \mathbf{x}_v. \end{aligned} \tag{2.29}$$

Substituting (2.16) and $d\mathbf{e}_3 = \mathbf{e}_{3u}\, du + \mathbf{e}_{3v}\, dv$ in the second equation of (2.20) and using (2.29), (2.13), and (2.28); we obtain

$$\begin{aligned} L &= -\mathbf{e}_{3u} \cdot \mathbf{x}_u = \mathbf{e}_3 \cdot \mathbf{x}_{uu} = \frac{|\mathbf{x}_u, \mathbf{x}_v, \mathbf{x}_{uu}|}{\sqrt{EG-F^2}}, \\ M &= -\mathbf{e}_{3u} \cdot \mathbf{x}_v = -\mathbf{e}_{3v} \cdot \mathbf{x}_u = \mathbf{e}_3 \cdot \mathbf{x}_{uv} = \frac{|\mathbf{x}_u, \mathbf{x}_v, \mathbf{x}_{uv}|}{\sqrt{EG-F^2}}, \\ N &= -\mathbf{e}_{3v} \cdot \mathbf{x}_v = \mathbf{e}_3 \cdot \mathbf{x}_{vv} = \frac{|\mathbf{x}_u, \mathbf{x}_v, \mathbf{x}_{vv}|}{\sqrt{EG-F^2}}, \end{aligned} \tag{2.30}$$

where $|\mathbf{u}, \mathbf{v}, \mathbf{w}|$ denotes the determinant of the three vectors $\mathbf{u}, \mathbf{v}, \mathbf{w}$ in E^3 [as (3.2.6) in Chapter 1].

2.16. Examples. 1. Let $P \subset E^3$ be a plane passing though the point $\mathbf{p}_0 = (x_1^0, x_2^0, x_3^0)$ and containing the orthogonal unit vectors $\mathbf{w}_1 = (a_1, a_2, a_3)$, $\mathbf{w}_2 = (b_1, b_2, b_3)$, and let a coordinate system for P be given by

$$\mathbf{x}(u, v) = \mathbf{p}_0 + u\mathbf{w}_1 + v\mathbf{w}_2, \qquad (u, v) \in E^2.$$

Then $\mathbf{x}_u = \mathbf{w}_1$ and $\mathbf{x}_v = \mathbf{w}_2$; therefore the first fundamental form at a general point $\mathbf{x}(u, v)$ of P is

$$ds^2 = du^2 + dv^2. \tag{2.31}$$

2. Let S be the right cylinder over the circle $x_1^2+x_2^2=1$, $x_3=0$ given by the parametrization $\mathbf{x}$: $U\to E^3$ where

$$\mathbf{x}(u,v)=(\cos u,\sin u,v),$$
$$U=\{(u,v)\in E^2|0<u<2\pi,\ -\infty<v<\infty\}.$$

Then

$$\mathbf{x}_u=(-\sin u,\cos u,0),\qquad \mathbf{x}_v=(0,0,1),$$

and therefore the first fundamental form at the point $\mathbf{x}(u,v)$ of S is also (2.31).

3. The first fundamental form of the sphere given by (1.8) is

$$ds^2=a^2\,d\theta^2+a^2\sin^2\theta\,d\phi^2, \tag{2.32}$$

and that of the surface of revolution given by (1.12) is

$$ds^2=(1+f'^2)\,du^2+u^2\,dv^2, \tag{2.33}$$

where $f'=df/du$.

Exercises

1. Prove that the definition in part 2 of Remarks 2.2 does not depend on the choice of the parametrizations $\mathbf{x}_1$ and $\mathbf{x}_2$.
2. A necessary and sufficient condition that two curves on a surface S be orthogonal to each other at a point $\mathbf{p}$ of intersection is that the directions dv/du, dv_1/du_1 of the two curves at $\mathbf{p}$ satisfy the equation
$$E\,du\,du_1+F(du\,dv_1+du_1\,dv)+G\,du\,dv_1=0,$$
where E, F, G are the coefficients of the first fundamental form of S.
3. Show that on a surface S with parameters u and v the orthogonal trajectories of the curves $a\,du+b\,dv=0$ are given by
$$(Eb-Fa)\,du+(Fb-Ga)\,dv=0,$$
where E, F, G are the coefficients of the first fundamental form of S.
4. Prove that a necessary and sufficient condition that two one-parameter families of curves on a surface with parameters u and v given by
$$a\,du^2+2b\,du\,dv+c\,dv^2=0 \tag{2.34}$$
be orthogonal is that
$$Ec-2Fb+Ga=0.$$

5. Let θ be the angle at the point (u,v) between the two directions given by (2.34). Show that

$$\tan\theta = \frac{2\sqrt{EG-F^2}\,\sqrt{b^2-ac}}{Ec-2Fb+Ga}.$$

6. Assume the parametric u- and v-curves on a surface S to be orthogonal. Show that the differential equation of the curves on S bisecting the angles of the parametric curves is

$$E\,du^2 - G\,dv^2 = 0.$$

7. Find the equation of the tangent plane of the elliptic paraboloid $\mathbf{x}(u,v)=(au\cos v, bu\sin v, u^2)$, $u>0, 0<v<2\pi$, at the point $\mathbf{x}(u_0,v_0)$.

***8.** Show that $(f_{x_1}(\bar{x}_1,\bar{x}_2,\bar{x}_3), f_{x_2}(\bar{x}_1,\bar{x}_2,\bar{x}_3), f_{x_3}(\bar{x}_1,\bar{x}_2,\bar{x}_3))$ is a normal vector of a surface given by $f(x_1,x_2,x_3)=0$ at the point $(\bar{x}_1,\bar{x}_2,\bar{x}_3)$, where $f_{x_i}=\partial f/\partial x_i$.

9. Find the equation of the tangent plane of a surface, which is the graph of a differentiable function $x_3=f(x_1,x_2)$, at the point $\mathbf{p}=(\bar{x}_1,\bar{x}_2,\bar{x}_3)$.

10. Show that

$$(f_{x_2}g_{x_3}-f_{x_3}g_{x_2}, f_{x_3}g_{x_1}-f_{x_1}g_{x_3}, f_{x_1}g_{x_2}-f_{x_2}g_{x_1})$$

is a tangent vector at a point of the curve of intersection of two surfaces given, respectively, by

$$f(x_1,x_2,x_3)=0, \qquad g(x_1,x_2,x_3)=0.$$

11. Find the equation of the normal plane at the point $(1,1,1)$ of the curve

$$x_1x_2x_3=1, \qquad x_2^2=x_1.$$

12. Show that if all normals to a connected surface pass through a fixed point, the surface is contained in a sphere.

***13.** Show that if all the normals of a connected surface intersect a given straight line, the surface is a surface of revolution.

14. (a) Let $f\colon S\to E^1$ be a differentiable function on a surface S given by

$$f(\mathbf{p})=\|\mathbf{p}-\mathbf{p}_0\|, \qquad \mathbf{p}\in S,\ \mathbf{p}_0\notin S.$$

Show that $\mathbf{p}\in S$ is a critical point of f if and only if the line joining $\mathbf{p}$ to $\mathbf{p}_0$ is normal to S at $\mathbf{p}$.

(b) Let $h: S \to E^1$ be given by $h(\mathbf{p}) = \mathbf{p} \cdot \mathbf{v}$, where $\mathbf{v} \in E^3$ is a unit vector. Show that $\mathbf{p} \in S$ is a critical point of h if and only if $\mathbf{v}$ is a normal vector of S at $\mathbf{p}$.

***15.** Let a surface S be given by $f(x_1, x_2, x_3) = c$ where $(x_1, x_2, x_3) \in E^3$ and c is constant. Show that the *gradient* vector field of f, grad $f = (\partial f/\partial x_1, \partial f/\partial x_2, \partial f/\partial x_3)$, is a nonvanishing normal vector field on S, and verify this geometrically for the sphere $f(x_1, x_2, x_3) \equiv \Sigma_{i=1}^3 x_i^2 = r^2$.

16. Let $\mathbf{x}: U \subset E^2 \to S$ be a parametrization of a surface S, and let $(u, v) \in U$.

(a) If x is the derivative mapping of x (see Definition 4.2.1 of Chapter 1), show that

$$(\mathrm{U}_1) = \mathbf{x}_u, \quad \mathbf{x}\,(\mathrm{U}_2) = \mathbf{x}_v,$$

where U_1, U_2 are the unit vector fields in the positive u, v directions, respectively.

(b) If f is a differentiable function on S, prove that

$$\mathbf{x}_u[f] = \frac{\partial}{\partial u}(f(\mathbf{x})), \qquad \mathbf{x}_v[f] = \frac{\partial}{\partial v}(f(\mathbf{x})).$$

17. Prove (2.7) and (2.9) through (2.12).

18. Find the first and second fundamental forms of the following surfaces:

(a) The sphere given by (1.8).
(b) The surface of revolution given by (1.12).
(c) The torus given by

$$x_1 = u\cos v, \qquad x_2 = u\sin v, \qquad x_3 = \sqrt{b^2 - (u-a)^2}\,, \qquad a > b.$$

(d) The helicoid given by

$$x_1 = u\cos v, \qquad x_2 = u\sin v, \qquad x_3 = av + f(u), \qquad a = \text{const.}$$

19. For each of the following surfaces, describe the image of the Gauss mapping on the unit sphere (use either normal):

(a) Cylinder: $x_1^2 + x_2^2 = r^2.$
(b) Plane: $x_1 + x_2 + x_3 = 0.$
(c) Sphere: $(x_1 - 1)^2 + x_2^2 + (x_3 + 2)^2 = 1.$
(d) Cone: $x_3 = \sqrt{x_1^2 + x_2^2}\,.$

20. Let $g: T \to S^2$ be the Gauss mapping of the torus T (as in Example 1.14) derived from the outward unit normal. What are the image curves under g of the meridians and parallels of T^2 (for the definitions, see Example 1.10). What points of S^2 are the image of exactly two points of T?

21. Find the normal curvature of the curve $\mathbf{x}(t^2, t)$ at $t=1$ on the surface

$$\mathbf{x}(u,v)=(u,v,u^2+v^2).$$

3. FORM OF A SURFACE IN A NEIGHBORHOOD OF A POINT

Let $\mathbf{x}$: $U \subset E^2 \to S$ be a parametrization of a surface S, and let $(u, v) \in U$. On S consider a point $\mathbf{p}$ at which for simplicity we assume that $u=v=0$. Then in a neighborhood of the point $\mathbf{p}$ we have the following Taylor's expansion (by Theorem 2.2.2 of Chapter 1)

$$\mathbf{x}(u,v)=\mathbf{x}(0,0)+\mathbf{x}_u^0 u+\mathbf{x}_v^0 v+\frac{1}{2}(\mathbf{x}_{uu}^0 u^2+2\mathbf{x}_{uv}^0 uv+\mathbf{x}_{vv}^0 v^2)+(3), \quad (3.1)$$

where the subscripts u, v denote the partial derivatives, the superscript zero denotes the value at $\mathbf{p}$, and (3) denotes the terms of degree $\geqslant 3$ in u and v. Let d be an oriented distance of a neighboring point $\mathbf{x}(u, v)$ of $\mathbf{p}$ on S to the tangent plane of S at $\mathbf{p}$. Then by (2.30) and (3.1) we obtain

$$\begin{aligned} d&=[\mathbf{x}(u,v)-\mathbf{x}(0,0)]\cdot\mathbf{e}_3^0 \\ &=\frac{1}{2}(\mathbf{x}_{uu}^0 u^2+2\mathbf{x}_{uv}^0 uv+\mathbf{x}_{vv}^0 v^2)\cdot\mathbf{e}_3^0+(3) \qquad (3.2) \\ &=\frac{1}{2}(L_0u^2+2M_0uv+N_0v^2)+(3), \end{aligned}$$

where the subscript zero denotes the value at $\mathbf{p}$. Thus the form of S in the neighborhood of $\mathbf{p}$ depends on the behavior of L_0, M_0, N_0.

3.1. Definition. The point $\mathbf{p}$ is *elliptic*, *hyperbolic*, or *parabolic* according to whether $L_0N_0-M_0^2>0$, <0, or $=0$.

Since

$$L_0u^2+2M_0uv+N_0v^2=L_0\left(u+\frac{M_0}{L_0}v\right)^2+\frac{v^2}{L_0}(L_0N_0-M_0^2),$$

in a neighborhood of an elliptic point $\mathbf{p}$ the distance d is of the same sign for all u and v, so that the surface S lies wholly on one side of its tangent plane at $\mathbf{p}$.

Similarly, at a hyperbolic point $\mathbf{p}$ the surface S is of the form of a saddle, that is, lies on both sides of its tangent plane at $\mathbf{p}$. At a parabolic point $\mathbf{p}$

with L_0, M_0, N_0 not all zero, the surface S lies entirely on one side of its tangent plane at $\mathbf{p}$. In particular, a parabolic point $\mathbf{p}$ is called a *planar point* if $L_0 = M_0 = N_0 = 0$. At such a planar point the form of the surface is not defined by the process above because nothing can be said by using the sign of the distance function $d(u, v)$, which now is zero. However, we can have Lemma 3.2.

3.2. Lemma. *A surface having only planar points is a plane.*

A proof of Lemma 3.2 is given in Section 7.

To further study the form of a surface S in a neighborhood of a point $\mathbf{p}$, we take a positively oriented frame (see Definition 5.4.1 of Chapter 1) $\mathbf{p}\mathbf{e}_1^0\mathbf{e}_2^0\mathbf{e}_3^0$ such that $\mathbf{e}_3^0$ is the normal vector of S at $\mathbf{p}$, and define the coordinates x_1, x_2, x_3 of a point $\mathbf{x}(u, v)$ on S relative to this frame by

$$\mathbf{x}(u,v) - \mathbf{x}(0,0) = x_1\mathbf{e}_1^0 + x_2\mathbf{e}_2^0 + x_3\mathbf{e}_3^0. \tag{3.3}$$

Furthermore, we take x_1, x_2 to be the parameters u, v. Then from (3.2) it follows that d now is the x_3-coordinate, and the equation of S is

$$x_3 = \frac{1}{2}\left(L_0x_1^2 + 2M_0x_1x_2 + N_0x_2^2\right) + (3), \tag{3.4}$$

where (3) denotes the terms of degree $\geqslant 3$ in x_1 and x_2. By differentiating (3.3) with respect to x_1 and x_2, respectively, and noticing that $(\partial x_3/\partial x_1)_0 = (\partial x_3/\partial x_2)_0 = 0$, we obtain, at $\mathbf{p}$,

$$(\mathbf{x}_u)_0 = \mathbf{e}_1^0, \qquad (\mathbf{x}_v)_0 = \mathbf{e}_2^0, \qquad E_0 = G_0 = 1, \qquad F_0 = 0. \tag{3.5}$$

A unit tangent vector $\mathbf{t}$ of S at $\mathbf{p}$ can be written as

$$\mathbf{t} = \left(\frac{d\mathbf{x}}{ds}\right)_0 = \mathbf{e}_1^0\frac{du}{ds} + \mathbf{e}_2^0\frac{dv}{ds}, \tag{3.6}$$

where s is the arc length of a curve on S, to which $\mathbf{t}$ is tangent. If ϕ denotes the angle that $\mathbf{t}$ makes with e_1^0-axis, we have

$$\cos\phi = \mathbf{t}\cdot\mathbf{e}_1^0 = \frac{du}{ds}, \qquad \sin\phi = \mathbf{t}\cdot\mathbf{e}_2^0 = \frac{dv}{ds}. \tag{3.7}$$

Then by (2.26) the normal curvature in the direction $\mathbf{t}$ is given by the formula

$$\begin{aligned}\kappa_N(\phi) &= L_0\left(\frac{du}{ds}\right)^2 + 2M_0\frac{du}{ds}\frac{dv}{ds} + N_0\left(\frac{dv}{ds}\right)^2 \\ &= L_0\cos^2\phi + 2M_0\cos\phi\sin\phi + N_0\sin^2\phi.\end{aligned} \tag{3.8}$$

This leads to the following geometric interpretation.

3.3. Lemma. *In the tangent plane of a surface S at a point* **p** *let us consider the conics*

$$L_0x_1^2+2M_0x_1x_2+N_0x_2^2=\pm 1, \qquad x_3=0, \tag{3.9}$$

where the sign on the right-hand side of the first equation is so chosen that the conics will be real. Then the absolute value of the radius of the normal curvature of S at **p** *in a direction is equal to the square of the length of the position vector, with reference to* **p**, *of the conic* (3.9) *in that direction.*

Proof. Let the position vector, with reference to **p**, of the conic (3.9) be given by $(x_1, x_2, 0)$ with

$$x_1=r\cos\phi, \qquad x_2=r\sin\phi, \tag{3.10}$$

so that r is the length of the vector. Then substitution of (3.10) in (3.8) gives the required result immediately.

3.4. Definition. The conic (3.9) is called the *Dupin indicatrix* of the surface S at the point **p**. At a nonparabolic point **p** if the conic (3.9) is not a circle, the directions of the principal axes of (3.9) are completely determined and are called the *principal directions* of S at **p**.

The Dupin indicatrix is an ellipse at an elliptic point, a pair of conjugate hyperbolas at a hyperbolic point, and a pair of straight lines at a non-parabolic point. At a nonparabolic point **p** we can choose $\mathbf{e}_1^0$, $\mathbf{e}_2^0$ to be in the directions of the principal axes of (3.9). By this choice of $\mathbf{e}_1^0$, $\mathbf{e}_2^0$, (3.9) takes the normal form so that $M_0=0$, and the normal curvatures in the directions of e_1^0, e_2^0, for which $\phi=0$, $\pi/2$, are equal to L_0, N_0, respectively, by (3.8), which hence becomes

$$\kappa_N(\phi)=\kappa_1\cos^2\phi+\kappa_2\sin^2\phi, \tag{3.11}$$

where κ_1 and κ_2 are the normal curvatures in the principal directions.

3.5. Definition. Equation 3.11 is Euler's formula; κ_1, κ_2 are the *principal curvatures*, and $1/\kappa_1$, $1/\kappa_2$ the *radii of principal curvatures* of the surface S at the point **p**.

3.6. Definition. At a hyperbolic point **p** on a surface S, the Dupin indicatrix has two asymptotes whose directions are called the *asymptotic directions* of S at **p**. On a surface S a point at which the normal curvatures of S are the same for all directions is called an *umbilical point* of S.

From (2.26) it follows that the conditions for a point on a surface to be an umbilical point are $L/E=M/F=N/G$ at the point. Thus an umbilical point is either planar or elliptic. Furthermore, at an elliptic umbilical point **p**, by taking $\mathbf{e}_1^0$, $\mathbf{e}_2^0$ to be in the directions of the principal axes of the Dupin indicatrix and noting that the normal curvature at **p** is the same for all directions, we have $M_0=0$ and $L_0=N_0$; therefore by (3.9) the Dupin indicatrix at **p** is a circle.

The behavior of a surface S in a neighborhood of a point **p** can also be described by studying the properties of the *normal sections* of S at **p**, which are defined to be the curves of intersection of S by planes through the normal of S at **p**.

3.7. Lemma. *A normal section C of a surface S at a point* **p** *has* **p** *as a point of inflexion only if the plane of C passes through an asymptotic direction of S at* **p**; *otherwise the normal section C has a nonzero curvature at* **p**.

Proof. Let the equation of the plane of the normal section C be

$$x_2=\lambda x_1. \tag{3.12}$$

Substituting (3.12) in (3.4) we see that C is given by (3.12) and

$$x_3=\frac{1}{2}(L_0+2M_0\lambda+N_0\lambda^2)x_1^2+(3), \tag{3.13}$$

where (3) denotes the terms of degree $\geqslant$ in λ. Thus C has **p** as a point of inflexion only if

$$L_0+2M_0\lambda+N_0\lambda^2=0, \tag{3.14}$$

which, together with (3.12), implies that

$$L_0x_1^2+2M_0x_1x_2+N_0x_2^2=0, \tag{3.15}$$

so that the plane (3.12) of C passes through an asymptotic direction of S at **p**, since those asymptotic directions are given by (3.15) and $x_3=0$.

On the other hand, if the plane (3.12) of C does not pass through an asymptotic direction of S at **p**, that is, if (3.14) does not hold, then from (3.13) we can easily see that the curvature of C at **p** is nonzero.

3.8. Lemma. *The locus of the centers of curvature of the normal sections of a surface S at a point* **p**: (*i*) *is a finite line segment lying on one side of the normal* $\mathbf{e}_3$ *of S at* **p** *if* **p** *is elliptic*, (*ii*) *consists of two infinite line segments on both sides of* $\mathbf{e}_3$ *if* **p** *is hyperbolic*, (*iii*) *is an infinite line segment on one side of* $\mathbf{e}_3$ *if* **p** *is nonplanar parabolic* (*Fig. 3.19*).

Proof. The normal curvature of S at **p** is a function of a direction in the tangent plane of S at **p**. By Corollary 1.5.10 of Chapter 1, this function attains a maximum and a minimum. Since $\mathbf{e}_3$ is the principal normal of a

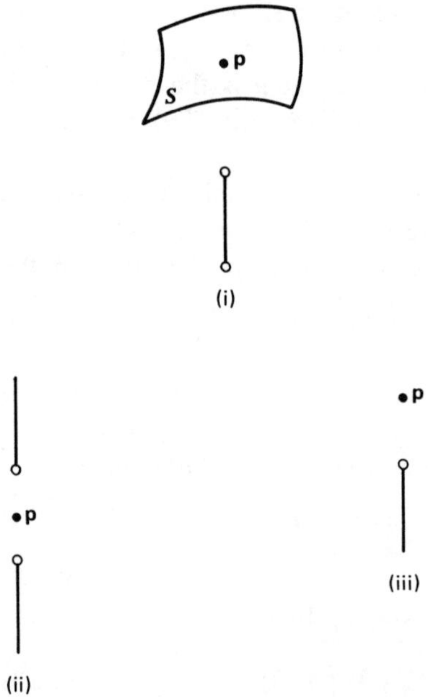

Figure 3.19

normal section C of S at $\mathbf{p}$, the center of curvature of C at $\mathbf{p}$ is on $\mathbf{e}_3$; hence the lemma follows immediately from Definition 1.4.2.5 of Chapter 2.

3.9. Examples of Points.

1. *Elliptic points.* Every point on an ellipsoid is elliptic.
2. *Hyperbolic points.* Every point on a hyperboloid of one sheet is hyperbolic.
3. *Umbilical points.* Every point on a sphere is an umbilical point, and on an ellipsoid with unequal axes there are four umbilical points (see Exercise 3).
4. *Parabolic points.* The point $(0,0,0)$ on the surface

$$x_3 = x_1^2 + \alpha x_2^2 \tag{3.16}$$

is a parabolic point.

5. *Planar points.* Every point of a plane is a planar point.

A "monkey saddle" (see Fig. 3.20) given by

$$x_1 = u, \qquad x_2 = v, \qquad x_3 = u^3 - 4v^2u \tag{3.17}$$

has the point $(0,0,0)$ as a planar point. The surface has three hills and three downward slopes; it is so named because a saddle for a monkey needs two downward slopes for the legs and one for the tail.

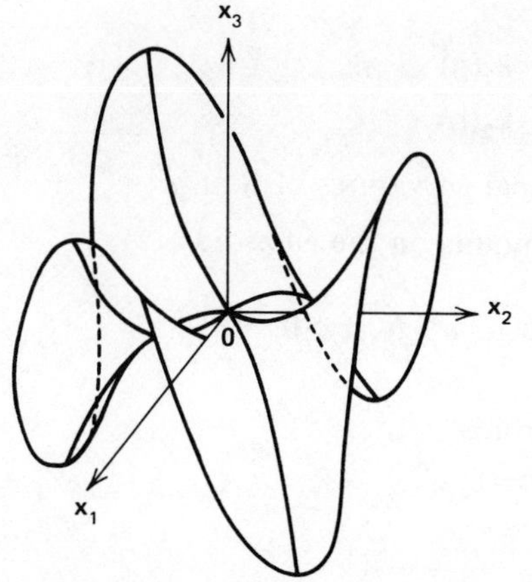

Figure 3.20

All types of points are illustrated on a torus such as the one given by (1.18). The outside points (obtained by rotating BCA) are elliptic; the inside points (BDA) are hyperbolic; the circles obtained by rotating B and A consist of only parabolic points (Fig. 3.21).

Exercises

***1.** Let C be the curve of intersection of two surfaces S and $\bar{S}$. Show that the curvature κ of C at a point $\mathbf{p}$ is given by

$$\kappa^2 \sin^2\theta = \kappa_N^2 + \bar{\kappa}_N^2 - 2\kappa_N\bar{\kappa}_N \cos\theta,$$

where κ_N and $\bar{\kappa}_N$ are the normal curvatures at $\mathbf{p}$, along the tangent line to C, of S and $\bar{S}$, respectively, and θ is the angle between the normals of S and $\bar{S}$ at $\mathbf{p}$.

***2.** Show that there are exactly two asymptotic directions of a surface S at a hyperbolic point $\mathbf{p}$, which are bisected by the principal directions

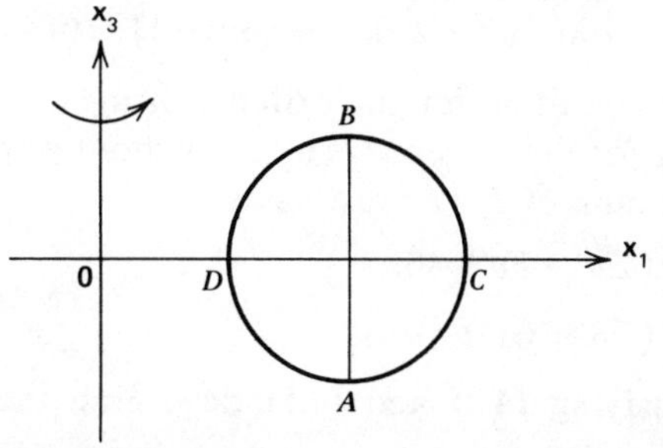

Figure 3.21

at angle θ such that

$$\tan^2\theta = -\frac{\kappa_1(\mathbf{p})}{\kappa_2(\mathbf{p})},$$

where $\kappa_1(\mathbf{p})$ and $\kappa_2(\mathbf{p})$ are the principal curvatures of S at $\mathbf{p}$.

3. Show that there are four umbilical points on the ellipsoid

$$\frac{x_1^2}{a^2} + \frac{x_2^2}{b^2} + \frac{x_3^2}{c^2} = 1, \qquad a > b > c > 0.$$

4. Find the umbilical points of the surface

$$x_1 x_2 x_3 = 8.$$

4. PRINCIPAL CURVATURES, ASYMPTOTIC CURVES, AND CONJUGATE DIRECTIONS

Let $\mathbf{x}$: $U \subset E^2 \to S$ be a parametrization of a surface S, and let $(u, v) \in U$. The normal curvature of S at a fixed point is a function of a tangent direction (or simply just direction) of S at the fixed point. It is therefore natural to study the directions for each of which the normal curvature takes an extremum, and to study the values of these extrema.

Putting

$$u' = \frac{du}{ds}, \qquad v' = \frac{dv}{ds}, \tag{4.1}$$

where s is the arc length of a curve on S, from (2.26) we obtain

$$\kappa_N = Lu'^2 + 2Mu'v' + Nv'^2, \tag{4.2}$$

$$Eu'^2 + 2Fu'v' + Gv'^2 = 1. \tag{4.3}$$

The problem is therefore to determine an extremum of $\kappa_N(u', v')$, where u', v' are related by (4.3). According to Lagrange's method of multipliers we introduce a new variable λ and put

$$F(u', v') = Lu'^2 + 2Mu'v' + Nv'^2 - \lambda(Eu'^2 + 2Fu'v' + Gv'^2 - 1). \tag{4.4}$$

By Theorem 2.4.1 of Chapter 1, the condition for the extrema of κ_N with the relation (4.3) is the same as that for the extrema $F(u', v')$ without any relation between u', v'. For an extremum of F we must have

$$\begin{aligned} Lu' + Mv' - \lambda(Eu' + Fv') &= 0, \\ Mu' + Nv' - \lambda(Fu' + Gv') &= 0. \end{aligned} \tag{4.5}$$

Therefore the values of u', v' satisfying (4.3) and (4.5) determine the extrema of κ_N. Multiplying (4.5) by u' and v', respectively, and adding the resulting equations we know that such a extremum of κ_N is equal to

$$Lu'^2+2Mu'v'+Nv'^2=\lambda(Eu'^2+2Fu'v'+Gv'^2)=\lambda.$$

Eliminating u', v' from (4.5) gives

$$\begin{vmatrix} L-\lambda E & M-\lambda F \\ M-\lambda F & N-\lambda G \end{vmatrix}=0, \tag{4.6}$$

which defines the two extrema of κ_N. Eliminating λ from (4.5), we obtain

$$\begin{vmatrix} L\,du+M\,dv & E\,du+F\,dv \\ M\,du+N\,dv & F\,du+G\,dv \end{vmatrix}=0, \tag{4.7}$$

which gives the directions where the extrema of κ_N are attained.

4.1. Definition. The roots of (4.6) are the *principal curvatures*, and the roots of (4.7) the *principal directions* of the surface S at a point $\mathbf{p}(u,v)$.

From (3.11) it follows that κ_1 and κ_2 are the extrema of the normal curvature at the point $\mathbf{p}$, so that Definition 4.1 of the principal curvatures and principal directions is equivalent to Definition 3.5. Furthermore, by the properties of the principal axes of a central conic we see that the two principal directions of a surface at a nonparabolic point are real and orthogonal to each other.

Equations 4.6 and 4.7 can also be written, respectively, as

$$(EG-F^2)\lambda^2-(EN-2FM+GL)\lambda+LN-M^2=0, \tag{4.8}$$

$$(EM-FL)\,du^2+(EN-GL)\,du\,dv+(FN-GM)\,dv^2=0. \tag{4.9}$$

From (4.8) the two elementary symmetric functions of the principal curvatures κ_1, κ_2 have the expressions:

$$H=\frac{1}{2}(\kappa_1+\kappa_2)=\frac{1}{2}\,\frac{EN-2FM+GL}{EG-F^2},$$
$$K=\kappa_1\kappa_2=\frac{LN-M^2}{EG-F^2}. \tag{4.10}$$

4.2. Definition. H and K are, respectively, the *mean curvature* and the *Gaussian curvature* of a surface S at a point $\mathbf{p}(u,v)$.

From Definition 3.1 and (4.10) it follows that a point on a surface is elliptic, hyperbolic, or parabolic according as $K>0$, <0, or $=0$ at the point.

4.3. Lemma. *The two principal curvatures κ_1, κ_2 can be expressed in terms of H, K as follows:*

$$\kappa_1, \kappa_2 = H \pm \sqrt{H^2 - K}\,. \tag{4.11}$$

Proof. Equation 4.11 can be proved by directly solving (4.10) simultaneously for κ_1, κ_2 and noting that

$$H^2 - K = \frac{1}{4}(\kappa_1 - \kappa_2)^2 \geqslant 0. \tag{4.12}$$

By (4.11) or (4.12) we immediately have Corollary 4.4.

4.4. Corollary. *A point on a surface is an umbilical point if and only if at the point*

$$H^2 - K = 0. \tag{4.13}$$

A plane or sphere consists entirely of umbilical points. We shall prove the converse of this result.

4.5. Theorem. *A connected surface S consisting entirely of umbilical points is contained in either a plane or a sphere.*

Proof. Let $\mathbf{p} \in S$, and let $\mathbf{x}(u, v)$ be a parametrization of S at $\mathbf{p}$ such that the coordinate neighborhood U is connected. The hypothesis implies that the normal curvature at each point $\mathbf{q} \in U$ is independent of the direction, so that at $\mathbf{q}$

$$\frac{L}{E} = \frac{M}{F} = \frac{N}{G} = \rho \text{(say)}. \tag{4.14}$$

Before we go on, we need the following elementary lemma.

4.6. Lemma. *In E^3 the two vectors*

$$\mathbf{a} = a_1\mathbf{x}_u + a_2\mathbf{x}_v + a_3\mathbf{e}_3, \qquad \mathbf{b} = b_1\mathbf{x}_u + b_2\mathbf{x}_v + b_3\mathbf{e}_3 \tag{4.15}$$

are equal if and only if

$$\mathbf{a}\cdot\mathbf{x}_u = \mathbf{b}\cdot\mathbf{x}_u, \qquad \mathbf{a}\cdot\mathbf{x}_v = \mathbf{b}\cdot\mathbf{x}_v, \qquad \mathbf{a}\cdot\mathbf{e}_3 = \mathbf{b}\cdot\mathbf{e}_3. \tag{4.16}$$

Proof. The "only if" part is obvious. For the "if" part we first note that the third equation of (4.16) implies

$$a_3 = b_3. \tag{4.17}$$

By means of (2.28), from the first two equations of (4.16) we then obtain

$$\begin{aligned}(a_1-b_1)E+(a_2-b_2)F=0,\\(a_1-b_1)F+(a_2-b_2)G=0.\end{aligned} \tag{4.18}$$

Since $EG-F^2>0$, (4.18) holds only if $a_1=b_1$, $a_2=b_2$, which together with (4.17) shows that $\mathbf{a}=\mathbf{b}$. Q.E.D.

Now we continue with the proof of Theorem 4.5. By Lemma 4.6; (2.30), and (2.28), we can easily show that (4.14) implies that at $\mathbf{q}$

$$e_{3u}=-\rho x_u, \qquad e_{3v}=-\rho x_v. \tag{4.19}$$

Differentiating the first and second equations of (4.19) with respect to v and u, respectively, and subtracting one of the two resulting equations from the other, we obtain at all $\mathbf{q}\in U$

$$\rho_v \mathbf{x}_u-\rho_u \mathbf{x}_v=0. \tag{4.20}$$

Since $\mathbf{x}_u$ and $\mathbf{x}_v$ are linearly independent, (4.20) holds only when $\rho_u=\rho_v=0$ for all $\mathbf{q}\in U$. Since U is connected, ρ is constant in U.

Suppose first that $\rho=0$ at all $\mathbf{q}\in U$. Then from (4.19), $\mathbf{e}_3$ is constant in U. Integrating $\mathbf{e}_3\cdot d\mathbf{x}=0$ gives $\mathbf{e}_3\cdot\mathbf{x}=$constant or $ax_1+bx_2+cx_3=d$, where $\mathbf{x}=(x_1,x_2,x_3)$. Thus U is contained in a plane.

Next suppose that $\rho\neq0$ at all $\mathbf{q}\in U$. Then by (4.19) we have at all $\mathbf{q}\in U$

$$d\left(\mathbf{x}+\frac{1}{\rho}\mathbf{e}_3\right)=0,$$

which shows that the point $\mathbf{z}=\mathbf{x}+\dfrac{1}{\rho}\mathbf{e}_3$ is fixed. Since

$$(\mathbf{x}-\mathbf{z})^2=\frac{1}{\rho^2}\mathbf{e}_3^2=\frac{1}{\rho^2},$$

U is contained in a sphere with center $\mathbf{z}$ and radius $1/\rho$. This completes the proof of the theorem in the local sense, that is, for a neighborhood of a point $\mathbf{p}\in S$.

To complete the proof we first notice that given any other point $\mathbf{r}\in S$, due to the connectedness of S there exists a continuous curve α: $[0,1]\to S$ with $\alpha(0)=\mathbf{p}$, $\alpha(1)=\mathbf{r}$. For each point $\alpha(t)\in S$ of this curve, there exists a neighborhood U_t in S contained in a plane or a sphere and such that $\alpha^{-1}(U_t)$ is an open interval of $[0,1]$. The union $\cup\alpha^{-1}(U_t)$, $t\in[0,1]$, covers the closed interval $[0,1]$, so by the Heine-Borel theorem (Theorem 1.5.5 of Chapter 1) the interval $[0,1]$ is covered by finitely many elements of the family $\{\alpha^{-1}(U_t)\}$. Thus the curve $\alpha([0,1])$ is covered by a finite number of the neighborhoods U_t. If one of these neighborhoods is contained in a plane or sphere, all the other neighborhoods are in the same plane or sphere. Since the point $\mathbf{r}$ is arbitrary, the whole S is contained in this plane or sphere.

4.7. Definition. On a surface S a curve C is a *line of curvature* if the tangent of C at every point $\mathbf{p}$ is a principal direction of S at $\mathbf{p}$.

The lines of curvature are therefore defined by the differential equation (4.7) or (4.9). From (4.5) it follows that along a line of curvature

$$\begin{aligned} L\,du + M\,dv - k(E\,du + F\,dv) &= 0, \\ M\,du + N\,dv - k(F\,du + G\,dv) &= 0, \end{aligned} \tag{4.21}$$

where k is the principal curvature in the direction of the line of curvature. By using Lemma 4.6, the equation $d\mathbf{e}_3 = \mathbf{e}_{3u}\,du + \mathbf{e}_{3v}\,dv$, and (2.16), (2.28), and (2.30), from (4.21), we can easily verify

$$d\mathbf{e}_3 + k\,d\mathbf{x} = 0, \tag{4.22}$$

since the inner products of the vector on the left-hand side of (4.22) with the vectors $\mathbf{x}_u, \mathbf{x}_v, \mathbf{e}_3$ are zero. Equation 4.22 is called the *equation of Rodriques*.

4.8. Lemma. *Necessary and sufficient conditions for the parametric curves $u = const$ and $v = const$ of a surface at a point $\mathbf{x}(u, v)$ to be the lines of curvature at the point are*

$$F = M = 0. \tag{4.23}$$

Proof. Assume that the parametric curves to be the lines of curvature. Since the lines of curvature are orthogonal, $F = 0$, and (4.9) then becomes

$$EM\,du^2 + (EN - GL)\,du\,dv - GM\,dv^2 = 0.$$

For this equation to be $du\,dv = 0$ we have

$$EM = 0, \qquad GM = 0, \qquad EN - GL \neq 0,$$

which imply immediately that $M = 0$.

Now suppose that $F = M = 0$. Then (4.9) becomes $du\,dv = 0$, which means that the parametric curves are the line of curvature.

4.9. Definition. On a surface S a curve C is an *asymptotic curve* if the tangent of C at every point $\mathbf{p}$ is an aysmptotic direction of S at $\mathbf{p}$.

From Lemma 3.7 and (2.26) it follows that if C is an asymptotic curve of a surface S, the normal curvature of S for the tangent to C at each point $\mathbf{p}$ is zero; therefore at each $\mathbf{p}$ the second fundamental form II of S vanishes, that is,

$$L\,du^2+2M\,du\,dv+N\,dv^2=0. \tag{4.24}$$

Hence (4.24) *is the differential equation of the asymptotic curves of the surface S.*

An asymptotic curve on a surface can be also defined in two other ways as follows. By Lemma 3.7 we obtain Lemma 4.10.

4.10. Lemma. *The tangent at each point* $\mathbf{p}$ *of an asymptotic curve of a surface S has a contact of three points with S at* $\mathbf{p}$.

Furthermore, we have Lemma 4.11.

4.11. Lemma. *If the osculating plane of a curve C on a surface S at each point* $\mathbf{p}$ *coincides with the tangent plane of S at* $\mathbf{p}$, *then C is an asymptotic curve of S.*

Proof. Let $\mathbf{x}(u,v)$ be a parametrization of the surface S, and let

$$u=u(t), \qquad v=v(t) \tag{4.25}$$

be the coordinate functions of the curve C with respect to this parametrization $\mathbf{x}$. Then from the unit tangent vector $\mathbf{x}'=\mathbf{x}_u u'+\mathbf{x}_v v'$ of C at a general point $\mathbf{p}$ we obtain the principal normal vector of C at $\mathbf{p}$:

$$\mathbf{x}''=\mathbf{x}_{uu}u'^2+2\mathbf{x}_{uv}u'v'+\mathbf{x}_{vv}v'^2+\mathbf{x}_u u''+\mathbf{x}_v v'', \tag{4.26}$$

where the primes denote the derivatives with respect to the arc length s of C. By the assumption, at each point $\mathbf{p}$ the normal vector $\mathbf{e}_3$ of the surface S is orthogonal to the osculating plane of C, therefore to the principal normal vector (4.26). Thus

$$\mathbf{e}_3\cdot\mathbf{x}''=\mathbf{e}_3\cdot\mathbf{x}_{uu}u'^2+2(\mathbf{e}_3\cdot\mathbf{x}_{uv})u'v'+\mathbf{e}_3\cdot\mathbf{x}_{vv}v'^2=0,$$

which together with (2.30) implies (4.24). Hence C is an asymptotic curve of S. Q.E.D.

Since any plane through a straight line may be considered to be an osculating plane of the line, and the osculating plane at a point of a plane curve is the plane of the curve, Corollary 4.12 follows immediately from Lemma 4.11.

4.12. Corollary. *Every straight line on a surface is an asymptotic curve, and every curve on a plane is an asymptotic curve.*

4.13. Definition. At a point on a surface two directions $du:dv$ and $\delta u:\delta v$ are *conjugate* to each other if they satisfy

$$L\,du\,\delta u+M(du\,\delta v+dv\,\delta u)+N\,dv\,\delta v=0. \tag{4.27}$$

Using (4.24), (4.9), and the elementary relations between the roots and coefficients of a quadratic equation, we can easily obtain Lemma 4.14.

4.14. Lemma. *An asymptotic direction is self-conjugate, and at a non-parabolic point the principal directions are conjugate.*

4.15. Definition. A *net* of curves on a surface S is a pair of one-parameter families of curves on S such that through each point $\mathbf{p}$ of S there passes just one curve of each family, the two tangents of the curves at $\mathbf{p}$ being distinct.

4.16. Definition. A *conjugate net* of curves on a surface S is a net of curves such that the two tangents of the curves at each point are conjugate.

Exercises

* **1.** Find the principal curvatures of a surface of revolution.

2. Show that there are no umbilical points of a surface with negative Gaussian curvature K, and that umbilical points are planar if $K \leqslant 0$.

3. Show that the sum of the normal curvatures of a surface S in two orthogonal directions at a point $\mathbf{p}$ is always equal to $2H$, where H is the mean curvature of S at $\mathbf{p}$.

4. Show that

$$H(\mathbf{p})=\frac{1}{2\pi}\int_0^{2\pi}\kappa_N(\phi)\,d\phi,$$

where $\kappa_N(\phi)$ is the normal curvature, as in (3.11).

5. Give an analytic proof as well as a geometric proof that the sphere of radius a has Gaussian curvature $K=1/a^2$ and mean curvature $H=1/a$. For the analytic proof use the parametrization (1.8) of the sphere.

6. Let T be a torus with parametrization (1.18). Show that the Gaussian curvature of T is given by

$$K=\frac{\cos u}{r(R+r\cos u)}.$$

Thus K has its maximum value $1/r(R+r)$ on the outer equator $(u=0)$ and its minimum value $-1/r(R-r)$ on the inner equator $(u=\pi)$, and $K=0$ on the top and bottom circles $(u=\pm\pi/2)$.

7. Use part a of Exercise 3, Section 2 and part b of Exercise 2, Section 3, to show that the mean curvature H of a surface S vanishes at a nonplanar point $\mathbf{p}$ if and only if there exist two orthogonal asymptotic directions at $\mathbf{p}$.

8. A *tractrix* is a plane curve with the following property: the segment of its tangent between the point of contact $\mathbf{p}$ and some fixed straight line in the plane (the *asymptote* of the tractrix) is of constant length a. If the fixed line is taken to the $\mathbf{x}_1$-axis, the tractrix (Fig. 3.22) is given by $(x_1(t), x_2(t))$, where

$$x_1 = a\cos t + a\ln\tan\frac{t}{2}, \qquad x_2 = a\sin t, \qquad 0 \leqslant t \leqslant \pi. \tag{4.28}$$

A *pseudosphere* (Fig. 3.23) is the surface of revolution obtained by revolving a tractrix about its asymptote. The pseudosphere with respect to the tractrix (4.28) is given by $\mathbf{x}(u,v) = (x_1, x_2, x_3)$ where

$$x_1 = a\sin u\cos v,\ x_2 = a\sin u\sin v,\ x_3 = a\left(\cos u + \ln\tan\frac{u}{2}\right). \tag{4.29}$$

Show that the Gaussian curvature of the pseudosphere (4.29) is $-1/a^2$.

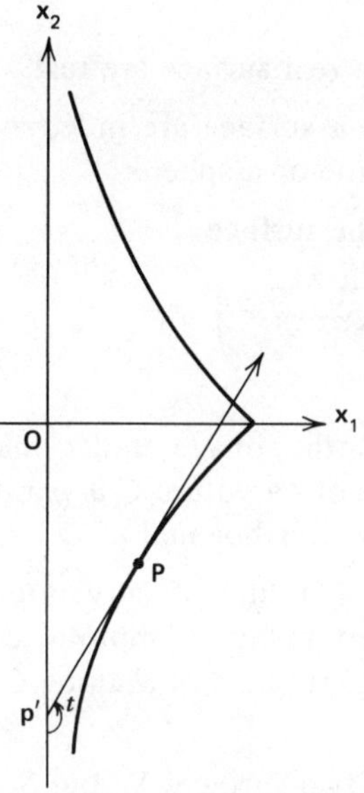

Figure 3.22

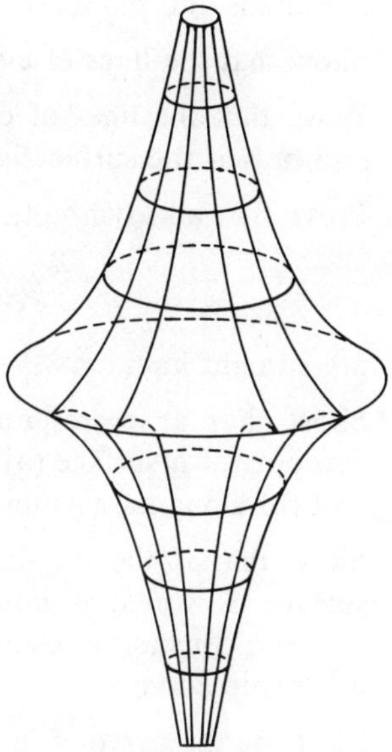

Figure 3.23

9. Let a surface S be the graph of a differentiable function $x_3 = f(x_1, x_2)$, and write

$$f_{x_1} = p, \qquad f_{x_2} = q, \qquad r = f_{x_1x_1}, \qquad s = f_{x_1x_2}, \qquad t = f_{x_2x_2},$$

where the subscript letters denote partial derivatives.

(a) Prove that the principal curvatures of S are the roots of the equation

$$(1+p^2+q^2)^2\lambda^2 - \left[(1+q^2)r - 2pqs + (1+p^2)t\right]\sqrt{1+p^2+q^2}\,\lambda + (rt - s^2) = 0.$$

(b) Prove that the Gaussian and mean curvatures of S are

$$K = \frac{rt - s^2}{(1+p^2+q^2)^2}, \qquad H = \frac{(1+q^2)r - 2pqs + (1+p^2)t}{2(1+p^2+q^2)^{3/2}}.$$

10. Use Exercise 9 to find the Gaussian curvature K and the mean curvature H of the surface

$$2x_3 = 6x_1^2 - 5x_1x_2 - 6x_2^2$$

at the origin $(0,0,0)$.

11. Prove Lemma 4.10.

12. Show that the lines of curvature of a surface of revolution are the meridians and parallels.

13. Show that the lines of curvature on a real surface are real.

14. Prove that the lines of curvature on a surface are indeterminate if and only if the surface is either a plane or a sphere.

15. Prove that all asymptotic curves of the surface

$$x(u,v) = \left(u, v, \frac{u^2}{4} - v^2\right)$$

are straight lines.

16. Show that at corresponding points the image under the Gauss mapping of a surface (a) along a line of curvature C is parallel to C and (b) along an asymptotic curve D is orthogonal to D.

17. Show that if the osculating plane of a line of curvature C of a surface S, which is nowhere tangent to an asymptotic direction, makes a constant angle with the tangent plane of S along C, then C is a plane curve.

***18.** Let C be the curve of intersection of two surfaces S_1 and S_2, and let θ be the angle between S_1 and S_2 at any point $\mathbf{p}$ on C. Suppose that

C is a line of curvature of S_1. Show that θ is constant if and only if C is also a line of curvature of S_2 (F. Joachimstahl).

19. Show that if a line of curvature C of a surface S is in a plane π or on a sphere Σ, along C the surface S makes a constant angle with π or Σ.

***20.** Show that if all lines of curvature of a surface are planar, the planes of the lines of curvature of one family are orthogonal to those of the other family, and the planes of each family are parallel to each other.

21. Can a line of curvature ever be an asymptotic curve?

22. Prove that the torsion of an asymptotic curve of a surface S at a point $\mathbf{p}$ is equal to $\sqrt{-K}$, where K is the Gaussian curvature of S at $\mathbf{p}$ (E. Beltrami–A. Enneper).

23. Prove that every net of curves in a plane is a conjugate net.

24. Prove Lemma 4.14.

25. Prove that a necessary and sufficient condition that a net given by the differential equation

$$a\,du^2 + 2b\,du\,dv + c\,dv^2 = 0$$

be a conjugate net is

$$aN - 2bM + cL = 0.$$

***26.** Show that at a nonplanar umbilical point of a surface, any two orthogonal directions are conjugate.

27. Show that the net of the lines of curvature on a surface is the only orthogonal conjugate net on the surface.

5. MAPPINGS OF SURFACES

We discussed mappings and isometries for Euclidean spaces in Sections 4 and 5, respectively, of Chapter 1. Here we study isometries (or isometric mappings) and conformal mappings of surfaces.

5.1. Definition. An *isometry* $f: S \to \bar{S}$ of two surfaces $S, \bar{S}$ in E^3 is a bijective differentiable mapping that preserves the first fundamental form; when this is the case, the two surfaces $S, \bar{S}$ are said to be *isometric*.

Lemma 5.2 follows immediately from Definition 2.11.

5.2. Lemma. *An isometry of $f: S \to \bar{S}$ of two surfaces $S, \bar{S}$ in E^3 preserves the length of every curve on S.*

5.3. Definition. A mapping f: $V \to \bar{S}$ of a neighborhood V of a point $\mathbf{p}$ on a surface S into a surface $\bar{S}$ is a *local isometry* at $\mathbf{p}$ if there exists a neighborhood $\bar{V}$ of $f(\mathbf{p}) \in \bar{S}$ such that f: $V \to \bar{V}$ is an isometry. If there exists a local isometry of S into $\bar{S}$ at every point $\mathbf{p} \in S$, the surface S is said to be *locally isometric* to $\bar{S}$. Surfaces S and $\bar{S}$ are *locally isometric* if S is locally isometric to $\bar{S}$, and $\bar{S}$ is locally isometric to S.

It is obvious that if $f: S \to \bar{S}$ is a diffeomorphism and a local isometry at every point $\mathbf{p} \in S$, then f is an isometry (globally). However two locally isometric surfaces may not be (globally) isometric, as shown in the following example.

5.4. Example. Let f be a mapping of the coordinate neighborhood of the cylinder S in part 2 of Examples 2.16 into the plane $P = \mathbf{x}(E^2)$ in part 1 of Examples 2.16 defined by

$$f = \mathbf{x} \cdot \bar{\mathbf{x}}^{-1}: S \to P,$$

(the parametrization $\mathbf{x}$ of S in part 2 of Examples 2.16 is written as $\bar{\mathbf{x}}$ here). From Examples 2.16 it follows that S and P have the same first fundamental form given by (2.31) at each pair of corresponding points under f, therefore that f is a local isometry, which cannot be extended to the entire cylinder S because S is not even homeomorphic to a plane. The latter can be seen intuitively by noting that any simple closed curve in a plane can be shrunk continuously into a point without leaving the plane, but a parallel (i.e., a curve with $v = \text{const}$) of the cylinder does not have this property.

5.5. Example. A one-sheeted cone (minus the vertex and a generator) is locally isometric to a plane. This can be seen geometrically by cutting the cone along the generator and unrolling it onto a piece of a plane. In this case the "unrolling" is a bending of the surface without stretching or shrinking; therefore it is an isometry. This is also clear, perhaps even clearer, from the inverse process of the unrolling: a piece of paper in the form of a circular sector can be bent around a cone; the radius of the sector becomes the length of the generator (Fig. 3.24).

5.6. Definition. A *conformal* mapping f: $S \to \bar{S}$ of two surfaces $S, \bar{S}$ in E^3 is a bijective differentiable mapping that preserves the angle between any two intersecting curves on the surface S. A mapping f: $V \to \bar{S}$ of a

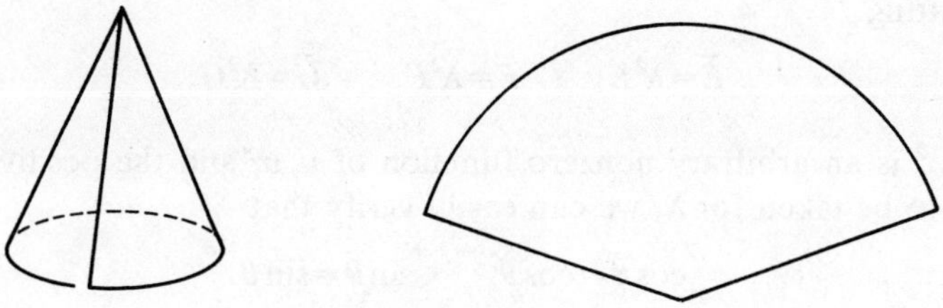

Figure 3.24

neighborhood V of a point $\mathbf{p}$ on a surface S into another surface $\bar{S}$ is a *local conformal* mapping at $\mathbf{p}$ if there exists a neighborhood $\bar{V}$ of $f(\mathbf{p}) \in \bar{S}$ such that $f: V \to \bar{V}$ is a conformal mapping. If there exists a local conformal mapping at each $\mathbf{p} \in S$, the surface S is *locally conformal* to the surface $\bar{S}$.

5.7. Lemma. *A mapping $f: S \to \bar{S}$ of two surfaces $S, \bar{S}$ is locally conformal at a point $\mathbf{p} \in S$ if and only if the first fundamental forms of $S, \bar{S}$ at $\mathbf{p}$, $f(\mathbf{p})$, respectively, are proportional.*

Proof. Let $\mathbf{x}(u,v)$ be a parametrization of the surface S, and $f\mathbf{x}(u,v) = \bar{\mathbf{x}}(u,v)$ be that of $\bar{S}$. Let C_1, C_2 be two curves on the surface S intersecting at a point $\mathbf{p} = \mathbf{x}(u,v)$ given by the coordinate functions, respectively,

$$u = u(s),\ v = v(s); \qquad u = u_1(s_1),\ v = v_1(s_1), \tag{5.1}$$

where s, s_1 are the arc lengths of C_1, C_2 respectively. Then the unit tangent vectors of C_1, C_2 at $\mathbf{p}$ are, respectively,

$$\mathbf{t} = \mathbf{x}_u \frac{du}{ds} + \mathbf{x}_v \frac{dv}{ds}, \qquad \mathbf{t}_1 = \mathbf{x}_u \frac{du_1}{ds_1} + \mathbf{x}_v \frac{dv_1}{ds_1}. \tag{5.2}$$

From (2.28) the angle between $\mathbf{t}, \mathbf{t}_1$ is therefore given by

$$\cos\theta = \frac{1}{ds\, ds_1}\left[E\, du\, du_1 + F(du\, dv_1 + du_1\, dv) + G\, dv\, dv_1 \right], \tag{5.3}$$

provided that the sign of $\sin\theta$ is properly chosen.

To prove the "only if" part, let $\bar{\theta}$ be the angle between the curves corresponding to $C, \bar{C}$ under f at the corresponding point $f(\mathbf{p})$ on the surface $\bar{S}$. Then by replacing E, F, G in (5.3), respectively, by $\bar{E}, \bar{F}, \bar{G}$, the coefficients of the first fundamental form on $\bar{S}$, using

$$\sin\theta = -\frac{\sqrt{EG - F^2}}{ds\, ds_1}(du\, dv_1 - du_1\, dv),$$

and putting

$$\bar{E}=\lambda^2 E, \qquad \bar{F}=\lambda^2 F, \qquad \bar{G}=\lambda^2 G, \tag{5.4}$$

where λ^2 is an arbitrary nonzero function of u, v, and the positive square root is to be taken for λ, we can easily verify that

$$\cos\bar{\theta}=\cos\theta, \qquad \sin\bar{\theta}=\sin\theta. \tag{5.5}$$

Thus $\bar{\theta}=\theta$, and f is locally conformal at $\mathbf{p}$.

To prove the "if" part, let us observe that if the mapping f preserves all angles, it surely preserves right angles. Therefore from (5.3), the conditions of orthogonality,

$$\begin{aligned} E\,du\,du_1+F(du\,dv_1+du_1\,dv)+G\,dv\,dv_1&=0,\\ \bar{E}\,du\,du_1+\bar{F}(du\,dv_1+du_1\,dv)+\bar{G}\,dv\,dv_1&=0, \end{aligned} \tag{5.6}$$

must be equivalent for all directions dv/du, dv_1/du_1. Hence the conditions of the form (5.4) must be satisfied. Q.E.D.

It is easily seen that local conformality is an equivalence relation, so that if a surface S_1 is locally conformal to a surface S_2, and S_2 is locally conformal to a surface S_3, then S_1 is locally conformal to S_3.

The most important property of conformal mappings is given by the following theorem, which we shall not prove here.

5.8. Theorem. *Any two surfaces are locally conformal.*

The proof is based on the possibility of parametrizing a neighborhood of any point of a surface such that the coefficients of the first fundamental form are

$$E=\lambda^2(u,v)>0, \qquad F=0, \qquad G=\lambda^2(u,v). \tag{5.7}$$

This possibility was proved by A. Korn, L. Lichtenstein, and S. S. Chern [12].

Such a coordinate system and such parameters u, v are said to be *isothermal.* With an isothermal coordinate system, a surface is clearly locally conformal to a plane, therefore to any other surface by composition of mappings.

Exercises

1. Use the stereographic projection (cf. Exercise 11 of Section 1) to show that the sphere is locally conformal to a plane.

***2.** Consider the sphere S^2 given by (1.8). Show that the mapping f: $S^2 \to E^2$ defined by

$$f(a\sin\theta\cos\phi, a\sin\theta\sin\phi, a\cos\theta)=(x_1, x_2),$$

where $x_1=\phi$, $x_2=\ln\cot(\theta/2)$, is conformal. Such a mapping f is called the *Mercator projection*. The following additional properties of the Mercator projection are immediately evident. The meridians of longitude, $\phi=$const, become straight lines parallel to the $\mathbf{x}_1$-axis. The loxodromes

$$\phi\cot a=\ln\tan\frac{\theta}{2}+c \quad (c=\text{const}),$$

which are the curves on S^2 crossing the meridians of S^2 at a constant angle a, become the straight lines represented by

$$x_2+x_1\cot a-c=0.$$

These properties explain why the Mercator projection is so extensively used in making maps of the earth's surface, particularly for use by mariners.

3. Let S^2 be the sphere of radius a with center at the origin of E^3 [cf. (1.8)]. Describe the effect of the following mappings $f: S^2 \to S^2$ on the meridians and parallels of S^2:

(a) $f(\mathbf{x})=-\mathbf{x}$; (b) $f(x_1,x_2,x_3)=(x_3,x_1,x_2)$;

(c) $f(x_1,x_2,x_3)=\left(\dfrac{x_1+x_2}{\sqrt{2}}, \dfrac{x_1-x_2}{\sqrt{2}}, -x_3\right)$.

4. Surfaces of each of the following pairs are congruent under a certain mapping; find the mapping in each case:

(a) $x_1^2+(x_2-2)^2+(x_3+3)^2=4$; $\quad (x_1-1)^2+x_2^2+x_3^2=4$.

(b) $x_3=x_1x_2$; $\quad x_3=\frac{1}{2}(x_1^2-x_2^2)$.

(c) $x_1^2+x_2^2=1$; $\quad x_1=0,\ 0\leqslant x_2\leqslant 2\pi$.

5. Let f: $U\subset E^2\to E^3$ be given by

$$f(u,v)=(u\sin\alpha\cos v, u\sin\alpha\sin v, u\cos\alpha),$$

$$(u,v)\in U=\{(u,v)\in E^2; u>0\}, \quad \alpha=\text{const}\neq 0.$$

(a) Show that f is a local diffeomorphism of U onto a cone C (with the vertex at the origin and 2α as the angle of the vertex) minus the vertex.

(b) Is f a local isometry?

6. Let $f: E^2 \to E^2$ be given by $f(x, y) = (u(x, y), v(x, y))$, where u and v are differentiable functions satisfying the Cauchy-Riemann equations

$$u_x = v_y, \qquad u_y = -v_x,$$

the subscript letters denoting the partial derivatives. Show that f is a local conformal mapping from $E^2 - Q$ into E^2, where $Q = \{(x, y) \in E^2 \mid u_x^2 + u_y^2 = 0\}$.

7. A diffeomorphism $f: S \to \bar{S}$ of a surface S onto a surface $\bar{S}$ is said to be *area preserving* if the area of any region $R \subset S$ is equal to the area of $f(R)$. Prove that if f is area preserving and conformal, f is an isometry.

***8.** Let $S^2 = \{(x_1, x_2, x_3) \in E^3 \mid x_1^2 + x_2^2 + x_3^2 = 1\}$ be the unit sphere, and $C = \{(x_1, x_2, x_3) \in E^2 \mid x_1^2 + x_2^2 = 1\}$ be the circumscribed cylinder. Let

$$f: S^2 - \{(0,0,1) \cup (0,0,-1)\} = M \to C$$

be a mapping defined as follows. For each $\mathbf{p} \in M$, let $\mathbf{q}$ be the point of intersection of the $\mathbf{x}_3$-axis and the straight line passing through $\mathbf{p}$ and orthogonal to the $\mathbf{x}_3$-axis, and let l be the half-line starting from $\mathbf{q}$ and containing $\mathbf{p}$ (Fig. 3.25). Then we define $f(\mathbf{p}) = C \cap l$. Show that f is an area-preserving diffeomorphism.

9. Determine the function $g(v)$ so that the conoid with parametrization $\mathbf{x}(u, v) = (u \cos v, u \sin v, g(v))$ shall be isometric to some surface of revolution. [A conoid is a ruled surface (see Section 8) generated by a straight line that moves along a given curve, and is orthogonal to and intersects a fixed straight line l (Fig. 3.26).]

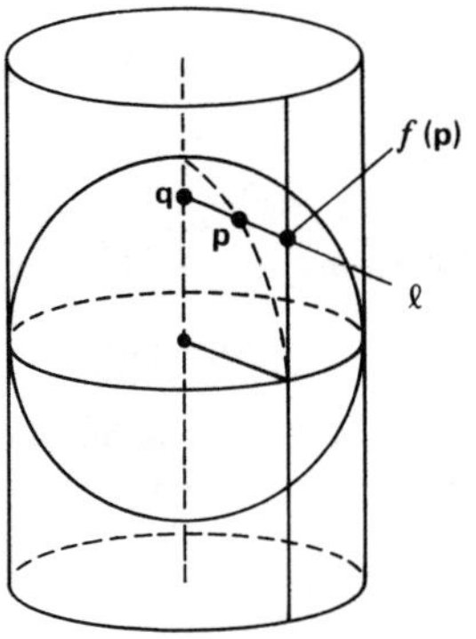

Figure 3.25

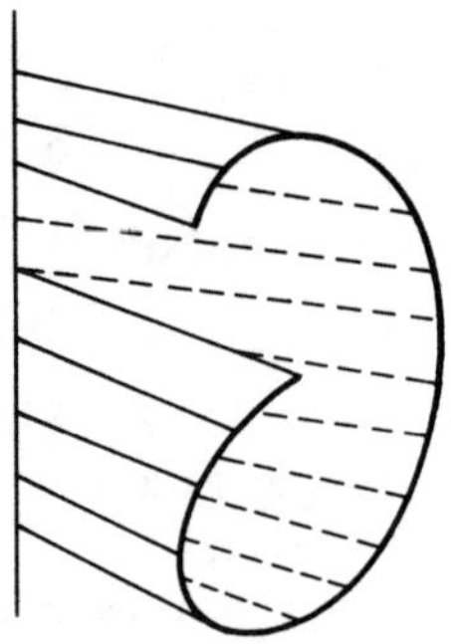

Figure 3.26

10. Prove that the surface of revolution

$$\mathbf{x}(u,v)=(u\cos v, u\sin v, \ln u)$$

and the screw surface

$$\mathbf{y}(u,v)=(u\cos v, u\sin v, v)$$

have the same Gaussian curvature K at each pair of corresponding points (u,v), the formula for K being

$$K=-\frac{1}{(1+u^2)^2};$$

but the two surfaces are not isometric.

11. Show that if a local isometry of two surfaces preserves the mean curvature, it also preserves the principal curvatures.

***12.** An *inversion* in a space E^3 is a mapping f: $E^3-\{(0,0,0)\}\to E^3$ defined by

$$f(\mathbf{x})=\mathbf{x}^*=\frac{\mathbf{x}}{\mathbf{x}\cdot\mathbf{x}}.$$

Show that an inversion (a) takes spheres into spheres, (b) is a conformal mapping of two surfaces, (c) preserves the lines of curvature of a surface, and (d) preserves the asymptotic curves of a surface.

6. TRIPLY ORTHOGONAL SYSTEMS, AND THE THEOREMS OF DUPIN AND LIOUVILLE

A parametrization $\mathbf{x}$ of a surface is, as we know, a function of two variables u, v. Now we generalize this notion of a surface by taking the parametrization $\mathbf{x}$ as a function of three variables u, v, w, and suppose that it possesses continuous second partial derivatives in u, v, w with a Jacobian $|\mathbf{x}_u, \mathbf{x}_v, \mathbf{x}_w| \neq 0$ in the domain D of the (u,v,w)-space under consideration. Then the variables u, v, w can be considered as *local coordinates* of a point in the space, and each of the equations

$$u=\text{const}, \qquad v=\text{const}, \qquad w=\text{const}$$

defines a surface.

6.1. Definition. These three families of surfaces, of which there is a member from each family through each point $\mathbf{p}\in D$, form a *triple system*, and a triple system is *orthogonal* if through $\mathbf{p}$ the three surfaces are mutually orthogonal.

Therefore the conditions for a triple system to be orthogonal are

$$\mathbf{x}_v \cdot \mathbf{x}_w = 0, \qquad \mathbf{x}_w \cdot \mathbf{x}_u = 0, \qquad \mathbf{x}_u \cdot \mathbf{x}_v = 0. \tag{6.1}$$

6.2. Example. Consider the coordinate surfaces of a spherical coordinate system defined by the equations

$$x_1 = r\cos u \cos v, \qquad x_2 = r\cos u \sin v, \qquad x_3 = r\sin u. \tag{6.2}$$

The three families of surfaces are (i) the spheres: $r=$constant, (ii) the cones of revolution: $u=$constant, (iii) the planes: $v=$constant.

6.3. Theorem (C. Dupin). *The surfaces of a triply orthogonal system intersect each other in lines of curvature.*

Proof. Differentiating (6.1) partially with respect to u, v, w, respectively, we obtain

$$\mathbf{x}_{uv} \cdot \mathbf{x}_w + \mathbf{x}_{wu} \cdot \mathbf{x}_v = 0,$$
$$\mathbf{x}_{vw} \cdot \mathbf{x}_u + \mathbf{x}_{uv} \cdot \mathbf{x}_w = 0,$$
$$\mathbf{x}_{wu} \cdot \mathbf{x}_v + \mathbf{x}_{vw} \cdot \mathbf{x}_u = 0,$$

or

$$\mathbf{x}_{vw} \cdot \mathbf{x}_u - \mathbf{x}_{wu} \cdot \mathbf{x}_v = 0, \qquad \mathbf{x}_{wu} \cdot \mathbf{x}_v + \mathbf{x}_{vw} \cdot \mathbf{x}_u = 0,$$

which give $\mathbf{x}_{vw} \cdot \mathbf{x}_u = 0$, $\mathbf{x}_{wu} \cdot \mathbf{x}_v = 0$, and therefore $\mathbf{x}_{uv} \cdot \mathbf{x}_w = 0$. Thus

$$\mathbf{x}_{vw} \cdot \mathbf{x}_u = 0, \qquad \mathbf{x}_{wu} \cdot \mathbf{x}_v = 0, \qquad \mathbf{x}_{uv} \cdot \mathbf{x}_w = 0.$$

From the last equation above and (6.1) it follows that $\mathbf{x}_u, \mathbf{x}_v, \mathbf{x}_{uv}$ all are orthogonal to $\mathbf{x}_w$, therefore lie on a plane. Thus $|\mathbf{x}_u, \mathbf{x}_v, \mathbf{x}_{uv}| = 0$. For a surface $w=$constant, this means that the coefficient M of the second fundamental form of the surface vanishes if u, v are taken as the parameters. Since the u- and v-curves are also orthogonal, the coefficient F of the first fundamental form of the surface also vanishes. Hence by Lemma 4.8, the parametric u- and v-curves are the lines of curvature of the surface $w=$constant. Since the same argument can be applied to the surfaces $u=$constant and $v=$constant, the theorem is proved.

6.4. Example. A nontrival example of a triply orthogonal system is given by the confocal quadrics. Limited to the central quadrics this system is defined by the equation

$$\frac{x_1^2}{t-a_1} + \frac{x_2^2}{t-a_2} + \frac{x_3^2}{t-a_3} = 1, \tag{6.3}$$

where t is a parameter, and $a_1 > a_2 > a_3 > 0$. If (x_1, x_2, x_3) are the coordinates of a given point distinct from the origin, then (6.3) is a cubic equation in t, namely,

$$f(t) \equiv (t-a_2)(t-a_3)x_1^2 + (t-a_1)(t-a_3)x_3^2$$
$$+(t-a_1)(t-a_2)x_3^2 - (t-a_1)(t-a_2)(t-a_3) = 0.$$

Since

$$f(a_1) = +, f(a_2) = -, f(a_3) = +, \lim_{t\to\infty} \frac{f(t)}{(t-a_1)(t-a_2)(t-a_3)} = -,$$

by the fundamental property of continuous functions (if $g(x)$ is a real continuous function for $x \in [a, b]$ and if $f(a)$ and $f(b)$ have opposite signs, the equation $g(x) = 0$ has at least one real root in $[a, b]$), (6.3) has three real roots t_1, t_2, t_3 satisfying

$$t_1 > a_1 > t_2 > a_2 > t_3 > a_3. \tag{6.4}$$

Therefore through the given point there pass three quadrics of the family: an ellipsoid corresponding to t_1, a hyperboloid of one sheet corresponding to t_2, and a hyperboloid of two sheets corresponding to t_3.

The tangent plane of the quadric (6.3) at a point $\mathbf{p}(\bar{x}_1, \bar{x}_2, \bar{x}_3)$ is given by the equation

$$\frac{\bar{x}_1 x_1}{t-a_1} + \frac{\bar{x}_2 x_2}{t-a_2} + \frac{\bar{x}_3 x_3}{t-a_3} = 1. \tag{6.5}$$

Let θ be the angle between the two tangent planes at $\mathbf{p}$ of the two quadrics (6.3) corresponding to $t = t_i, t_j$ where $i, j = 1,2,3$, and $i \neq j$. From (6.5) for $t = t_i, t_j$ it follows that $\cos\theta$ is proportional to

$$\sum_{k=1}^{3} \frac{\bar{x}_k^2}{(t_i - a_k)(t_j - a_k)}. \tag{6.6}$$

On the other hand, $(\bar{x}_1, \bar{x}_2, \bar{x}_3)$ satisfy (6.3) for $t = t_i, t_j$. Subtracting one of these two equations from the other shows immediately that (6.6) is zero. Thus the three quadrics (6.3) corresponding to $t = t_1, t_2, t_3$ are mutually orthogonal at each point of intersection, hence form a triply orthogonal system. Furthermore, (t_1, t_2, t_3) can be taken as local coordinates of a point in a domain not containing the origin and are called *elliptic coordinates*. Since every central quadric can be "embedded" in a confocal family, the following property of its lines of curvature is deduced immediately from Theorem 6.3.

6.5. Lemma. *In general, the lines of curvature of a central quadric are quartic curves (space curves of the fourth degree), which are the curves of intersection of the quadric with its confocal quadrics.*

Related to our present considerations is a theorem due to J. Liouville, which is concerned with the conformal mappings of a three-space E^3 into itself. In the plane the class of conformal mappings is defined by arbitrary analytic functions and is rather extensive. The following theorem shows that for E^3 the class is very restricted.

6.6. Theorem (J. Liouville). *Every conformal mapping in E^3 carries spheres or planes into spheres or planes.*

Proof. Let $\mathbf{x}=\mathbf{x}(u,v)$ define a sphere S. Then from (4.19) we have

$$\mathbf{e}_{3u}+\rho\mathbf{x}_u=0, \qquad \mathbf{e}_{3v}+\rho\mathbf{x}_v=0,$$

where $\mathbf{e}_3$ is the unit normal vector of S, and ρ is a constant. Suppose that the parametric u- and v-curves are orthogonal, and that

$$\bar{\mathbf{x}}(u,v,w)=\mathbf{x}(u,v)+w\mathbf{e}_3(u,v). \tag{6.7}$$

Then

$$\begin{aligned}
\bar{\mathbf{x}}_u&=\mathbf{x}_u+w\mathbf{e}_{3u}=(1-w\rho)\mathbf{x}_u,\\
\bar{\mathbf{x}}_v&=\mathbf{x}_v+w\mathbf{e}_{3v}=(1-w\rho)\mathbf{x}_v,\\
\bar{\mathbf{x}}_w&=\mathbf{e}_3.
\end{aligned}$$

In a domain of the (u,v,w)-space such that $w\neq 1/\rho$, $\bar{\mathbf{x}}(u,v,w)$ defines a triply orthogonal system that contains the given sphere S as a member corresponding to $w=0$. Geometrically the three families of surfaces are two families of cones described by the normals to the given sphere S, respectively, along the two families of the parametric curves with the common vertex at the center of S, and a family of concentric spheres containing S.

Under a conformal mapping a triply orthogonal system goes to a triply orthogonal system, and, by Theorem 6.3, S goes to a surface S' on which the parametric curves are the lines of curvature. Since the parametric curves are arbitrary orthogonal curves on S, it follows that any orthogonal curves on S' are the lines of curvature. But this is possible only when every point on S' is an umbilical point, that is, by Theorem 4.5 only when S' is a sphere. Thus Theorem 6.6 is proved.

Exercises

1. Show that each of the following two systems of surfaces is triply orthogonal:

(a) $x_1^2+x_2^2+x_3^2=u, \qquad x_2=vx_1, \qquad x_1^2+x_2^2=wx_3,$

(b) $x_1^2+x_2^2+x_3^2=ux_1, \quad x_1^2+x_2^2+x_3^2=vx_2, \quad x_1^2+x_2^2+x_3^2=wx_3.$

7. FUNDAMENTAL EQUATIONS

In the theory of curves, the formulas of fundamental importance are the Frenet formulas [(1.3.11) of Chapter 2], which allow us to express the derivatives of the vectors of a Frenet frame field as linear combinations of these vectors themselves. The corresponding problem in the theory of surfaces is to express $\mathbf{x}_{uu}, \mathbf{x}_{uv}, \mathbf{x}_{vv}, \mathbf{e}_{3u}, \mathbf{e}_{3v}$ as linear combinations of $\mathbf{x}_u, \mathbf{x}_v, \mathbf{e}_3$, where $\mathbf{x}$: $U \subset E^2 \to S$ is a parametrization of a surface S in E^3, $(u, v) \in U$, and $\mathbf{e}_3$ is the unit normal vector of S. Unlike the theory of curves, where the arc length is a natural parameter, it is in general not possible to choose invariant parameters on a surface. Therefore the coefficients of those linear combinations for a surface are not invariants, and their properties are more complicated. Nevertheless, they still play an important role in describing the geometrical properties of surfaces.

We first consider the partial derivatives of $\mathbf{e}_3$. Put

$$\mathbf{e}_{3u}=a\mathbf{x}_u+b\mathbf{x}_v, \quad \mathbf{e}_{3v}=c\mathbf{x}_u+d\mathbf{x}_v.$$

Taking the inner products of each of these two equations with $\mathbf{x}_u, \mathbf{x}_v$, we get

$$\begin{aligned} -L&=aE+bF, & -M&=cE+dF,\\ -M&=aF+bG, & -N&=cF+dG. \end{aligned}$$

Solve these equations simultaneously for the coefficients a, b, c, d; the result is

$$\begin{aligned} \mathbf{e}_{3u}&=\frac{(FM-GL)\mathbf{x}_u+(FL-EM)\mathbf{x}_v}{EG-F^2},\\ \mathbf{e}_{3v}&=\frac{(FN-GM)\mathbf{x}_u+(FM-EN)\mathbf{x}_v}{EG-F^2}, \end{aligned} \tag{7.1}$$

which are called the *Weingarten formulas*.

Several consequences can be derived from (7.1). At first, $\mathbf{e}_{3u}=k\mathbf{e}_{3v}$ if and only if $LN-M^2=0$. Second, there is a linear relation among the three fundamental forms

$$K\mathrm{I}-2H\mathrm{II}+\mathrm{III}=0, \tag{7.2}$$

which can be obtained by direct computation and using (7.1). However, the computation can be simplified by choosing the parameters u and v such that $F=M=0$, since (7.2) is independent of such choice.

It also follows from (7.1) that the plane is the only surface consisting entirely of planar points (Lemma 3.2). For the conditions $L=M=N=0$ imply that $\mathbf{e}_3$ is a constant vector, so that the surface is a plane (see the proof of Theorem 4.5).

As another application of (7.1) we shall study the surfaces on which the Gaussian curvature $K=LN-M^2$ vanishes identically. We suppose that the neighborhood of our surface under consideration contains no planar point. Then the differential equation (4.24) of the asymptotic curves can be written as $(\sqrt{L}\,du+\sqrt{N}\,dv)^2=0$, the left-hand side of which is not identically zero. By choosing these coincident asymptotic curves as the u-curves, we have $L=M=0$. From (7.1) it follows that $\mathbf{e}_{3u}=0$, so that $\mathbf{e}_3$ is a function of v alone. Integrating the equation $(\mathbf{x}\cdot\mathbf{e}_3)_u=\mathbf{x}_u\cdot\mathbf{e}_3=0$, we get

$$\mathbf{e}_3(v)\cdot\mathbf{x}=p(v), \tag{7.3}$$

where $p(v)$ is a function of v. Differentiation of (7.3) with respect to v gives

$$\mathbf{e}_{3v}\cdot\mathbf{x}=p'(v). \tag{7.4}$$

The vector $\mathbf{e}_{3v}$ is orthogonal to $\mathbf{e}_3$, but is not zero. This implies that $\mathbf{e}_3$ and $\mathbf{e}_{3v}$ are linearly independent, so that along each u-curve, (7.3) and (7.4) are the equations of the two distinct planes, that is, each u-curve is the line of intersection of the two planes (7.3) and (7.4). Therefore the asymptotic curves are straight lines. Since along each u-curve, $\mathbf{e}_3$ is constant and orthogonal to the tangent plane of our surface, our surface thus has the property that *it is generated by a family of straight lines such that it has the same tangent plane along each of these generators*. Such a surface is called a *developable surface* or simply a *developable*.

To find the formulas for $\mathbf{x}_{uu},\mathbf{x}_{uv},\mathbf{x}_{vv}$, we first notice that

$$L=\mathbf{x}_{uu}\cdot\mathbf{e}_3,\qquad M=\mathbf{x}_{uv}\cdot\mathbf{e}_3,\qquad N=\mathbf{x}_{vv}\cdot\mathbf{e}_3,$$

given by (2.30). Then we can write

$$\begin{aligned}\mathbf{x}_{uu}&=\Gamma_1{}^1{}_1\mathbf{x}_u+\Gamma_1{}^2{}_1\mathbf{x}_v+L\mathbf{e}_3,\\ \mathbf{x}_{uv}&=\Gamma_1{}^1{}_2\mathbf{x}_u+\Gamma_1{}^2{}_2\mathbf{x}_v+M\mathbf{e}_3,\\ \mathbf{x}_{vv}&=\Gamma_2{}^1{}_2\mathbf{x}_u+\Gamma_2{}^2{}_2\mathbf{x}_v+N\mathbf{e}_3,\end{aligned} \tag{7.5}$$

where the coefficients Γ's are to be determined.

To be able to write our formulas in a convenient form, we shall introduce some new notation. We put

$$u^1 = u, \qquad u^2 = v, \qquad \mathbf{e}_1 = \mathbf{x}_u, \qquad \mathbf{e}_2 = \mathbf{x}_v, \tag{7.6}$$

and make the convention that all Latin indices run from 1 to 2. Since the symmetric matrix formed by the elements

$$g_{ij} = \mathbf{e}_i \cdot \mathbf{e}_j \tag{7.7}$$

is nonsingular, we introduce the elements g^{ij} of the inverse of the matrix according to the equations

$$\sum_j g_{ij} g^{jk} = \delta_i^k. \tag{7.8}$$

These quantities g^{ij} and g_{ij} can be used to introduce the processes of raising or lowering indices. Associated with a quantity with a number of indices, we can define quantities with indices raised or lowered. For instance, from A_{ijk} we define

$$A_{i\cdot k}^{\ j} = \sum_l g^{jl} A_{ilk}, \qquad A^{ij}_{\cdot\cdot k} = \sum_{l,m} g^{il} g^{jm} A_{lmk}, \ etc. \tag{7.9}$$

Then we have

$$A_{ijk} = \sum_l g_{jl} A_{i\cdot k}^{\ l} = \sum_{l,m} g_{jl} g_{km} A_{i\cdot\cdot}^{\ lm}, \ etc. \tag{7.10}$$

After these preparations let us proceed to the determination of the coefficients in (7.5). In our notation they are $\Gamma_i{}^j{}_k$, which we can suppose to be symmetric in i, k by introducing $\Gamma_2{}^j{}_1$ with $\Gamma_2{}^j{}_1 = \Gamma_1{}^j{}_2$. We introduce the differential forms

$$\omega_{i\cdot}^{\ j} = \sum_k \Gamma_i{}^j{}_k \, du^k. \tag{7.11}$$

According to the rules above, the indices of $\omega_{i\cdot}^{\ j}$ can be raised or lowered. Thus we have

$$\omega_{ij} = \sum_k g_{jk} \omega_{i\cdot}^{\ k}, \qquad \omega_{i\cdot}^{\ j} = \sum_k g^{jk} \omega_{ik}. \tag{7.12}$$

By means of these differential forms $\omega_{i\cdot}^{\ j}$ we can write (7.5) as

$$d\mathbf{e}_i = \sum_j \omega_{i\cdot}^{\ j} \mathbf{e}_j + \omega_i{}^3 \mathbf{e}_3, \tag{7.13}$$

where

$$\omega_1{}^3 = L\, du^1 + M\, du^2, \qquad \omega_2{}^3 = M\, du^1 + N\, du^2. \tag{7.14}$$

Differentiating (7.7) and using (7.13) and (7.12) we obtain

$$dg_{ij}=\omega_{ij}+\omega_{ji}. \tag{7.15}$$

But

$$dg_{ij}=\sum_k \frac{\partial g_{ij}}{\partial u^k}\,du^k,$$

and from (7.11),

$$\omega_{ij}=\sum_k \Gamma_{ijk}\,du^k.$$

Substituting these two equations in (7.15) we thus obtain

$$\Gamma_{ijk}+\Gamma_{jik}=\frac{\partial g_{ij}}{\partial u^k}. \tag{7.16}$$

Since $\Gamma_i{}^j{}_k$ is symmetric in i, k, the same is true of Γ_{ijk}:

$$\Gamma_{ijk}=\Gamma_{kji}. \tag{7.17}$$

Equations 7.16 and 7.17 are sufficient to determine Γ_{ijk}. In fact, permuting (7.16) cyclically in i, j, k we obtain

$$\Gamma_{jki}+\Gamma_{kji}=\frac{\partial g_{jk}}{\partial u^i}, \qquad \Gamma_{kij}+\Gamma_{ikj}=\frac{\partial g_{ki}}{\partial u^j},$$

which together with (7.16) give

$$\Gamma_{ijk}=\Gamma_{kji}=\frac{1}{2}\left(\frac{\partial g_{ij}}{\partial u^k}+\frac{\partial g_{jk}}{\partial u^i}-\frac{\partial g_{ik}}{\partial u^j}\right). \tag{7.18}$$

From Γ_{ijk} we can determine $\Gamma_i{}^j{}_k$ as follows:

$$\Gamma_i{}^j{}_k=\Gamma_k{}^j{}_i=\sum_l g^{jl}\Gamma_{ilk}. \tag{7.19}$$

7.1. Definition. The functions Γ_{ijk} and $\Gamma_i{}^j{}_k$ first introduced by E. B. Christoffel are the *Christoffel symbols of the first and second kinds*, respectively, and (7.5), (7.18), and (7.19) are the *equations of Gauss*.

It is important to notice that Γ_{ijk} and $\Gamma_i{}^j{}_k$ depend only on the first fundamental form.

Written in terms of the notation of Gauss, the Christoffel symbols of the second kind have the expressions:

$$\Gamma_1{}^1{}_1 = \frac{GE_u - 2FF_u + FE_v}{2(EG-F^2)}, \qquad \Gamma_1{}^2{}_1 = \frac{2EF_u - EE_v - FE_u}{2(EG-F^2)},$$

$$\Gamma_1{}^1{}_2 = \frac{GE_v - FG_u}{2(EG-F^2)}, \qquad \Gamma_1{}^2{}_2 = \frac{EG_u - FE_v}{2(EG-F^2)}, \tag{7.20}$$

$$\Gamma_2{}^1{}_2 = \frac{2GF_v - GG_u - FG_v}{2(EG-F^2)}, \qquad \Gamma_2{}^2{}_2 = \frac{EG_v - 2FF_v + FG_u}{2(EG-F^2)}.$$

From (7.1) and (7.5) it is possible to express higher partial derivatives of $\mathbf{x}$ and $\mathbf{e}_3$ with respect to u, v as linear combinations of $\mathbf{x}_u, \mathbf{x}_v$ and $\mathbf{e}_3$. Some of these derivatives can be calculated in two different ways and are independent of the order of differentiation. Thus we have

$$(\mathbf{x}_{uu})_v = (\mathbf{x}_{uv})_u, \qquad (\mathbf{x}_{uv})_v = (\mathbf{x}_{vv})_u, \qquad (\mathbf{e}_{3u})_v = (\mathbf{e}_{3v})_u.$$

By putting

$$\mathbf{x}_{uuv} - \mathbf{x}_{uvu} = \alpha_1 \mathbf{x}_u + \beta_1 \mathbf{x}_v + \gamma_1 \mathbf{e}_3,$$
$$\mathbf{x}_{uvv} - \mathbf{x}_{vvu} = \alpha_2 \mathbf{x}_u + \beta_2 \mathbf{x}_v + \gamma_2 \mathbf{e}_3,$$
$$\mathbf{e}_{3uv} - \mathbf{e}_{3vu} = \alpha_3 \mathbf{x}_u + \beta_3 \mathbf{x}_v + \gamma_3 \mathbf{e}_3,$$

where the order of differentiation is from the left to the right, all coefficients on the right-hand side must then vanish. By a calculation, which we shall not give here, we find that the conditions $\alpha_1 = \beta_1 = 0$ can be written as

$$F\frac{LN-M^2}{EG-F^2} = (\Gamma_1{}^1{}_2)_u - (\Gamma_1{}^1{}_1)_v + \Gamma_1{}^2{}_2\Gamma_1{}^1{}_2 - \Gamma_1{}^2{}_1\Gamma_2{}^1{}_2, \tag{7.21}$$

$$-E\frac{LN-M^2}{EG-F^2} = (\Gamma_1{}^2{}_2)_u - (\Gamma_1{}^2{}_1)_v + \Gamma_1{}^1{}_2\Gamma_1{}^2{}_1 - \Gamma_1{}^1{}_1\Gamma_1{}^2{}_2 + \Gamma_1{}^2{}_2\Gamma_1{}^2{}_2 - \Gamma_1{}^2{}_1\Gamma_2{}^2{}_2. \tag{7.22}$$

Equation 7.22 is called the *Gauss equation*.

The conditions $\alpha_2 = \beta_2 = 0$ can be obtained from (7.21) and (7.22) by interchanging $1, u, E, L$ and $2, v, G, N$, respectively, and keeping F, M.

From each of the conditions $\alpha_1 = \beta_1 = \alpha_2 = \beta_2 = 0$ we obtain the following important theorem of Gauss.

7.2. Theorem. *The Gaussian curvature of a surface depends only on the first fundamental form of the surface.*

Furthermore, for the Gaussian curvature K we can have the following formula, which is symmetric with respect to E, F, G and u, v:

$$K = -\frac{1}{4W^4}\begin{vmatrix} E & E_u & E_v \\ F & F_u & F_v \\ G & G_u & G_v \end{vmatrix} - \frac{1}{2W}\left(\frac{\partial}{\partial v}\frac{E_v - F_u}{W} - \frac{\partial}{\partial u}\frac{F_v - G_u}{W}\right), \quad (7.23)$$

where $W = \sqrt{EG - F^2}$.

The conditions $\gamma_1 = \gamma_2 = \alpha_3 = \beta_3 = 0$ give two other equations, which can be written as

$$\begin{aligned} \frac{\partial L}{\partial v} - \frac{\partial M}{\partial u} &= L\Gamma_1{}^1{}_2 + M(\Gamma_1{}^2{}_2 - \Gamma_1{}^1{}_1) - N\Gamma_1{}^2{}_1, \\ \frac{\partial M}{\partial v} - \frac{\partial N}{\partial u} &= L\Gamma_2{}^1{}_2 + M(\Gamma_2{}^2{}_2 - \Gamma_1{}^1{}_2) - N\Gamma_1{}^2{}_2, \end{aligned} \quad (7.24)$$

and are known as the *Mainardi-Codazzi* equations. According to E. Study, (7.24) can also be written in the form:

$$\begin{matrix} (EG - 2FF + GE)(L_v - M_u) \\ -(EN - 2FM + GL)(E_v - F_u) \end{matrix} + \begin{vmatrix} E & E_u & L \\ F & F_u & M \\ G & G_u & N \end{vmatrix} = 0,$$

$$\begin{matrix} (EG - 2FF + GE)(M_v - N_u) \\ -(EN - 2FM + GL)(F_v - G_u) \end{matrix} + \begin{vmatrix} E & E_v & L \\ F & F_v & M \\ G & G_v & N \end{vmatrix} = 0.$$

Finally, the condition $\gamma_3 = 0$ is identically satisfied.

A system of linear partial differential equations is said to be *completely integrable* if its integrability conditions are satisfied. The system consisting of (7.5) and (7.1), starting with the parametrization $\mathbf{x}(u, v)$ of a surface S, is such a system whose integrability conditions are (7.22) and (7.24).

The method we have used in developing the local theory of surfaces is not the only method that could be used. Another method of defining a surface gives the following fundamental theorem in the theory.

7.3. Theorem (O. Bonnet). *Let E, F, G, and L, M, N be functions of classes C^2 and C^1, respectively, defined in an open set $V \subset E^2$ with $E > 0$ and $G > 0$. Suppose that $EG - F^2 > 0$ and that the six functions satisfy the Gauss and Mainardi-Codazzi equations* (7.22) *and* (7.24). *Then for every $\mathbf{q} \in V$ there exist a neighborhood $U \subset V$ of $\mathbf{q}$ and a C^3 homeomorphism $\mathbf{x}$: $U \to \mathbf{x}(U) \subset E^3$ such that the regular surface $\mathbf{x}(U) \subset E^3$ has E, F, G and L, M, N as coeffi-*

cients of the first and second fundamental forms, respectively. Furthermore, if U is connected and

$$\mathbf{x}^*: U \to \mathbf{x}^*(U) \subset E^3$$

is another C^3 homeomorphism satisfying the same conditions, there exists an orientation-preserving isometry (see Definition 5.4.6 of Chapter 1) f in E^3 such that $\mathbf{x}^* = f\mathbf{x}$.

Proof. A proof of the first part of the theorem is given in Appendix 2, and a proof of the second part in Section 4 of Chapter 4.

Exercises

1. Prove (7.2).

2. Show that

$$\frac{\partial g_{ij}}{\partial u^k} - \frac{\partial g_{jk}}{\partial u^i} = \Gamma_{jik} - \Gamma_{ikj}.$$

***3.** Show that

$$\sum_{l=1}^{2} g_{hl} \frac{\partial}{\partial u^j} \Gamma_i{}^l{}_k = \frac{\partial}{\partial u^j} \Gamma_{ihk} - \sum_{m=1}^{2} \Gamma_i{}^m{}_k (\Gamma_{mhj} + \Gamma_{hmj}).$$

4. Show that in terms of the *Riemann symbols of the second kind* defined by

$$R^l{}_{ijk} = \frac{\partial}{\partial u^j} \Gamma_i{}^l{}_k - \frac{\partial}{\partial u^k} \Gamma_i{}^l{}_j + \sum_{m=1}^{2} \Gamma_i{}^m{}_k \Gamma_m{}^l{}_j - \sum_{m=1}^{2} \Gamma_i{}^m{}_j \Gamma_m{}^l{}_k,$$

(7.21) and (7.22) can be written as

$$R^i{}_{121} = g^{i2}(LN - M^2), \qquad i = 1,2.$$

5. Show that in terms of the *Riemann symbols of the first kind* defined by

$$R_{hijk} = \sum_{m=1}^{2} g_{hl} R^l{}_{ijk},$$

the Gaussian curvature K can be written as

$$K = \frac{R_{1212}}{EG - F^2}.$$

*6. Show that

$$R_{hijk}=\frac{\partial}{\partial u^j}\Gamma_{ihk}-\frac{\partial}{\partial u^k}\Gamma_{ihj}+\sum_{m=1}^{2}\Gamma_i{}^m{}_j\Gamma_{hmk}-\sum_{m=1}^{2}\Gamma_i{}^m{}_k\Gamma_{hmj}, \quad (7.25)$$

$$R_{hijk}=\frac{1}{2}\left(\frac{\partial^2 g_{hk}}{\partial u^i\partial u^j}+\frac{\partial^2 g_{ij}}{\partial u^h\partial u^k}-\frac{\partial^2 g_{hi}}{\partial u^i\partial u^k}-\frac{\partial^2 g_{ik}}{\partial u^h\partial u^j}\right) \quad (7.26)$$

$$+\sum_{l,m=1}^{2} g^{lm}(\Gamma_{imj}\Gamma_{hlk}-\Gamma_{imk}\Gamma_{hlj}).$$

7. Show that Riemann symbols R_{hijk} satisfy the following identities:

$$R_{hijk}=-R_{hikj},$$
$$R_{hijk}=-R_{ihjk},$$
$$R_{hijk}=R_{jkhi},$$
$$R_{hijk}+R_{hjki}+R_{hkij}=0.$$

8. Show that $R^i{}_{ijk}=0$.

8. RULED SURFACES AND MINIMAL SURFACES

In this section we study surfaces satisfying certain simple analytic or geometric conditions. In the last section we considered surfaces with zero Gaussian curvature under the further restriction that there are no planar points, and found that such a surface is generated by a family of straight lines such that the tangent plane is the same along each of these generators. These surfaces have been called *developable surfaces*, whereas a surface generated by a family of straight lines without the other restriction is called a *ruled surface*. A parametrization of a ruled surface is given by

$$\mathbf{x}(s,t)=\mathbf{y}(s)+t\mathbf{z}(s), \quad (8.1)$$

so that the surface is regarded as formed by the lines with direction vector $\mathbf{z}(s)$ along the curve $\mathbf{y}(s)$. A line with direction vector $\mathbf{z}(s)$ and the curve $\mathbf{y}(s)$ are known as a *ruling* and the *directrix of the ruled surface*, respectively.

The condition that the tangent plane of the surface stay constant along the generators is not a property of an arbitrary ruled surface. For example, on the hyperboloid of one sheet there is twisting of the tangent planes as a generator is traversed.

8.1. Theorem. *The ruled surface* (8.1) *is developable if and only if the determinant*

$$|\mathbf{y}', \mathbf{z}, \mathbf{z}'| = 0, \tag{8.2}$$

where the prime denotes the derivative with respect to s.

Proof. To prove this, we want to find the analytic condition that the ruled surface has the same tangent plane along each one of its generators. Let us choose s to be the arc length of the curve $\mathbf{y}(s)$ so that $\mathbf{y}'^2 = 1$, and let us assume that $\mathbf{z}^2 = 1$. Then from (8.1) we have the partial derivatives of $\mathbf{x}$ with respect to s and t, respectively:

$$\mathbf{x}_s = \mathbf{y}' + t\mathbf{z}', \qquad \mathbf{x}_t = \mathbf{z}.$$

If the tangent plane of the surface is constant along each generator, the tangent vectors $\mathbf{x}_t, \mathbf{x}_s(t_1), \mathbf{x}_s(t_2)$ for all t_1 and t_2, where $\mathbf{x}_s(t_i) = \mathbf{y}' + t_i\mathbf{z}'$, $i = 1, 2$, are coplanar, and therefore $|\mathbf{x}_t, \mathbf{x}_s(t_1), \mathbf{x}_s(t_2)| = 0$ for all t_1 and t_2. This is equivalent to $(t_1 - t_2)|\mathbf{z}, \mathbf{y}', \mathbf{z}'| = 0$. Thus if we choose $t_1 \neq t_2$, we have (8.2). Condition 8.2 implies that there exist functions $\alpha(s)$, $\beta(s)$, and $\gamma(s)$, not all zero, such that

$$\alpha\mathbf{y}' + \beta\mathbf{z} + \gamma\mathbf{z}' = 0, \tag{8.3}$$

from which we can classify two possible cases of developable surfaces.

Case I: $\alpha = 0$. This implies that $\beta\mathbf{z} + \gamma\mathbf{z}' = 0$. If we take the inner product of this equation with $\mathbf{z}$, we have $\beta = -\gamma\mathbf{z}' \cdot \mathbf{z} = 0$. Since $\alpha = 0$ and $\beta = 0$, we must have $\gamma \neq 0$, and condition 8.3 becomes $\mathbf{z}' = 0$, which gives that $\mathbf{z}$ is constant. Thus all the generating lines are parallel to the same direction $\mathbf{z}$, and we obtain either a plane or a cylinder.

Case II: $\alpha \neq 0$. Dividing condition 8.3 by α, we have

$$\mathbf{y}' = \lambda(s)\mathbf{z} + \mu(s)\mathbf{z}'.$$

We want to find the condition that these lines $\mathbf{y}'$ are tangent to some curve on the surface. For this purpose we consider a curve C described by $\mathbf{y}^* = \mathbf{y} + \rho(s)\mathbf{z}$ and intersecting all the generating lines of direction vector $\mathbf{z}$. Since

$$\frac{d\mathbf{y}^*}{ds} = \lambda\mathbf{z} + \mu'\mathbf{z} + \rho'\mathbf{z} + \rho\mathbf{z}',$$

the tangent vector of the curve C at $\mathbf{y}^*$ will be in the direction $\mathbf{z}$ only if $d\mathbf{y}^*/ds$ is a multiple of $\mathbf{z}$, say equal to $\zeta\mathbf{z}$. This implies that $\rho = -\mu$, and the curve C is given by $\mathbf{y}^* = \mathbf{y} - \mu\mathbf{z}$.

We are now led to consider two subcases:

(i) $\lambda - \mu' = 0$. In this case we have $d\mathbf{y}^*/ds = 0$, so that the curve C is the fixed point $\mathbf{y}^*$ through which all the generating lines pass. Thus the surface is a cone.

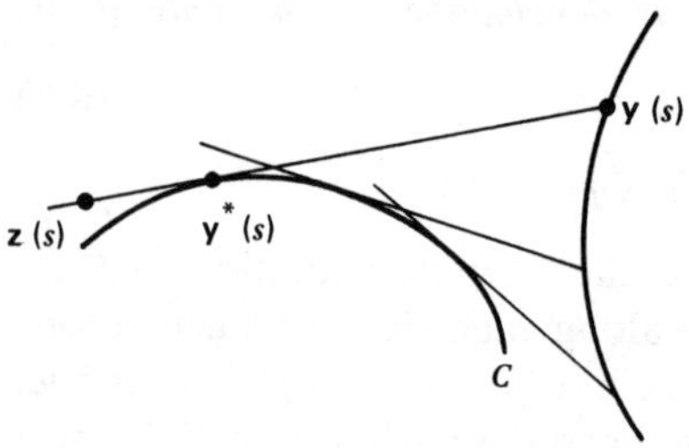

Figure 3.27

(ii) $\lambda-\mu'\neq 0$. In this case we have $d\mathbf{y}^*/ds\neq 0$, and the surface is the tangent surface of the curve C (Fig. 3.27), by the foregoing remarks.

Now we must note that Cases I and II are not exhaustive. If α has only isolated zeros, we can shrink our neighborhoods so that at any time we only consider regions on which $\alpha\neq 0$; but if α has a cluster of zeros, the problem becomes difficult.

Since a surface with zero Gaussian curvature is a developable surface, the proof of Theorem 8.1 is complete by the first part of the following theorem.

8.2. Theorem. *The plane, cylinder, cone, and tangent surfaces of curves all have a metric with zero Gaussian curvature and are locally isometric to the plane.*

Proof. In the case of the cone and cylinder we can cut them along a generator and unroll them onto the plane (cf. Example 5.5), and the inverse of this mapping induces the metric of the plane on the cone and cylinder, and has of course zero Gaussian curvature. The problem is thus reduced to considering the tangent surface of a curve. We first show that the surface generated by the tangent lines to a curve has zero Gaussian curvature.

Let $\mathbf{y}(u)$ denote the position vector of a curve C. Then the surface S generated by the tangent lines of the curve C is given by

$$\mathbf{x}(u,v)=\mathbf{y}(u)+v\mathbf{y}'(u), \tag{8.4}$$

where the prime denotes the derivative with respect to u. It suffices to show that $LN-M^2=0$ where L, M, N are the coefficients of the second fundamental form of the surface S, since up to a nonzero factor $LN-M^2$ is equal to the Gaussian curvature. From (2.30) we must evaluate

$$c^2L=|\mathbf{x}_u,\mathbf{x}_v,\mathbf{x}_{uu}|,\ c^2M=|\mathbf{x}_u,\mathbf{x}_v,\mathbf{x}_{uv}|,\ c^2N=|\mathbf{x}_u,\mathbf{x}_v,\mathbf{x}_{vv}|,$$

where $c\neq 0$. Now $\mathbf{x}_u=\mathbf{y}'+v\mathbf{y}''$, $\mathbf{x}_v=\mathbf{y}'$, $\mathbf{x}_{uv}=\mathbf{y}''$, and $\mathbf{x}_{vv}=0$, which implies that $N=0$. Thus it suffices to shown that $M=0$, which is obviously true, since $c^2M=|\mathbf{y}'+v\mathbf{y}'',\mathbf{y}',\mathbf{y}''|$.

Next, we shall show that the tangent surface of a curve is locally isometric to the plane.

Let us choose the parameter u of the curve C to be the arc length so that $\mathbf{y}'(u)^2=1$. Then $\mathbf{y}''^2=\kappa^2$, $\kappa \geqslant 0$, where κ is the curvature of the curve C. From (8.4) we thus obtain the first fundamental form of the surface S:

$$\mathrm{I}=(1+v^2\kappa^2)\,du^2+2\,du\,dv+dv^2. \tag{8.5}$$

Now in the following we introduce a metric into a neighborhood on the plane with the same first fundamental form as (8.5). A plane curve is completely determined by its unit tangent and normal vectors $\mathbf{t}$ and $\mathbf{n}$ satisfying the equations

$$\frac{d\mathbf{t}}{ds}=\kappa\mathbf{n}, \qquad \frac{d\mathbf{n}}{ds}=-\kappa\mathbf{t} \tag{8.6}$$

with the initial conditions

$$\mathbf{t}(0)=\mathbf{t}_0, \qquad \mathbf{n}(0)=\mathbf{n}_0 \tag{8.7}$$

for given t_0 and n_0. In fact, since for a given κ, (8.6) is a system of ordinary linear homogeneous differential equations, there exists a local solution $\mathbf{t}(s),\mathbf{n}(s)$ satisfying the given initial conditions (8.7), and once $\mathbf{t}(s)=d\mathbf{y}^*(s)/ds$ is determined, $\mathbf{y}^*(s)$ can be found by an integration. Hence there exists a plane curve $\mathbf{y}^*(s)$ that has the same curvature function $\kappa(s)$ as that of the space curve $\mathbf{y}(u)$.

Now we look at the family of the tangent lines of this curve $\mathbf{y}^*(s)$, which can be given by

$$\mathbf{x}^*(s,t)=\mathbf{y}^*(s)+t\mathbf{y}^{*\prime}(s). \tag{8.8}$$

In a small neighborhood these tangent lines cover an open set U of the plane, and (8.8) is a parametrization of the set U. From (8.4), (8.5) and (8.8) it follows immediately that the first fundamental form of U is

$$\mathrm{I}^*=(1+t^2\kappa^2)\,ds^2+2\,ds\,dt+dt^2. \tag{8.9}$$

Hence the surface S and the set U with metrics given by (8.5) and (8.9) are locally isometric.

In the remainder of this section we consider the class of surfaces called *minimal surfaces*, for each of which the mean curvature H vanishes everywhere. The designation "minimal" is motivated by the following theorem related to the classical problem of Plateau, which is to determine a surface of minimal area bounded by a given closed curve in E^3.

8.3. Theorem. *A necessary condition that a surface S with a boundary curve C solve Plateau's problem for C is that S be minimal.*

Proof. We utilize some rudimentary techniques of the calculus of variation. Let $\mathbf{x}: U \subset E^2 \to E^3$ be a parametrization of the surface S, and let $(u, v) \in U$. Then a small *normal variation* of the surface S with respect to a differentiable function λ on S, which vanishes on C, is a mapping (Fig. 3.28) $\mathbf{x}^*: U \times (-\varepsilon, \varepsilon) \to E^3$ given by

$$\mathbf{x}^*(u, v, t) = \mathbf{x}(u, v) + t\lambda(u, v)\mathbf{e}_3(u, v), \tag{8.10}$$

where ε is small, $t \in (-\varepsilon, \varepsilon)$, and $\mathbf{e}_3$ is the unit normal vector of S. For each fixed $t \in (-\varepsilon, \varepsilon)$, the mapping $\mathbf{x}^t: U \to E^3$ given by

$$\mathbf{x}^t(u, v) = \mathbf{x}^*(u, v, t) \tag{8.11}$$

with

$$\begin{aligned} \mathbf{x}_u^t &= \mathbf{x}_u + t\lambda_u \mathbf{e}_3 + t\lambda \mathbf{e}_{3u}, \\ \mathbf{x}_v^t &= \mathbf{x}_v + t\lambda_v \mathbf{e}_3 + t\lambda \mathbf{e}_{3v}, \end{aligned} \tag{8.12}$$

where the subscripts u and v denote the partial derivatives, is a parametrization of a surface.

Next we assume the given surface S to have the minimal area among this family of surfaces $\mathbf{x}^t(u, v)$, and deduce necessary conditions. The area of S is given by

$$\begin{aligned} A &= \int_S\!\!\int \|\mathbf{x}_u \times \mathbf{x}_v\| \, du \, dv \\ &= \int_S\!\!\int \sqrt{EG - F^2} \, du \, dv. \end{aligned} \tag{8.13}$$

[The definition of the area A and the differential of (8.13) can be found in

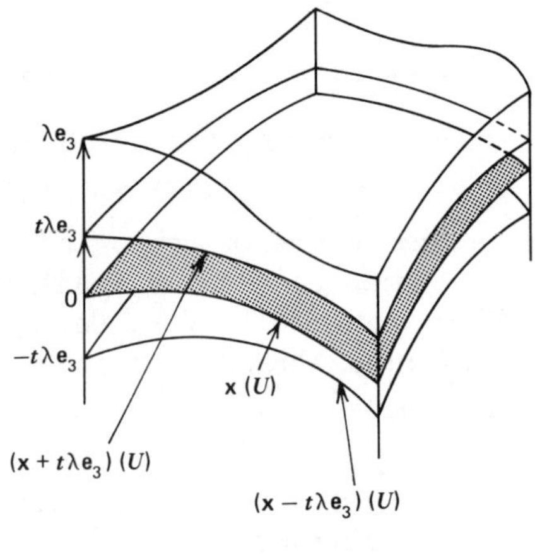

Figure 3.28

most calculus books and are not given here.] By using the Weingarten equations (7.1) we obtain, from (8.12),

$$\mathbf{x}'_u=\mathbf{x}_u+t\lambda_u\mathbf{e}_3+\frac{t\lambda}{EG-F^2}[(FM-GL)\mathbf{x}_u+(FL-EM)\mathbf{x}_v],$$
$$\mathbf{x}'_v=\mathbf{x}_v+t\lambda_v\mathbf{e}_3+\frac{t\lambda}{EG-F^2}[(FN-GM)\mathbf{x}_u+(FM-EN)\mathbf{x}_v]. \tag{8.14}$$

Let E', F', G' be the coefficients of the first fundamental form of the surface $\mathbf{x}'(u,v)$. Then from (8.14) and (2.28) it follows that

$$E'=E-2t\lambda L+(t^2), \tag{8.15}$$

where (t^2) denotes the terms of degree $\geqslant 2$ in t. By interchanging u and v, we obtain

$$G'=G-2t\lambda N+(t^2). \tag{8.16}$$

Similarly,

$$F'=F-2t\lambda M+(t^2). \tag{8.17}$$

From (8.15), (8.16), and (8.17) it thus follows that

$$E'G'-(F')^2=EG-F^2-2t\lambda(EN+GL-2FM)+(t^2). \tag{8.18}$$

Substitution of (4.10) in (8.18) gives

$$E'G'-(F')^2=(EG-F^2)(1-4t\lambda H+(t^2)),$$

which, by the binomial expansion for a square root, leads to

$$\sqrt{E'G'-(F')^2}=\sqrt{EG-F^2}\,(1-2t\lambda H+(t^2)). \tag{8.19}$$

Therefore the area of the surface $\mathbf{x}'(u,v)$ is

$$A'=\int_S\int\sqrt{E'G'-(F')^2}\,du\,dv \tag{8.20}$$
$$=A-2t\int_S\int\lambda H\sqrt{EG-F^2}\,du\,dv+(t^2).$$

For the original surface S to have the minimal area, the derivative of the area A' with respect to t must vanish for all λ, which satisfies the boundary condition. This means that

$$\delta A\equiv 2\int_S\int\lambda H\sqrt{EG-F^2}\,du\,dv=0 \tag{8.21}$$

for all λ. δA is called the *first variation* of the area A. By an argument analogous to that of Lemma 7.1 of Chapter 1, from (8.21) it follows that a necessary and sufficient condition is that

$$H\sqrt{EG-F^2}=0.$$

Since $EG-F^2$ is strictly positive, we must have $H=0$, and the theorem is proved.

8.4. Remark. A minimal surface may be made experimentally by dipping a simple closed curve made out of wire into some soap solution. The resulting soap film bounded by the wire frame is a minimal surface. However not all soap films are minimal surfaces according to our definition of surfaces.

8.5. Example. The catenoid given by

$$\begin{aligned}\mathbf{x}(u,v)&=(a\cosh v\cos u,\, a\cosh v\sin u,\, bv),\\ &0<u<2\pi,\ -\infty<v<\infty.\end{aligned} \tag{8.22}$$

This is the surface of revolution generated by revolving the catenary $x_1=b\cosh(x_3/a)$ about the $\mathbf{x}_3$-axis (cf. Fig. 3.13). It is easy to verify that the catenoid is a minimal surface.

A plane is trivially a minimal surface. Furthermore, we have Theorem 8.6.

8.6. Theorem. *Catenoids are the only nonplanar surfaces of revolution that are minimal.*

Proof. A nonplanar surface of revolution with $\mathbf{x}_3$ as the axis of revolution has a parametrization given by (see Example 1.10)

$$\mathbf{x}(u,v)=(u\cos v,\, u\sin v,\, f(u)), \tag{8.23}$$

where $f'(u)=df(u)/du\neq 0$. Therefore

$$\mathbf{x}_u=(\cos v,\sin v, f'(u)),\qquad \mathbf{x}_v=(-u\sin v, u\cos v, 0),$$

from which we have $E=1+f'^2$, $F=0, G=u^2$. By (4.10) the condition for a surface to be minimal is

$$EN+GL-2FM=0. \tag{8.24}$$

But since $F=0$, we need only to compute L and N, for which we need the second derivatives $\mathbf{x}_{uu}=(0,0,f'')$ and $\mathbf{x}_{vv}=(-u\cos v, -u\sin v, 0)$. Thus we have

$$|\mathbf{x}_u,\mathbf{x}_v,\mathbf{x}_{uu}|=uf'',\qquad |\mathbf{x}_u,\mathbf{x}_v,\mathbf{x}_{vv}|=u^2f'',$$

and (8.24) becomes, in consequence of (2.30),

$$(1+f'^2)f'+uf''=0. \tag{8.25}$$

Multiplying (8.25) by f' and putting $f'^2=y$ we obtain

$$\frac{dy}{y(1+y)}+\frac{2du}{u}=0. \tag{8.26}$$

Integration of (8.26) gives

$$yu^2=a(1+y), \tag{8.27}$$

where a is constant. Thus $f'^2=a/(u^2-a)$ or

$$f'=\sqrt{a/(u^2-a)}\,. \tag{8.28}$$

By integrating (8.28) we obtain

$$f=b\cosh^{-1}\left(\frac{u}{\sqrt{a}}\right), \tag{8.29}$$

where b is constant. Substituting (8.29) in (8.23) and comparing the resulting equation with (8.22), we see that our surface is a catenoid. Q.E.D.

Minimal surfaces have various interesting and important properties, two of which are contained in Lemma 8.7 and Corollary 8.10.

8.7. Lemma. *The Gauss mapping of a surface S is conformal if and only if S is either a sphere or a minimal surface.*

Proof. Let $g: S\to S^2$ be the Gauss mapping (see Definition 2.12) of S onto the unit sphere S^2. At first suppose that S is a sphere. Without loss of generality we may assume the center of S to be at the origin so that $\mathbf{x}=r\mathbf{e}_3$, where $\mathbf{x}$ is the parametrization, r the radius, and $\mathbf{e}_3$ the unit normal vector of S. Then $d\mathbf{x}=r\,d\mathbf{e}_3$, and therefore $\mathrm{I}=r^2\,\mathrm{III}$; that is, the first and third fundamental forms of S are proportional. Thus by Lemmas 2.13 and 5.7 the Gauss mapping g is conformal.

Next we notice that since $H=\frac{1}{2}(\kappa_1+\kappa_2)$ from (4.10), $H=0$ implies that $\kappa_1=-\kappa_2$ and therefore that $K=\kappa_1\kappa_2=-\kappa_1^2\leqslant 0$. Thus on a minimal surface the Gaussian curvature $K\leqslant 0$, so that there are only hyperbolic or parabolic points and no elliptic points. Now suppose that S is a minimal surface. Then $H=0$, and by the identity (7.2): $K\mathrm{I}-2H\mathrm{II}+\mathrm{III}=0$ among the three fundamental forms, we therefore have $\mathrm{III}=-K\mathrm{I}$, which implies that the Gauss mapping g is conformal. In this case the mapping g further preserves the sense, since $-K\geqslant 0$.

Now, to prove the converse, we suppose that $\mathrm{III}=c\mathrm{I}$ and $H\neq 0$, where c is a nonzero constant. By the identity (7.2) we then have $\mathrm{II}/\mathrm{I}=\frac{1}{2}(K+c)/H$, which shows that every point of the surface S is an umbilical point. Hence by Theorem 4.5, S is a sphere.

8.8. Lemma. *Let* $\mathbf{x}=\mathbf{x}(u,v)$ *be a parametrization of a surface S, and assume* $\mathbf{x}$ *to be isothermal. Then*

$$\mathbf{x}_{uu}+\mathbf{x}_{vv}=2\lambda^2\mathbf{H}, \tag{8.30}$$

where $\lambda^2=\mathbf{x}_u^2=\mathbf{x}_v^2$, *and* $\mathbf{H}=H\mathbf{e}_3$ *is called the mean curvature vector of S.*

Proof. Since $\mathbf{x}$ is isothermal, we have (5.7), that is,

$$\mathbf{x}_u^2=\mathbf{x}_v^2, \qquad \mathbf{x}_u\cdot\mathbf{x}_v=0, \tag{8.31}$$

and therefore H is reduced to

$$H=\frac{L+N}{2\lambda^2}. \tag{8.32}$$

Differentiating (8.31) gives

$$\mathbf{x}_{uu}\cdot\mathbf{x}_u=\mathbf{x}_{uv}\cdot\mathbf{x}_v=-\mathbf{x}_u\cdot\mathbf{x}_{vv},$$

which implies that

$$(\mathbf{x}_{uu}+\mathbf{x}_{vv})\cdot\mathbf{x}_u=0.$$

Similarly,

$$(\mathbf{x}_{uu}+\mathbf{x}_{vv})\cdot\mathbf{x}_v=0.$$

Thus $\mathbf{x}_{uu}+\mathbf{x}_{vv}$ is parallel to $\mathbf{e}_3$, and therefore can be written as

$$\mathbf{x}_{uu}+\mathbf{x}_{vv}=a\mathbf{e}_3. \tag{8.33}$$

Taking the inner product of (8.33) with $\mathbf{e}_3$ and using (2.30) and (8.32) we obtain

$$a=L+N=2\lambda^2H. \tag{8.34}$$

Substitution of (8.34) in (8.33) gives (8.30) immediately.

8.9. Definition. The *Laplacian* Δf of a C^2 function $f\colon U\subset E^2\to E^1$ is defined by

$$\Delta f=\frac{\partial^2 f}{\partial u^2}+\frac{\partial^2 f}{\partial v^2}, \qquad (u,v)\in U. \tag{8.35}$$

f is *harmonic* in U if $\Delta f=0$.

From Lemma 8.8 and Definition 8.9 we obtain the following corollary immediately.

8.10. Corollary. *Let* $\mathbf{x}(u,v)=(x_1(u,v),x_2(u,v),x_3(u,v))$ *be an isothermal parametrization of a surface S. Then S is minimal if and only if its coordinate functions* x_1, x_2, x_3 *are harmonic.*

Exercises

1. Prove that the element of area of the image of a surface S under the Gauss mapping is $K\,dA$, where K and dA are, respectively, the Gaussian curvature and the element of area of S.

2. The surface

$$x_1 = u^2 + 2uv, \qquad x_2 = u + v, \qquad x_3 = u^3 + 3u^2v$$

is a ruled surface since the v-curves are straight lines. Check to see whether this surface is developable.

3. Show that a curve C on a surface S is a line of curvature in case the ruled surface of normals of S at points of C is a developable surface.

***4.** Prove that each of the following surfaces is doubly ruled, that is, has two families of rulings:

(a) Hyperboloid of one sheet: $\dfrac{x_1^2}{a^2} + \dfrac{x_2^2}{b^2} - \dfrac{x_3^2}{c^2} = 1.$

(b) Saddle surface: $x_3 = kx_1x_2$, for nonzero constant k.

(c) Hyperbolic paraboloid: $x_3 = \dfrac{x_1^2}{a^2} - \dfrac{x_2^2}{b^2}.$

5. Show that each of the following surfaces is a minimum surface:

*(a) $e^{x_3} = \dfrac{\cos x_2}{\cos x_1}$ (H. F. Scherk).

(b) $\mathbf{x}(u,v) = \left(u - \dfrac{u^3}{3} + uv^2,\ v - \dfrac{v^3}{3} + vu^2,\ u^2 - v^2\right), \qquad (u,v) \in E^2$

(A. Enneper).

6. Determine the functions $f(x_1)$ and $F(x_2)$ so that the surface of the type

$$x_3 = f(x_1) + F(x_2)$$

may be a minimal surface.

***7.** Prove that the helicoid (Fig. 3.29)

$$\mathbf{x}(u,v) = (v\cos u,\ v\sin u,\ au),$$

$$0 < u < 2\pi, \ -\infty < v < \infty, \qquad a\text{: constant},$$

is the only minimal surface, other than the plane, that is also a ruled surface.

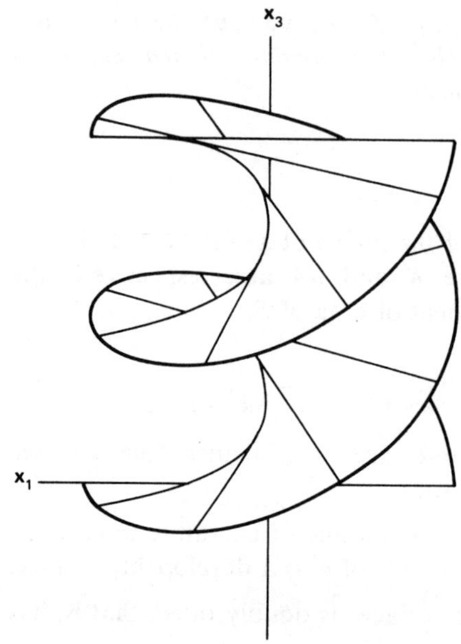

Figure 3.29

9. LEVI-CIVITA PARALLELISM

Let $\mathbf{x}(u, v)$ be a parametrization of a surface S. To simplify our notation we write $u^1 = u$, $u^2 = v$, and

$$\mathbf{x}_{u^i} = \mathbf{x}_i, \qquad \mathbf{x}_{u^i u^j} = \mathbf{x}_{ij}, \qquad 1 \leqslant i, j \leqslant 2,$$

and adopt the convention that subscripts denote differentiation only when applied to vectors. From (7.5) we have the differential forms

$$d\mathbf{x}_i = \sum_{j=1}^{2} \mathbf{x}_{ij} du^j = \sum_{j,k=1}^{2} \Gamma_{ij}^k \mathbf{x}_k du^j + \sum_{j=1}^{2} b_{ij} \mathbf{e}_3 du^j, \tag{9.1}$$

where $b_{11} = L$, $b_{12} = M$, $b_{22} = N$.

Now we consider a tangent vector field $\mathbf{v}$ of the surface S, which is a function assigning a tangent vector to each point of S (see Definition 2.5). Since any tangent vector of S at the point $\mathbf{p}(\equiv \mathbf{x}(u, v))$ can be represented as a linear combination of $\mathbf{x}_1$ and $\mathbf{x}_2$, $\mathbf{v}$ is represented by

$$\mathbf{v} = \sum_{i=1}^{2} y^i \mathbf{x}_i. \tag{9.2}$$

By (9.1) we obtain

$$d\mathbf{v}=\sum_{i=1}^{2}\left[(dy^i)\mathbf{x}_i+y^i\sum_{j,k=1}^{2}\Gamma_{ij}^k\mathbf{x}_k\,du^j+y^i\sum_{j=1}^{2}b_{ij}\mathbf{e}_3\,du^j\right]. \tag{9.3}$$

9.1. Definition. The *covariant differential* $\nabla\mathbf{v}$ of a tangent vector $\mathbf{v}$ of a surface S at a point $\mathbf{p}$ is the orthogonal projection of the differential $d\mathbf{v}$ from the normal $\mathbf{e}_3$ of S at $\mathbf{p}$ onto the tangent plane of S at $\mathbf{p}$.

Analytically, by just omitting the terms containing $\mathbf{e}_3$ from (9.3), we have

$$\begin{aligned}\nabla\mathbf{v}&=\sum_{i=1}^{2}\left(dy^i+\sum_{j,k=1}^{2}y^k\Gamma_{kj}^i\,du^j\right)\mathbf{x}_i\\&=\sum_{i=1}^{2}\left[(\nabla y^i)\mathbf{x}_i+y^i\nabla\mathbf{x}_i\right],\end{aligned} \tag{9.4}$$

or the relations

$$\begin{aligned}\nabla\mathbf{v}&\equiv d\mathbf{v}, \qquad \mathrm{mod}\ \mathbf{e}_3,\\(\nabla\mathbf{v})\cdot\mathbf{e}_3&=0.\end{aligned} \tag{9.5}$$

Since Γ_{ij}^k depend only on the first fundamental form, the following fundamental theorem is obvious.

9.2. Theorem. *The covariant differential* $\nabla\mathbf{v}$ *of a tangent vector* $\mathbf{v}$ *of a surface* S *at a point* $\mathbf{p}$ *depends only on the first fundamental form of the surface* S.

9.3. Definition. A tangent vector field $\mathbf{v}$ of a surface S is *parallel in the sense of Levi-Civita* along a curve C on S if $\nabla\mathbf{v}=0$ along C.

Since $\mathbf{x}_1, \mathbf{x}_2$ and du^1, du^2 are, respectively, linearly independent, it follows from (9.4) that $\nabla\mathbf{v}=0$ if and only if

$$\frac{\partial y^i}{\partial u^j}+\sum_{k=1}^{2}y^k\Gamma_{kj}^i=0, \qquad i,j=1,2. \tag{9.6}$$

Equations 9.6 form a system of four differential equations of the first order in two independent variables u^1, u^2 and two dependent variables y^1, y^2. Since the number of the equations is greater than the number of the unknown functions, such a system (9.6) is said to be *overdetermined* and in general we cannot get a solution on a surface, so we restrict our parallelism along a curve on the surface.

Along a curve given by $u^i = u^i(t)$ on the surface S, from (9.5) it follows that $\nabla \mathbf{v} = 0$ if and only if

$$\frac{dy^i}{dt} + \sum_{j,k=1}^{2} y^k \Gamma^i_{kj} \frac{du^j}{dt} = 0, \qquad i = 1,2, \tag{9.7}$$

which form a system of two linear homogeneous equations in y^i and dy^i/dt, $i = 1,2$. By the existence and uniqueness theorems for ordinary differential equations, equations 9.7 have unique solutions $y^i(t)$, when the initial values $y^i(t_0)$ are given. Thus we have Theorem 9.4.

9.4. Theorem. *On a surface S there exists a unique parallel tangent vector field* $\mathbf{v}(t)$ *along a curve* $\mathbf{x}(t)$ *such that* $\mathbf{v}(t_0)$ *is a given tangent vector; in other words, every tangent vector* $\mathbf{v}(t_0)$ *of the surface S at the point* $\mathbf{x}(t_0)$ *can be parallelly translated along a curve* $\mathbf{x}(t)$ *on S, and the vector* $\mathbf{v}(t_1)$ *is called the parallel translate of* $\mathbf{v}(t_0)$ *along the curve* $\mathbf{x}(t)$ *at the point* $\mathbf{x}(t_1)$.

9.5. Example. If the surface $\mathbf{x}(u, v)$ is the Euclidean plane, then $\mathrm{I} = ds^2 = (du^1)^2 + (du^2)^2$ and $\Gamma^i_{jk} = 0$ for all i, j, k. Thus from (9.5) it follows that $\nabla \mathbf{v} = 0$ if and only if $dy^i = 0$ or the y^i are constant, for all i. Hence the Levi-Civita parallelism reduces to the usual one in the case of the Euclidean plane.

9.6. Definition. A geometric property of a surface is *intrinsic* if it depends only on the first fundamental form of the surface.

Geometrically, intrinsic properties of a surface are invariant under a bending of the surface without stretching or shrinking. The simplest example of such properties is the arc length of a curve on the surface. We have also shown that the Gaussian curvature (Theorem 7.2) and the covariant differentiation (Theorem 9.2) are intrinsic. It is possible to have two distinct surfaces with the same first fundamental form—for instance, a developable surface and a plane (Theorem 8.2). On the other hand, the first fundamental form can be defined abstractly. As an example we consider the upper half-plane $\{(x, y) | y > 0\}$, for which the first fundamental form is usually given by $ds^2 = dx^2 + dy^2$ but can also be defined by $ds^2 = (dx^2 + dy^2)/y^2$. This introduces a new concept of the arc length on the upper plane, and the half-plane with this new first fundamental form has its Gaussian curvature equal to -1 and is called the *Poincaré half-plane*. We discuss this half-plane further in Section 2 of Chapter 4.

9.7. Definition. A geometric property of a surface that does not completely depend on the first fundamental form of the surface is *extrinsic*.

As an example, the second fundamental form of a surface is extrinsic; it depends on the embedding of the surface in the Euclidean space, since it is defined in terms of the normal vector of the surface.

The result that covariant differentiation is intrinsic gives the important fact that any two surfaces with the same first fundamental form have the same parallelism. Since the cone and the plane admit the same first fundamental form, we can obtain the parallelism on the cone by unrolling it onto the plane, translating parallelly in the Euclidean sense and then rolling the plane back to the cone.

9.8. Lemma. *The Levi-Civita parallelism on a surface S preserves the inner product of two tangent vector fields* **v** *and* **w** *on S and, in particular, the angle* θ *between* **v** *and* **w** *is constant if they are of constant lengths.*

Proof. Since $\mathbf{v}\cdot\mathbf{e}_3=\mathbf{w}\cdot\mathbf{e}_3=0$, we have

$$d(\mathbf{v}\cdot\mathbf{w})=d\mathbf{v}\cdot\mathbf{w}+\mathbf{v}\cdot d\mathbf{w}=\nabla\mathbf{v}\cdot\mathbf{w}+\mathbf{v}\cdot\nabla\mathbf{w}. \tag{9.8}$$

If **v** and **w** are parallel along any curve C on S, then $\nabla\mathbf{v}=\nabla\mathbf{w}=0$, and therefore from (9.8) it follows that $d(\mathbf{v}\cdot\mathbf{w})=0$, which means that $\mathbf{v}\cdot\mathbf{w}$ is constant along the curve C. The particular case is obvious since in this case $\mathbf{v}\cdot\mathbf{w}$ is equal to $\cos\theta$ except for a constant factor. Q.E.D.

Since $EG-F^2>0$, from (9.3) we can also write the condition of parallel translation (9.7) in the form

$$\frac{d\mathbf{v}}{dt}\cdot\mathbf{x}_i=0, \qquad i=1,2. \tag{9.9}$$

Let Σ_1 and Σ_2 be two surfaces tangent to each other along a curve C, so that they have the same tangent plane and unit normal vector $\mathbf{e}_3$ at every point of C. From (9.9) or from the definition that the Levi-Civita parallelism of a tangent vector field **v** along C is the orthogonal projection of the derivative $d\mathbf{v}$ of **v** along C from the normal vector $\mathbf{e}_3$ onto the common tangent plane, it follows that the parallelism is the same with respect to both Σ_1 and Σ_2. To find a geometric realization of the parallelism along a curve C on a surface S, we take the developable surface S_0 enveloped by the tangent planes of S along C. Since S and S_0 are tangent to each other along C, the parallelism along C is the same with respect to them both. Since S_0 is isometric to the plane and the parallelism depends only on the first fundamental form, we can unroll S_0 onto the plane, perform the parallel translation there, and roll the plane back to S_0. This gives a geometric interpretation of the Levi-Civita parallelism on a surface.

9.9. Theorem. *If the parallelism on a surface is independent of the path chosen, the surface is developable.*

Proof. Let $\mathbf{p},\mathbf{q}$ be two points on a surface S, and C a curve on S joining $\mathbf{p}$ and $\mathbf{q}$. When a tangent vector $\mathbf{v}$ of S at $\mathbf{p}$ is translated parallelly along C, we obtain a vector $\mathbf{w}$ at $\mathbf{q}$. If C' is another curve on S joining $\mathbf{p}$ and $\mathbf{q}$, then the vector $\mathbf{w}'$ obtained at $\mathbf{q}$ by translating $\mathbf{v}$ parallelly along C' will in general be distinct from $\mathbf{w}$. Now suppose that $\mathbf{w}=\mathbf{w}'$ for any tangent vector $\mathbf{v}$, any two points $\mathbf{p},\mathbf{q}$ and all curves joining $\mathbf{p}$ and $\mathbf{q}$ (in a certain neighborhood). Then for $\mathbf{v}$ given by (9.2), y^i can be determined as functions of u^1, u^2 such that (9.7) is satisfied for all curves $u^i=u^i(t)$. Under this assumption we have

$$\sum_{j=1}^{2}\left(\frac{\partial y^i}{\partial u^j}+\sum_{k=1}^{2} y^k\Gamma_{kj}^{\,i}\right)\frac{du^j}{dt}=0, \qquad i=1,2, \tag{9.10}$$

and therefore (9.6), since equations 9.10 hold for arbitrary functions $u^j(t)$. By computing the second partial derivatives of y^i from (9.6), and comparing the resulting equations, we obtain

$$\begin{aligned}0&=\frac{\partial^2 y^i}{\partial u^j\partial u^l}-\frac{\partial^2 y^i}{\partial u^l\partial u^j}\\&=\sum_{k=1}^{2}\left[\frac{\partial\Gamma_{kl}^{\,i}}{\partial u^j}-\frac{\partial\Gamma_{kj}^{\,i}}{\partial u^l}+\sum_{m=1}^{2}\left(\Gamma_{kl}^{m}\Gamma_{mj}^{\,i}-\Gamma_{kj}^{m}\Gamma_{ml}^{\,i}\right)y^k\right],\end{aligned}$$

which implies, since y^k have arbitrary initial values,

$$\frac{\partial\Gamma_{kl}^{\,i}}{\partial u^j}-\frac{\partial\Gamma_{kj}^{\,i}}{\partial u^l}+\sum_{m=1}^{2}\left(\Gamma_{kl}^{m}\Gamma_{mj}^{\,i}-\Gamma_{kj}^{m}\Gamma_{ml}^{\,i}\right)=0 \qquad \text{for all} \quad i,j,k,l. \tag{9.11}$$

From (7.22) it follows that the Gaussian curvature K of the surface S vanishes. Hence S is a developable surface.

Remark. The converse of Theorem 9.9 is also true, since the surface is then isometric to the plane on which the usual Euclidean parallelism is of course independent of path.

Exercises

***1.** Let C be a parallel of colatitude given by $\theta=\theta_0$ (const) on a sphere S with parametrization (1.8):

$$\mathbf{x}(\theta,\phi)=(a\sin\theta\cos\phi,\, a\sin\theta\sin\phi,\, a\cos\theta).$$

Show that the parallel translate of the unit vector $(\partial\mathbf{x}/\partial\theta)(\theta_0,0)$ along

C is

$$\mathbf{v}(\phi)=\left[\cos((\cos\theta_0)\phi)\right]\mathbf{x}_1-\frac{\sin((\cos\theta_0)\phi)}{\sin\theta_0}\mathbf{x}_2,$$

where $\mathbf{x}_1=\partial\mathbf{x}/\partial\theta$, $\mathbf{x}_2=\partial\mathbf{x}/\partial\phi$, and therefore deduce that $\mathbf{v}(0)=\mathbf{v}(2\pi)$ if and only if C is the equator of S.

10. GEODESICS

As another notion that can be introduced in the intrinsic geometry of surfaces, which is the study of intrinsic properties of surfaces, we shall define the *geodesic curvature* of a curve C on a surface S by using the notion of parallelism. The curvature of a plane curve at a point $\mathbf{p}(s)$ can be defined as $\kappa=dt/ds$, where s is the arc length, and t the angle which the oriented tangent of C at $\mathbf{p}$ makes with a fixed direction (Theorem 1.3.3 of Chapter 2). By generalizing this idea we have the following definition.

10.1. Definition. Consider a family of unit tangent vectors $\mathbf{y}(s)$ of a surface S, which are parallel along a curve C on S where s is the arc length of C. Let t be the angle between $\mathbf{y}(s)$ and the unit tangent vector $\mathbf{x}'(s)$ of C, where the prime denotes the derivative with respect to s. Then the rate of change of t with respect to s

$$\kappa_g=\frac{dt}{ds} \tag{10.1}$$

is called the *geodesic curvature* at a point $\mathbf{p}$ of C on S, and a curve on a surface with zero geodesic curvature everywhere is a *geodesic* of the surface.

10.2. Remarks. 1. Let $\Delta\theta$ be the angle between the tangent vector $\mathbf{x}'(s+\Delta s)$ and the parallel translate of the vector $\mathbf{x}'(s)$ along C at $\mathbf{x}(s+\Delta s)$. Since the angle between two tangent vectors of constant lengths remains unchanged under a parallel translation, it follows that

$$\kappa_g=\lim_{\Delta s\to 0}\frac{\Delta\theta}{\Delta s}, \tag{10.2}$$

which shows that κ_g is independent of the choice of $\mathbf{y}(s)$.

2. Geometrically, the tangent vectors of a geodesic are parallel to themselves along the geodesic. When the enveloping developable surface of a geodesic is unrolled onto a plane, the geodesic goes to a straight line.

We shall give some more properties of geodesic curvature and geodesics.

Let C be an oriented curve on a surface S with arc length s. From the unit tangent vector $\mathbf{x}'(s)$ of C and the unit normal vector $\mathbf{e}_3$ of S we construct a positively oriented frame $\mathbf{x}\mathbf{z}_1\mathbf{z}_2\mathbf{e}_3$ (see Definition 1.2.9 of Chapter 2) such that $\mathbf{z}_1=\mathbf{x}'$. Therefore $\mathbf{z}_2$ is the unit vector in the tangent plane of S, which is orthogonal to $\mathbf{x}'$ and satisfies $|\mathbf{z}_1,\mathbf{z}_2,\mathbf{e}_3|=+1$. Let $\mathbf{y}_1(s)$ be a family of unit tangent vectors of S parallel along C, and $\mathbf{y}_2(s)$ the family of unit tangent vector of S such that $\mathbf{x}\mathbf{y}_1\mathbf{y}_2\mathbf{e}_3$ is a positively oriented frame at each point $\mathbf{x}(s)$. Then $\mathbf{y}_1$ and $\mathbf{y}_2$ are obtained from $\mathbf{z}_1,\mathbf{z}_2$ by a rotation about $\mathbf{e}_3$ (see Fig. 3.30), and we have

$$\begin{aligned}\mathbf{y}_1&=\mathbf{z}_1\cos t-\mathbf{z}_2\sin t,\\ \mathbf{y}_2&=\mathbf{z}_1\sin t+\mathbf{z}_2\cos t,\end{aligned}\tag{10.3}$$

where t is the angle between $\mathbf{x}'$ and $\mathbf{y}_1$.

By taking s for t, $\mathbf{y}_1$ for $\mathbf{v}$, and $\mathbf{y}_i$ for $\mathbf{x}_i$ in (9.9), we obtain the condition for $\mathbf{y}_1(s)$ to be parallel along C:

$$\frac{d\mathbf{y}_1}{ds}\cdot\mathbf{y}_2=0;\tag{10.4}$$

the other condition $(d\mathbf{y}_1/ds)\cdot\mathbf{y}_1=0$ holds automatically, since $\mathbf{y}_1$ is a unit vector. Making use of (10.3) and the relations $\mathbf{z}_i^2=1$ for $i=1,2$ and $\mathbf{z}_1\cdot\mathbf{z}_2=0$, we can easily obtain

$$\frac{d\mathbf{y}_1}{ds}\cdot\mathbf{y}_2=-\frac{dt}{ds}+\frac{d\mathbf{z}_1}{ds}\cdot\mathbf{z}_2.\tag{10.5}$$

From (10.1), (10.4), and (10.5) we thus obtain the following formula for the geodesic curvature:

$$\kappa_g=\frac{d\mathbf{z}_1}{ds}\cdot\mathbf{z}_2=\left|\frac{d\mathbf{z}_1}{ds},\mathbf{e}_3,\mathbf{z}_1\right|=\left|\frac{d\mathbf{x}}{ds},\frac{d^2\mathbf{x}}{ds^2},\mathbf{e}_3\right|,\tag{10.6}$$

where we have used (5.1.16) of Chapter 1 for the second equality.

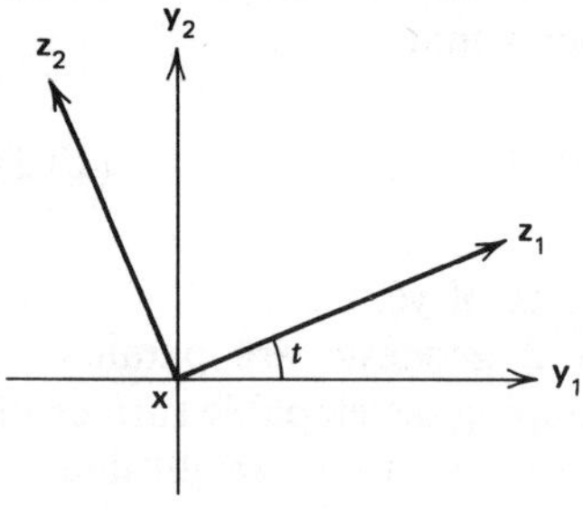

Figure 3.30

Let $\mathbf{n}$ and κ denote the unit principal normal and curvature of C, respectively. Then we have, from the Frenet formula,

$$\frac{d\mathbf{z}_1}{ds}=\kappa\mathbf{n},$$

and therefore, from (10.6),

$$\kappa_g=\kappa(\mathbf{n}\cdot\mathbf{z}_2)=\kappa\cos\left(\frac{\pi}{2}-\theta\right)=\kappa\sin\theta, \tag{10.7}$$

where θ is the angle between the principal normal $\mathbf{n}$ and the surface normal $\mathbf{e}_3$.

Thus we have the following characterization of geodesics.

10.3. Theorem. *A geodesic of a surface is either a straight line or a curve whose osculating plane passes through the surface normal.*

Proof. From (10.6) or (10.7) it follows immediately that $\kappa_g=0$ if either $\kappa=0$ or $\theta=0$. Q.E.D.

Now we continue to consider a curve C with arc length s on a surface S given by $\mathbf{x}(u^1,u^2)$. The unit tangent vector of C at a point $\mathbf{x}(s)$ can be written as

$$\frac{d\mathbf{x}}{ds}=\sum_{i=1}^{2}y^i\frac{\partial\mathbf{x}}{\partial u^i},$$

where

$$y^i=\frac{du^i}{ds}. \tag{10.8}$$

If the tangents of C are parallel along C, then C is a geodesic and y^i satisfy (9.7). Therefore substituting (10.8) in (9.7) with t replaced by s, we obtain the following differential equations defining the geodesic C:

$$\frac{d^2u^i}{ds^2}+\sum_{j,k=1}^{2}\Gamma^i_{jk}\frac{du^j}{ds}\frac{du^k}{ds}=0,\qquad i=1,2. \tag{10.9}$$

10.4. Lemma. *On a surface S through a point and tangent to a direction through the point, there passes exactly one geodesic.*

Proof. Since (10.9) is a pair of ordinary differential equations of the second order, by the existence and uniqueness theorems for such differential equations, (10.9) has uniqueness solutions $u^1(s),u^2(s)$ when the initial values $u^i(s_0)$, $(du^i/ds)(s_0)$, $i=1,2$, are given. Thus on S through a given point $u^i(s_0)$ and tangent to a given direction $(du^i/ds)(s_0)$ there passes exactly one geodesic. This geodesic is differentiable by the existence theorem and is regular by Lemma 9.8, which implies that the parallelism preserves the length of a tangent vector of s so that

$$\left(\frac{du^1}{ds}\right)^2+\left(\frac{du^2}{ds}\right)^2\neq 0$$

holds all along the geodesic whenever it holds at the initial point $u^i(s_0)$.

Q.E.D.

We shall define a system of parameters bearing a simple relationship to geodesics. We take a curve C_0 on a surface S given by $\mathbf{x}(u,v)$, and construct geodesics of S through the points of C_0 and orthogonal to C_0. Take these geodesics as u-curves and their orthogonal trajectories as v-curves. We restrict our discussion to a neighborhood in which this coordinate system is valid. Since the parametric curves are orthogonal we have $F=0$. Denoting the arc length on u-curve by s we find, along u-curves,

$$\mathbf{x}_s=\mathbf{x}_u\frac{du}{ds}=\frac{\mathbf{x}_u}{\sqrt{E}},\qquad \mathbf{x}_{ss}=\frac{\mathbf{x}_{su}}{\sqrt{E}}=\frac{\mathbf{x}_{uu}}{E}-\frac{E_u\mathbf{x}_u}{2E^2},$$

where the subscripts s and u denote partial derivatives. From (7.20) we have $\Gamma^2_{11}=-\frac{1}{2}E_v/G$; therefore from (10.6) and (7.5) we see that along u-curves

$$\kappa_g=-\frac{E_v}{2GE^{3/2}}|\mathbf{x}_u,\mathbf{x}_v,\mathbf{e}_3|=-\frac{E_v}{2E\sqrt{G}}=-\frac{(\sqrt{E}\,)_v}{\sqrt{EG}}=0,$$

which implies that E is a function of u alone. If we introduce a new parameter $\int\sqrt{E}\,du$ and denote it again by u, the first fundamental form of the surface S becomes

$$ds^2=du^2+G(u,v)\,dv^2. \tag{10.10}$$

10.5. Definition. The new u-curves above form a *field of geodesics*, and the v-curves their *geodesic parallels*.

10.6. Lemma. *The geodesics of a field are cut by any two geodesic parallels in arcs of the same length. Conversely, if any two v-curves cut equal lengths on their orthogonal trajectories, the trajectories are geodesics.*

Proof. From (10.10), the arc length on a u-curve is given by

$$s=\int_{u_0}^{u_1} du=u_1-u_0.$$

Thus the first part of the lemma is proved.

Conversely, if any two v-curves $u=u_0$ and $u=u_1$ cut equal lengths on their orthogonal trajectories, then

$$s=\int_{u=u_0}^{u=u_1}\sqrt{du^2+G\,dv^2}$$

is constant independent of v; therefore $G\,dv^2=0$, which means that the trajectories are geodesic u-curves, since $G\neq 0$. Q.E.D.

The geodesics are curves of the shortest length in the following sense.

10.7. Lemma. *A geodesic is the shortest curve between two points among all curves joining these two points and lying in a geodesic field.*

Proof. Let $\mathbf{p}_0(u_0,v_0)$, $\mathbf{p}_1(u_1,v_1)$ be two points, and $v=v(u)$ any curve joining $\mathbf{p}_0,\mathbf{p}_1$ on a portion of the surface, where our coordinate system is valid. Then the length between $\mathbf{p}_0,\mathbf{p}_1$ along this curve is

$$s=\int_{u_0}^{u_1}\sqrt{1+G(dv/du)^2}\;du \geqslant \int_{u_0}^{u_1} du=u_1-u_0.$$

Hence the lemma is proved. Q.E.D.

Similar to the coordinate system above we can define the so-called *geodesic polar coordinates*, which are convenient for the study of the intrinsic geometry of a surface in a neighborhood of a point. Through a point $\mathbf{0}$ of S given by $\mathbf{x}(u,v)$ we construct the geodesics and their orthogonal trajectories. By considerations analogous to the foregoing, we see that the first fundamental form can be written as

$$ds^2=dr^2+G(r,\phi)\,d\phi^2, \tag{10.11}$$

where r is the arc length along the geodesics through $\mathbf{0}$, and ϕ can be chosen to be the angle the geodesic makes with a fixed direction at $\mathbf{0}$.

To prove (10.11) we must show that $r=$const is orthogonal to $\phi=$const, or that $\mathbf{x}_r\cdot\mathbf{x}_\phi=0$. To this end, consider

$$(\mathbf{x}_r\cdot\mathbf{x}_\phi)_r=\mathbf{x}_{rr}\cdot\mathbf{x}_\phi+\mathbf{x}_r\cdot\mathbf{x}_{\phi r}.$$

Since $\mathbf{x}_r\cdot\mathbf{x}_r=1$, we have $\mathbf{x}_r\cdot\mathbf{x}_{r\phi}=0$. But as $\mathbf{x}_r$ is the unit tangent vector of the geodesic, it may be considered as a vector of a parallel field of unit vectors so that $\mathbf{x}_{rr}\cdot\mathbf{x}_\phi=0$, by (9.9). It therefore follows that $(\mathbf{x}_r\cdot\mathbf{x}_\phi)_r=0$, or

$\mathbf{x}_r \cdot \mathbf{x}_\phi$ is constant along a geodesic. Since the point $\mathbf{0}$ is determined by $\mathbf{x}(r,\phi)$ for $r=0$ independent of ϕ, $\mathbf{x}_\phi=0$ at $\mathbf{0}$. Therefore $\mathbf{x}_r \cdot \mathbf{x}_\phi=0$ holds at $\mathbf{0}$, hence everywhere along the geodesic.

The ϕ-curve is a geodesic parallel. The parameters r, ϕ generalize ordinary polar coordinates and are called *geodesic polar coordinates*. The point $\mathbf{0}$ itself is not regular with respect to the coordinate system.

Closely related to the coordinates r, ϕ are the *normal coordinates* x, y defined by

$$x=r\cos\phi, \qquad y=r\sin\phi. \tag{10.12}$$

The normal coordinates x, y form an admissible coordinate system in a neighborhood of the point $\mathbf{0}$ including $\mathbf{0}$, that is, a coordinate system such that the Jacobian of the transformation between x, y and the given system of coordinates u, v is not zero in the neighborhood. In fact, for simplicity we take u, v to be the arc lengths of the two orthogonal parametric curves starting from $\mathbf{0}$. Then at $\mathbf{0}$

$$u=v=0, \qquad E=G=1, \qquad F=0. \tag{10.13}$$

Writing $\mathbf{x}^1, \mathbf{x}^2$; u^1, u^2, s for x, y; u, v, r, respectively, we have (see Fig. 3.31) $du^1/ds=\cos\phi$, $du^2/ds=\sin\phi$, and therefore, from (10.12),

$$x^i=s\left(\frac{du^i}{ds}\right)_0. \tag{10.14}$$

where the subscript 0 denotes the value at the point $\mathbf{0}$. By using (10.9) and Maclaurin's formula

$$u^i(s)=u^i(0)+\frac{du^i}{ds}(0)s+\frac{1}{2}\frac{d^2u^i}{ds^2}(0)s^2+\cdots,$$

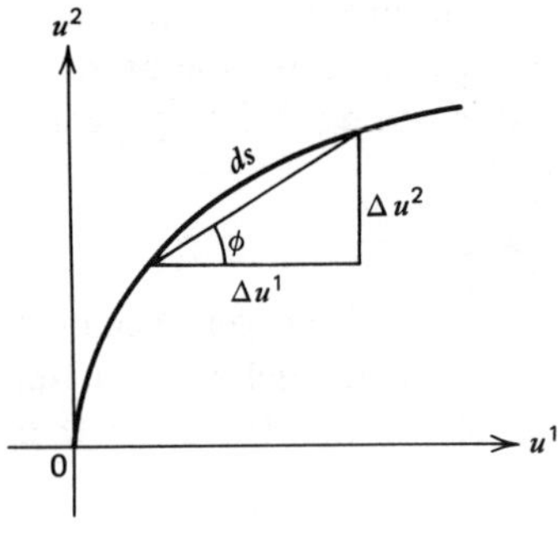

Figure 3.31

we obtain the relation between u^i and x^i:

$$u^i = s\left(\frac{du^i}{ds}\right)_0 - \frac{1}{2}s^2 \sum_{j,k=1}^{2} (\Gamma^i_{jk})_0 \left(\frac{du^j}{ds}\right)_0 \left(\frac{du^k}{ds}\right)_0 + \cdots$$

$$= x^i - \frac{1}{2} \sum_{j,k=1}^{2} (\Gamma^i_{jk})_0 x^j x^k + \cdots, \tag{10.15}$$

whose Jacobian at 0 is 1, since $x^i(0) = 0$.

From (10.12) we can express r, ϕ in terms of x, y:

$$r = \sqrt{x^2 + y^2}, \qquad \phi = \tan^{-1}\frac{y}{x}. \tag{10.16}$$

Since

$$dr = \frac{x\,dx + y\,dy}{r}, \qquad d\phi = \frac{x\,dy - y\,dx}{r^2},$$

the first fundamental form (10.11) becomes

$$ds^2 = g_{11}\,dx^2 + 2g_{12}\,dx\,dy + g_{22}\,dy^2, \tag{10.17}$$

where

$$g_{11} = \frac{x^2}{r^2} + G\frac{y^2}{r^4}, \qquad g_{12} = \frac{xy}{r^2}\left(1 - \frac{G}{r^2}\right), \qquad g_{22} = \frac{y^2}{r^2} + G\frac{x^2}{r^4}, \tag{10.18}$$

from which we may write

$$g_{11} - 1 = \frac{y^2}{r^2}\left(\frac{G}{r^2} - 1\right), \qquad g_{22} - 1 = \frac{x^2}{r^2}\left(\frac{G}{r^2} - 1\right), \tag{10.19}$$

so that

$$x^2(g_{11} - 1) = y^2(g_{22} - 1). \tag{10.20}$$

Since (10.20) must be true for all x, y, we have

$$g_{11} - 1 = \alpha y^2 + \cdots, \qquad g_{22} - 1 = \alpha x^2 + \cdots, \tag{10.21}$$

where α is a constant, and the dots denote terms of higher degrees in x and y; (10.21) is possible, since g_{11} and g_{22} are regular for $r = 0$. Substitution of (10.21) in (10.19) gives

$$\alpha y^2 + \cdots = \frac{y^2}{r^2}\left(\frac{G}{r^2} - 1\right), \qquad \alpha x^2 + \cdots = \frac{x^2}{r^2}\left(\frac{G}{r^2} - 1\right). \tag{10.22}$$

Thus when $(G/r^2 - 1)$ is expanded in powers of r by Maclaurin's formula, the first term must be αr^2. Hence

$$\frac{G}{r^2} - 1 = \alpha r^2 + \cdots, \tag{10.23}$$

or

$$G=r^2+\alpha r^4+\cdots, \tag{10.24}$$

and (10.17) becomes

$$ds^2=dx^2+dy^2+\alpha(y\,dx-x\,dy)^2+\cdots. \tag{10.25}$$

Since the normal coordinates x and y are determined up to a rotation about 0, the coefficients in (10.25), which are invariant under a rotation, are intrinsic invariants of the surface. This is the case of α; the relation of α to known invariants is easily found. In fact, with respect to the geodesic polar coordinates r, ϕ, we have $E=1$, $F=0$, $G=G(r,\phi)$, therefore, by (7.23),

$$K=-\frac{1}{\sqrt{G}}\frac{\partial^2\sqrt{G}}{\partial r^2}=\frac{G_r^2-2GG_{rr}}{4G^2}. \tag{10.26}$$

On the other hand, from (10.24) we obtain

$$\begin{aligned}\sqrt{G}&=r(1+\alpha r^2+\cdots)^{1/2}=r+\frac{\alpha}{2}r^3+\cdots,\\ \frac{\partial\sqrt{G}}{\partial r}&=1+\frac{3}{2}\alpha r^2+\cdots,\qquad \frac{\partial^2\sqrt{G}}{\partial r^2}3\alpha r+\cdots,\end{aligned} \tag{10.27}$$

so that

$$\sqrt{G}\,|_{r=0}=0,\qquad \frac{\partial\sqrt{G}}{\partial r}\Big|_{r=0}=1. \tag{10.28}$$

Thus (10.26) becomes

$$K=-\frac{3\alpha r+\cdots}{r+(\alpha/2)r^3+\cdots}.$$

Hence we have

$$K_0=\lim_{r\to 0}K=-3\alpha. \tag{10.29}$$

Using the element of arc in geodesic polar coordinates, we can derive some simple geometric interpretation of the Gaussian curvature. Call the curves $r=\text{const}$ *geodesic circles* with center $\mathbf{0}$ and radius r. By (10.11), (10.27), and (10.29) we obtain the perimeter of a geodesic circle of radius r:

$$\begin{aligned}L&=\int_{-\pi}^{\pi}\sqrt{G}\,d\phi=\int_{-\pi}^{\pi}\left(r-\frac{K_0}{6}r^3+\cdots\right)d\phi\\ &=2\pi r-\frac{\pi}{3}K_0r^3+\cdots.\end{aligned} \tag{10.30}$$

Hence

$$K_0=\lim_{r\to 0}\frac{3}{\pi}\frac{2\pi r-L}{r^3}. \tag{10.31}$$

Similarly, by (8.13) and (10.30) we obtain the area of a geodesic circle of radius r:

$$A=\int_0^r\int_{-\pi}^{\pi}\sqrt{G}\,d\phi\,dr=\int_0^r L\,dr=\pi r^2-\frac{\pi}{12}K_0r^4+\cdots,$$

from which follows immediately

$$K_0=\lim_{r\to 0}\frac{12}{\pi}\frac{\pi r^2-A}{r^4}. \tag{10.32}$$

10.8. Definition. The *geodesic torsion* τ_g at a point $\mathbf{p}$ of a curve C on a surface S is the torsion at $\mathbf{p}$ of the geodesic curve on S, which passes through $\mathbf{p}$ and is tangent to C.

To derive a formula for the geodesic torsion τ_g, let $\mathbf{t}$, $\mathbf{n}$ and $\mathbf{b}$ be, respectively, the unit tangent, principal normal, and binormal vectors of the curve C at $\mathbf{p}$, and let $\mathbf{t}_n$ be the unit vector orthogonal to both $\mathbf{t}$ and the unit normal vector $\mathbf{e}_3$ of the surface S such that both frames $\mathbf{tnb}$ and $\mathbf{t}\mathbf{t}_n\mathbf{e}_3$ are positively oriented (Definition 1.2.9 of Chapter 2). Then

$$|\mathbf{t},\mathbf{n},\mathbf{b}|=1,\qquad |\mathbf{t},\mathbf{t}_n,\mathbf{e}_3|=1. \tag{10.33}$$

Since the vectors $\mathbf{n},\mathbf{t}_n,\mathbf{b},\mathbf{e}_3$ are coplanar, we have (see Fig. 3.32)

$$\mathbf{b}\cdot\mathbf{e}_3=\cos\phi=\sin\theta. \tag{10.34}$$

Differentiating (10.34) with respect to the arc length s of the curve C, and using a Frenet formula for C we obtain

$$-\tau\mathbf{n}\cdot\mathbf{e}_3+\mathbf{b}\cdot\mathbf{e}_3'=\theta'\cos\theta, \tag{10.35}$$

where τ is the torsion of C at $\mathbf{p}$, and the prime denotes the derivative with respect to s. Since $\mathbf{n}\cdot\mathbf{e}_3=\cos\theta$, (10.35) is reduced to

$$(\theta'+\tau)\cos\theta=\mathbf{b}\cdot\mathbf{e}_3'. \tag{10.36}$$

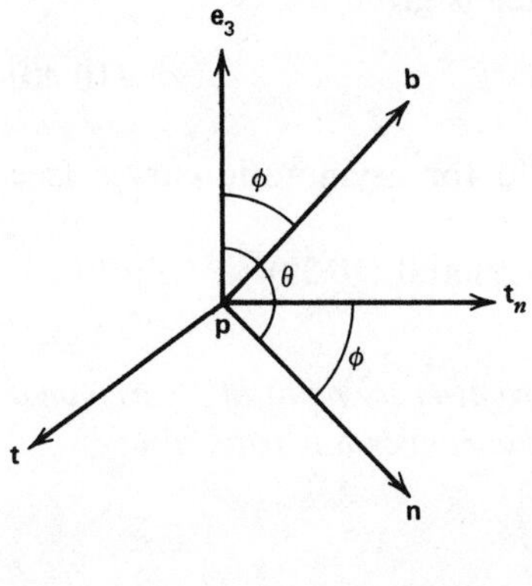

Figure 3.32

Using (5.1.16) of Chapter 1, (2.30), and the first equation of (10.33), and noticing that $|\mathbf{x}_u, \mathbf{x}_v, \mathbf{e}_3| = (\mathbf{x}_u \times \mathbf{x}_v)\cdot\mathbf{e}_3 = \sqrt{EG-F^2}$, we find

$$\begin{aligned}
\mathbf{b}\cdot\mathbf{e}_3' &= |\mathbf{t}, \mathbf{n}, \mathbf{e}_3'||\mathbf{x}_u, \mathbf{x}_v, \mathbf{e}_3| \Big/ \sqrt{EG-F^2} \\
&= \frac{1}{\sqrt{EG-F^2}} \begin{vmatrix} \mathbf{x}'\cdot\mathbf{x}_u & \mathbf{x}'\cdot\mathbf{x}_v & 0 \\ * & * & \mathbf{n}\cdot\mathbf{e}_3 \\ \mathbf{e}_3'\cdot\mathbf{x}_u & \mathbf{e}_3'\cdot\mathbf{x}_v & 0 \end{vmatrix} \qquad (10.37) \\
&= \frac{\cos\theta}{\sqrt{EG-F^2}\, ds^2} \begin{vmatrix} E\,du + F\,dv & F\,du + G\,dv \\ L\,du + M\,dv & M\,du + N\,dv \end{vmatrix}.
\end{aligned}$$

Substituting (10.37) in (10.36) gives

$$\theta' + \tau = \frac{(EM-FL)\,du^2 + (EN-GL)\,du\,dv + (FN-GM)\,dv^2}{\sqrt{EG-F^2}\,(E\,du^2 + 2F\,du\,dv + G\,dv^2)}. \qquad (10.38)$$

When θ is either 0 or π, from Theorem 10.3 the curve C is a geodesic, and thus we have Lemma 10.9.

10.9. Lemma. *The geodesic torsion at a point of a curve on a surface is given by*

$$\tau_g = \frac{(EM-FL)\,du^2 + (EN-GL)\,du\,dv + (FN-GM)\,dv^2}{\sqrt{EG-F^2}\,(E\,du^2 + 2F\,du\,dv + G\,dv^2)}. \qquad (10.39)$$

If $\cos\theta = 0$, then by Lemma 4.11 the curve C is an asymptotic curve. On the other hand, the relation between the torsion and the geodesic torsion at a point of a nonasymptotic curve on a surface is given by

$$\tau = \tau_g - \theta'. \qquad (10.40)$$

It should be noted that (10.40) is also valid for asymptotic curves (see Exercise 11 below).

Lemma 10.10 follows immediately from (4.9) and (10.39).

10.10. Lemma. *If the geodesic torsion is zero at every point of a curve on a surface, the curve is a line of curvature, and the converse is true, also.*

Exercises

***1.** Prove that

$$\kappa^2 = \kappa_N^2 + \kappa_g^2,$$

where κ, κ_N, and κ_g are, respectively, the curvature, normal curvature, and geodesic curvature at a point of a curve on a surface.

2. Show that the geodesic curvature at a point point $\mathbf{p}$ of an oriented curve C on a surface S is the curvature of the orthogonal projection of C onto the tangent plane of S at $\mathbf{p}$.

3. To which of the three types—lines of curvature, asymptotic curves, geodesics—do the following curves belong? (a) the top circle ($u = \pi/2$) on the torus with parametrization (1.18); (b) the outer equator ($u=0$) of the torus; (c) the $\mathbf{x}_1$-axis on the surface: $x_3 = x_1 x_2$.

4. Consider the following three properties of curves on a surface: (i) being an asymptotic curve, (ii) being a geodesic, and (iii) being a straight line.
(a) Show that properties i and ii together imply property iii.
(b) Do properties i and iii together imply property ii?
(c) Do properties ii and iii together imply property i?

5. Consider the following three properties of curves on a surface: (i) being a geodesic, (ii) being a line of curvature, and (iii) being a plane curve.
(a) Prove that properties i and ii together imply property iii.
(b) Prove that properties i and iii together imply property ii.
(c) Do properties ii and iii together imply property i?

***6.** Use differential equations to show that (a) the geodesics of a plane are straight lines, and (b) the great circles are geodesics of a sphere.

7. Show that the meridians (v=const) on a surface of revolution (1.12) are geodesics.

8. Find the equation in u and v of the geodesics of a surface whose first fundamental form is

$$ds^2 = v(du^2 + dv^2), \qquad v > 0.$$

9. Find the geodesics of a circular cylinder $x_1^2 + x_2^2 = a^2$.

10. Prove that a necessary and sufficient condition that the torsion and the geodesic torsion be equal at every point of a curve C on a surface S is that the angle θ between the principal normal of C and the normal of S be constant along C.

***11.** Taking the asymptotic curves as parametric on a nondevelopable surface S, to that $L=0$, $N=0$, $M\neq 0$, show by direct calculation that the torsions τ_1, τ_2 of the u- and v-curves, respectively, at a point $\mathbf{p}$ of S, are given by the formulas

$$\tau_1 = -\frac{M}{\sqrt{EG-F^2}}, \qquad \tau_2 = +\frac{M}{\sqrt{EG-F^2}}.$$

Prove that (10.39) gives these results in this case.

4

Global Theory of Surfaces

Sections 2 and 3 of Chapter 2 presented some global theorems on curves. Chapter 3 is a good basis for discussing some global theorems about surfaces. As before, the theorems deal with the relations between local and global (in general, topological) properties.

1. ORIENTATION OF SURFACES

In this section we discuss the orientability of a surface. Intuitively, since at each point $\mathbf{p}$ of a surface S there is a tangent plane $T_p(S)$, the choice of an orientation of $T_p(S)$ will induce an orientation of S in a neighborhood of $\mathbf{p}$ (as indicated by the dashed arrows in Fig. 4.1). If it is possible to make such a choice at each point $\mathbf{p}\in S$ so that in the intersection of any two neighborhoods the orientations coincide, then S is said to be *orientable*; otherwise, S is said to be *nonorientable*. Analytically, these ideas can be expressed as follows.

Let $\mathbf{x}(u,v)$ be a parametrization of a neighborhood of a point $\mathbf{p}$ of a surface S, and suppose that an orientation of the tangent plane $T_p(S)$ be given by the orientation of the associated ordered basis $\{\mathbf{x}_u,\mathbf{x}_v\}$. If $\mathbf{p}$ belongs to the coordinate neighborhood of another parametrization $\bar{\mathbf{x}}(\bar{u},\bar{v})$, the new basis $\{\bar{\mathbf{x}}_u,\bar{\mathbf{x}}_v\}$ is expressed in terms of the old one by

$$\begin{aligned}\bar{\mathbf{x}}_{\bar{u}}&=\mathbf{x}_u\frac{\partial u}{\partial \bar{u}}+\mathbf{x}_v\frac{\partial v}{\partial \bar{u}},\\ \bar{\mathbf{x}}_{\bar{v}}&=\mathbf{x}_u\frac{\partial u}{\partial \bar{v}}+\mathbf{x}_v\frac{\partial v}{\partial \bar{v}},\end{aligned}\tag{1.1}$$

where $u=u(\bar{u},\bar{v})$ and $v=v(\bar{u},\bar{v})$ are the expressions of the change of coordinates. Thus the bases $\{\mathbf{x}_u,\mathbf{x}_v\}$ and $\{\bar{\mathbf{x}}_{\bar{u}},\bar{\mathbf{x}}_{\bar{v}}\}$ determine the same

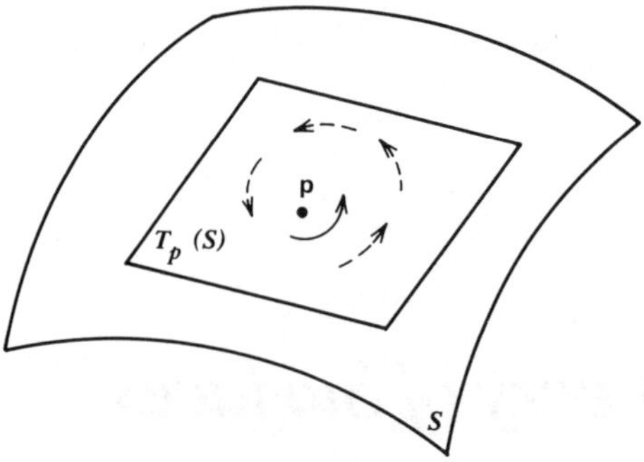

Figure 4.1

orientation of $T_p(S)$ if and only if the Jacobian determinant

$$\frac{\partial(u, v)}{\partial(\bar{u}, \bar{v})}$$

of the coordinate change is positive. Accordingly, we have Definition 1.1.

1.1. Definition. A surface S is said to be *orientable* if it can be covered by a family of coordinate neighborhoods such that if a point $\mathbf{p} \in S$ belongs to two neighborhoods of the family, the change of coordinates has a positive Jacobian determinant at $\mathbf{p}$. The choice of such a family is called an *orientation* of S. If such a choice is not possible, the surface is said to be *nonorientable*.

1.2. Examples. 1. A surface that is the graph of a C^3 function (cf. Lemma 1.5 of Chapter 3) is an orientable surface. In fact, every surface covered by one coordinate neighborhood is trivially orientable.

2. The sphere is an orientable surface. Instead of using a direct calculation we can show this by a general argument. The sphere can be covered by two coordinate neighborhoods (using stereographic projection; see Exercise 11 of Section 1, Chapter 3) with parameters (u, v) and $(\bar{u}, \bar{v})$ such that the intersection W of these two neighborhoods (the sphere minus two points) is a connected set. Take any point $\mathbf{p}$ in W. Then we can assume the Jacobian determinant of the coordinate change at $\mathbf{p}$ to be positive, since otherwise we could make it so by interchanging u and v in the first coordinate system. Since, in addition to this, the Jacobian determinant is continuous and different from zero in W, from the connectedness of W it follows (by Theorem 1.3.4 of Chapter 1) that the Jacobian determinant is positive everywhere in W. Thus by Definition 1.1 the sphere is orientable.

By the same argument we further can obtain the more general result: *A surface covered by two coordinate neighborhoods with connected intersection is orientable*.

Since Definition 1.1. is more or less analytic, we shall prove the following geometric criterion.

1.3. Theorem. *A surface $S \subset E^3$ is orientable if and only if there exists a differentiable unit normal vector field* $\mathbf{e}_3$ *on* S.

Proof. If S is orientable, we can cover it by a family of coordinate neighborhoods such that in the intersection of any two neighborhoods the change of coordinates has a positive Jacobian determinant. At the points $\mathbf{p}=\mathbf{x}(u,v)$ of each neighborhood, define $\mathbf{e}_3(\mathbf{p})=\mathbf{e}_3(u,v)$ by

$$\mathbf{e}_3(\mathbf{p})=\frac{\mathbf{x}_u\times\mathbf{x}_v}{\|\mathbf{x}_u\times\mathbf{x}_v\|} \tag{1.2}$$

(cf. (2.13) of Chapter 3). If $\mathbf{p}$ belongs to two neighborhoods with parameters (u,v) and $(\bar{u},\bar{v})$, then from (1.1) it follows that

$$\bar{\mathbf{x}}_{\bar{u}}\times\bar{\mathbf{x}}_{\bar{v}}=(\mathbf{x}_u\times\mathbf{x}_v)\frac{\partial(u,v)}{\partial(\bar{u},\bar{v})}, \tag{1.3}$$

so that $\mathbf{e}_3(u,v)$ and $\bar{\mathbf{e}}_3(\bar{u},\bar{v})$ coincide. Thus $\mathbf{e}_3(\mathbf{p})$ is well defined. Moreover, $\mathbf{e}_3(u,v)$ is differentiable (cf. the definition of a differentiable vector field in E^3 on *p.* 31) since its coordinates in E^3 are differentiable functions of (u,v) by (1.2).

Conversely, let $\mathbf{e}_3$ be a differentiable unit normal vector field on S, and consider a family of connected coordinate neighborhoods covering S. For the points $\mathbf{p}=\mathbf{x}(u,v)$ of each coordinate neighborhood $\mathbf{x}(U)$, $U\subset E^2$, take the inner product

$$\mathbf{e}_3(\mathbf{p})\cdot\frac{\mathbf{x}_u\times\mathbf{x}_v}{\|\mathbf{x}_u\times\mathbf{x}_v\|}=f(\mathbf{p})=\pm 1,$$

which is a continuous function on $\mathbf{x}(U)$. Since $\mathbf{x}(U)$ is connected, f is of constant sign, and we can assume $f=1$, for otherwise we could interchange u and v in the parametrization. Thus we can have (1.2) for all points $\mathbf{p}=\mathbf{x}(u,v)$ of each coordinate neighborhood $\mathbf{x}(U)$.

Continue this process for all coordinate neighborhoods, so that in the intersection of any two of them, say $\mathbf{x}(u,v)$ and $\bar{\mathbf{x}}(\bar{u},\bar{v})$, the Jacobian determinant $\partial(u,v)/\partial(\bar{u},\bar{v})$ must be positive; otherwise from (1.2), (1.3) we would have

$$\frac{\mathbf{x}_u \times \mathbf{x}_v}{\|\mathbf{x}_u \times \mathbf{x}_v\|} = \mathbf{e}_3(\mathbf{p}) = -\frac{\bar{\mathbf{x}}_{\bar{u}} \times \bar{\mathbf{x}}_{\bar{v}}}{\|\bar{\mathbf{x}}_{\bar{u}} \times \bar{\mathbf{x}}_{\bar{v}}\|} = -\mathbf{e}_3(\mathbf{p}),$$

which is a contradiction. Hence by Definition 1.1, S is orientable. Q.E.D.

We have seen (Examples 1.2) that a surface that is the graph of a C^3 function is orientable. Now we shall show that the inverse image of a regular value (for the definition of a regular value see Theorem 1.6 of Chapter 3) of a C^3 function is also an orientable surface.

1.4. Lemma. *If a surface S is given by $S = \{(x_1, x_2, x_3) \in E^3 \mid f(x_1, x_2, x_3) = a\}$, where $f\colon U \subset E^3 \to E^1$ is a C^3 function, and a is a regular value of f, then S is orientable.*

Proof. On S consider a curve $(x_1(t), x_2(t), x_3(t))$ with parameter $t \in I$ passing through a given point $\mathbf{p} = (\bar{x}_1, \bar{x}_2, \bar{x}_3) \in S$ for $t = t_0$ so that $x_i(t_0) = \bar{x}_i$ for $i = 1, 2, 3$. Then we have

$$f(x_1(t), x_2(t), x_3(t)) = a \tag{1.4}$$

for all $t \in I$. Differentiating (1.4) with respect to t and using the chain rule we obtain, at $t = t_0$,

$$f_{x_1}(\mathbf{p})\left(\frac{dx_1}{dt}\right)_{t_0} + f_{x_2}(\mathbf{p})\left(\frac{dx_2}{dt}\right)_{t_0} + f_{x_3}(\mathbf{p})\left(\frac{dx_3}{dt}\right)_{t_0} = 0.$$

This shows that the tangent vector to the curve at $t = t_0$ is orthogonal to the vector $(f_{x_1}, f_{x_2}, f_{x_3})$ at $\mathbf{p}$, which is a nonzero vector due to the regular value a of f. Since the curve and the point $\mathbf{p}$ are arbitrary, we have a differentiable unit normal vector field on S:

$$\mathbf{e}_3(x_1, x_2, x_3) = \left(\frac{f_{x_1}}{D}, \frac{f_{x_2}}{D}, \frac{f_{x_3}}{D}\right),$$

where $D = \sqrt{f_{x_1}^2 + f_{x_2}^2 + f_{x_3}^2}$. Hence by Theorem 1.3, S is orientable. Q.E.D.

Thus from Lemma 1.3, ellipsoids (Example 1.7, Chapter 3), hyperboloids of two sheets (Example 1.8, Chapter 3) (in fact, all quadric surfaces), cylinders (Example 1.9, Chapter 3), and surfaces of revolution (Example 1.10, Chapter 3) are orientable. However, nonorientable surfaces do exist in E^3; the simplest case is the Möbius band M (Example 1.15, Chapter 3). To show this, we suppose that there exists a differentiable unit normal vector field $\mathbf{e}_3$ on M with a parametrization $\mathbf{x}(u, v)$. Interchanging u and v if necessary, we can assume that

$$\mathbf{e}_3(\mathbf{p}) = \frac{\mathbf{x}_u \times \mathbf{x}_v}{\|\mathbf{x}_u \times \mathbf{x}_v\|}$$

for any point $\mathbf{p}$ in the coordinate neighborhood of $\mathbf{x}(u,v)$. Similarly, we assume that

$$\mathbf{e}_3(\mathbf{p}) = \frac{\bar{\mathbf{x}}_{\bar{u}} \times \bar{\mathbf{x}}_{\bar{v}}}{\|\bar{\mathbf{x}}_{\bar{u}} \times \bar{\mathbf{x}}_{\bar{v}}\|}$$

at all points of the coordinate neighborhood of another parametrization $\bar{\mathbf{x}}(u,v)$ of M. However, the intersection of the two coordinate neighborhoods has two connected components W_1 and W_2, and from (1.24) of Chapter 3 the Jacobian determinant of the change of coordinates must be -1 in either W_1 or W_2. If $\mathbf{p}$ is a point of that component, then $\mathbf{e}_3(\mathbf{p}) = -\mathbf{e}_3(\mathbf{p})$, which is a contradiction.

1.5. Remark. Orientation is definitely not a local property of a surface. Locally, every surface is diffeomorphic to an open set of the plane, hence orientable. Orientation is a global property in the sense that it involves the whole surface.

Concerning the orientation we have the following global theorem.

1.6. Theorem. *Every compact (Definition 1.5.4 of Chapter 1) surface in E^3 is orientable.*

Proof. A proof has been given by H. Samelson [49].

Exercises

1. Let S be a regular surface covered by coordinate neighborhoods V_1 and V_2. Assume that $V_1 \cap V_2$ has two connected components W_1, W_2, and that the Jacobian of the change of coordinates is positive in W_1 and negative in W_2. Show that S is nonorientable.
2. Let S_2 be an orientable regular surface, and $f\colon S_1 \to S_2$ be a C^3 mapping that is a local diffeomorphism of class C^3 at every point $\mathbf{p} \in S_1$. Prove that S_1 is orientable.
3. Let f be a differentiable real-valued function on a connected surface S. Show that (a) if $df = 0$, then f is constant; (b) if f is never zero, then either $f > 0$ or $f < 0$.

*4. Prove that a connected orientable surface has exactly two unit normal vector fields, which are negatives of each other.

5. Let $f\colon S_1 \to S_2$ be a C^3 differeomorphism of two regular surfaces.
   (a) Show that S_1 is orientable if and only if S_2 is orientable (thus *orientability is preserved by C^3 diffeomorphisms*).

(b) Let S_1 and S_2 be orientable and oriented. Prove that the diffeomorphism f induces an orientation on S_2. Use the antipodal mapping of the sphere† to show that this orientation may be distinct (cf. Exercise 4) from the initial one (thus *orientation may not be preserved by diffeomorphisms*; however, *if S_1 and S_2 are connected, a diffeomorphism either preserves or reverses the orientation*).

2. SURFACES OF CONSTANT GAUSSIAN CURVATURE

The surfaces of constant Gaussian curvature form an important class of surfaces. From Definition 5.3 and Theorem 7.2 of Chapter 3 we know that any two locally isometric surfaces have the same Gaussian curvature at each pair of corresponding points. However the converse of this result is not true in general, but is true for the special case in the following theorem.

2.1. Theorem (F. Minding). *Two surfaces of the same constant Gaussian curvature are locally isometric.*

Proof. Let S and $\bar{S}$ be two surfaces of the same constant Gaussian curvature K, and let f be a mapping between two neighborhoods of two arbitrary points $O, \bar{O}$ on $S, \bar{S}$ such that each pair of corresponding points has the same geodesic polar coordinates (r, ϕ) with respect to two arbitrary tangent directions of $S, \bar{S}$ at $O, \bar{O}$, respectively. Then by (10.11) and (10.28) of Chapter 3, the first fundamental forms of $S, \bar{S}$ at $O, \bar{O}$ are, respectively,

$$ds^2 = dr^2 + G\,d\phi^2, \qquad d\bar{s}^2 = dr^2 + \bar{G}\,d\phi^2, \tag{2.1}$$

where

$$G(0,\phi) = \bar{G}(0,\phi) = 0, \qquad \frac{\partial G(0,\phi)}{\partial r} = \frac{\partial \bar{G}(0,\phi)}{\partial r} = 1. \tag{2.2}$$

Furthermore, from (10.26) of Chapter 3 we have

$$K = -\frac{1}{\sqrt{G}}\frac{\partial^2 \sqrt{G}}{\partial r^2}, \qquad K = -\frac{1}{\sqrt{\bar{G}}}\frac{\partial^2 \sqrt{\bar{G}}}{\partial r^2}. \tag{2.3}$$

†Let $S^2 = \{(x_1, x_2, x_3) \in E^3 \mid x_1^2 + x_2^2 + x_3^2 = 1\}$ be the unit sphere. Then the antipodal mapping g: $S^2 \to S^2$ defined by $f(x_1, x_2, x_3) = (-x_1, -x_2, -x_3)$ is a diffeomorphism, since $f = f^{-1}$.

Thus $\sqrt{G}$ and $\sqrt{\bar{G}}$ are solutions fo the differential equation

$$\frac{\partial^2 z}{\partial r^2} = -Kz \tag{2.4}$$

with a constant coefficient K.

Since K does not depend on r, (2.4) can be treated as an ordinary differential equation, and a solution of the equation is uniquely determined by the initial values of z and $\partial z/\partial r$, which might also depend on ϕ as a parameter. But by (2.2), both $\sqrt{G}$ and $\sqrt{\bar{G}}$ satisfy the same initial conditions. Thus we have $\sqrt{G} = \sqrt{\bar{G}}$ everywhere, so that $ds^2 = d\bar{s}^2$. Hence f is a local isometry of S, $\bar{S}$. Q.E.D.

From Theorem 2.1 and the well-known properties of the sphere (Exercise 5 of Section 4, Chapter 3), plane (Theorem 8.2 of Chapter 3), and pseudosphere (Exercise 8 of Section 4, Chapter 3) we have the following two corollaries.

2.2. Corollary. *A surface of positive constant Gaussian curvature $K = a^{-2}$ is locally isometric to a sphere of radius a. A surface of zero Gaussian curvature is locally isometric to a plane. A surface of negative constant Gaussian curvature $K = -a^{-2}$ is locally isometric to the pseudosphere.*

2.3. Corollary. *Given two arbitrary points* $\mathbf{p}$ *and* $\mathbf{q}$ *of a surface S of constant Gaussian curvature and two arbitrary tangent directions of S at* $\mathbf{p}$ *and* $\mathbf{q}$, *respectively, there exists a local isometry of S into itself that takes* $\mathbf{p}$ *to* $\mathbf{q}$, *and the given direction at* $\mathbf{p}$ *to the given direction at* $\mathbf{q}$.

We have just discussed pieces of surfaces of constant Gaussian curvature. For compact surfaces of constant Gaussian curvature we first have Theorem 2.4.

2.4. Theorem (H. Liebmann). *Let S be a compact connected surface of constant Gaussian curvature K. Then S is a sphere of radius $1/\sqrt{K}$.*

Proof **(D. Hilbert).** First of all, the Gaussian curvature K must be positive. To show this, let x_3 be the oriented distance from a point of the surface S to an arbitrarily fixed horizontal plane in E^3. Since x_3 is a continuous function over S that is compact, by Corollary 1.5.10 of Chapter 1 on S there is a point $\mathbf{p}$ at which x_3 is maximum, so that there are no points on S that are on the other side of the tangent plane of S at $\mathbf{p}$. Thus $\mathbf{p}$ is not hyperbolic, and $K \geqslant 0$, since K is constant throughout the whole surface S. However, K cannot be zero, since if it were, S would be developable and not compact.

By Theorem 1.6 we assume that S is orientable and therefore two-sided in the sense that at each point of S the unit outward (or inward) normal is well defined. Using the family of outward normals of S, we can completely determine the second fundamental form (otherwise only up to a sign) and also the principal curvatures of S. At each point of S there are two principal curvatures which we denote by κ_1, κ_2. Let $\mathbf{p}$ be a point on S at which the principal curvatures attain a maximum, and denote the corresponding principal curvature by κ_1. Since $\kappa_1\kappa_2 = K = \text{const}$, κ_2 attains a minimum at $\mathbf{p}$. If $\kappa_1 = \kappa_2$ at $\mathbf{p}$, the two principal curvatures must be equal at every other point, so that the surface consists entirely of umbilical points and is therefore a sphere by Theorem 4.5 of Chapter 3. Suppose that this is not the case, so that $\kappa_1 > \kappa_2$ at $\mathbf{p}$. Then $\mathbf{p}$ is not an umbilical point, and by continuity there is a neighborhood U of $\mathbf{p}$ on S free from umbilical points. In U we can choose the lines of curvature as parametric curves, so that $F = M = 0$. Thus from (4.6), (7.23), and (7.25) of Chapter 3 we obtain the two principal curvatures, the Gauss equation, and the Mainardi-Codazzi equations, respectively:

$$\kappa_1 = \frac{L}{E}, \qquad \kappa_2 = \frac{N}{G}; \tag{2.5}$$

$$K = -\frac{1}{2\sqrt{EG}}\left[\frac{\partial}{\partial v}\left(\frac{E_v}{\sqrt{EG}}\right) + \frac{\partial}{\partial u}\left(\frac{G_u}{\sqrt{EG}}\right)\right]; \tag{2.6}$$

$$\begin{aligned} L_v &= \frac{E_v}{2}\left(\frac{L}{E} + \frac{N}{G}\right) = \frac{E_v}{2}(\kappa_1 + \kappa_2), \\ N_u &= \frac{G_u}{2}\left(\frac{L}{E} + \frac{N}{G}\right) = \frac{G_u}{2}(\kappa_1 + \kappa_2). \end{aligned} \tag{2.7}$$

Repeatedly differentiating (2.5) and using (2.7) give

$$\begin{aligned} \kappa_{1v} &= \frac{1}{E^2}(EL_v - LE_v) = \frac{E_v}{2E}(\kappa_2 - \kappa_1), \\ \kappa_{2u} &= \frac{1}{G^2}(GN_u - G_uN) = \frac{G_u}{2G}(\kappa_1 - \kappa_2); \end{aligned} \tag{2.8}$$

$$\begin{aligned} \kappa_{1vv} &= \frac{E_{vv}}{2E}(\kappa_2 - \kappa_1) + E_v(\cdots), \\ \kappa_{2uu} &= \frac{G_{uu}}{2G}(\kappa_1 - \kappa_2) + G_u(\cdots). \end{aligned} \tag{2.9}$$

Since at **p**, κ_1 is a maximum and κ_2 is a minimum, we have, at **p**,

$$\kappa_{1v}=0, \qquad \kappa_{2u}=0,$$
$$\kappa_{1vv}\leqslant 0, \qquad \kappa_{2uu}\geqslant 0, \tag{2.10}$$

which together with (2.8) and (2.9) imply that at **p**

$$E_v=0, \qquad G_u=0,$$
$$E_{vv}\geqslant 0, \qquad G_{uu}\geqslant 0. \tag{2.11}$$

Substituting (2.11) in (2.6), we find that at **p**

$$K=-\frac{1}{2EG}(E_{vv}+G_{uu})\leqslant 0, \tag{2.12}$$

which contradicts the fact that $K>0$. Hence Theorem 2.4 is proved.

2.5. Remarks. 1. Since the Gaussian curvature is an isometric invariant, and a continuous image of a compact connected set is compact and connected (Theorems 1.5.6 and 1.3.4 of Chapter 1), from Theorem 2.4 we obtain the rigidity of the sphere in the following sense: if $f\colon \Sigma\to S$ is an isometry of a sphere Σ onto a surface S, then S is a sphere. Therefore we can express Theorem 2.4 by saying that *a sphere cannot be bent*.

2. The compactness condition in Theorem 2.4 cannot be omitted, since there are many nonspherical surfaces in E^3 of constant (positive) Gaussian curvature by Corollary 2.2. In other words, a sphere with a hole in it can be bent.

3. From the proof of Theorem 2.4 it follows that a surface of constant negative Gaussian curvature in E^3 cannot be compact.

In Section 8 we shall introduce *complete surfaces*, the class of which is larger than that of compact surfaces. Concerning such surfaces of constant negative Gaussian curvature we have the following theorem.

2.6. Theorem (D. Hilbert). *There is no complete surface of constant negative Gaussian curvature in a Euclidean space E^3.*

Proof. The proof is omitted here, but can be found, for instance, in the books of J. J. Stoker [55, pp. 265–271] and M. P. do Carmo [8, pp. 446–453].

Having obtained some information on surfaces of constant Gaussian curvature from Theorems 2.1, 2.4, and 2.6, it is natural for us to investigate the abstract surfaces of constant Gaussian curvature, in particular, those of Gaussian curvature $K=-1$. A notable example of such a surface is the

Poincaré half-plane, which can be defined as follows (we mentioned this half-plane in the paragraph following Definition 9.6 of Chapter 3):

Take a system of coordinates u, v defined on p. 232 so that the first fundamental form is given by (10.10) on the same page. By substituting $E=1$, $F=0$ in (7.23) of Chapter 3 we obtain

$$K=-\frac{1}{\sqrt{G}}\frac{\partial^2\sqrt{G}}{\partial u^2}. \tag{2.13}$$

Then the condition that the Gaussian curvature K be -1 is

$$\frac{\partial^2\sqrt{G}}{\partial u^2}=\sqrt{G}, \tag{2.14}$$

from which we find

$$\sqrt{G}=A(v)e^{u}+B(v)e^{-u}. \tag{2.15}$$

In particular, when $B(v)=0$, $A(v)=1$, we have $G=e^{2u}$, and the first fundamental form (10.10) of Chapter 3 becomes

$$ds^2=du^2+e^{2u}dv^2, \tag{2.16}$$

with respect to which, of course, $K=-1$.

Putting

$$x=v, \qquad y=e^{-u}, \tag{2.17}$$

we obtain

$$ds^2=\frac{dx^2+dy^2}{y^2}, \tag{2.18}$$

which is valid in the half-plane $y>0$. The half-plane $y>0$ with the first fundamental form (2.18) is called the *Poincaré half-plane*.

Now let us determine the geodesics of the half-plane. Use $E=G=1/y^2$, $F=0$, $u=x, v=y$ in (7.20) of Chapter 3 to compute the Christoffel symbols; the nonzero ones are

$$\Gamma_1{}^2{}_1=\frac{1}{y}, \qquad \Gamma_1{}^1{}_2=\Gamma_2{}^2{}_2=-\frac{1}{y}. \tag{2.19}$$

Substituting (2.19) and

$$\frac{dy}{ds}=\frac{dy}{dx}\frac{dx}{ds}, \qquad \frac{d^2y}{ds^2}=\frac{dy}{dx}\frac{d^2x}{ds^2}+\left(\frac{dx}{ds}\right)^2\frac{d^2y}{dx^2}$$

in (10.9) of Chapter 3, we obtain the differential equation of the geodesics:

$$y\frac{d^2y}{dx^2}+\left(\frac{dy}{dx}\right)^2+1=0. \tag{2.20}$$

This differential equation can be integrated, giving

$$y^2\left[1+\left(\frac{dy}{dx}\right)^2\right]=a^2, \tag{2.21}$$

where a is a constant. An integration of (2.21) thus yields the equation of the geodesics:

$$(x-x_0)^2+y^2=a^2, \tag{2.22}$$

where x_0 is another constant. Hence *the geodesics of the Poincaré half-plane are the half-circle (in the Euclidean sense) with centers on the x-axis.*

By the identity mapping the Poincaré half-plane is mapped into the Euclidean plane. *This mapping is conformal*, for the angle θ between two tangent directions dx: dy and δx: δy is given by, in consequence of $E=G=1/y^2$, $F=0$,

$$\cos\theta=\frac{E\,dx\,\delta x+F(dx\,\delta y+\delta x\,dy)+G\,dy\,\delta y}{\sqrt{E\,dx^2+2F\,dxy+G\,dy^2}\,\sqrt{E\,\delta x^2+2F\delta x\,\delta y+G\,\delta y^2}}$$

$$=\frac{dx\,\delta x+dy\,\delta y}{\sqrt{dx^2+dy^2}\,\sqrt{\delta x^2+\delta y^2}}, \tag{2.23}$$

which is the same formula as for the Euclidean plane.

Exercises

1. Let S be a surface of revolution given by (1.12) of Chapter 3 with constant Gaussian curvature K and the first fundamental form given by (2.33) of Chapter 3. Prove the following by using the geodesic polar coordinates (u_1, v) in terms of which the first fundamental form is

$$ds^2=du_1^2+G\,dv^2,$$

and K is expressed by

$$\frac{d^2\sqrt{G}}{du_1^2}+K\sqrt{G}=0.$$

(a) For $K=1/a^2>0$,

$$u=c_1\cos\frac{u_1}{a}+c_2\sin\frac{u_1}{a},$$

$$x_3=\int\left[1-\frac{1}{a^2}\left(-c_1\sin\frac{u_1}{a}+c_2\cos\frac{u_1}{a}\right)^2\right]^{1/2}du_1.$$

In particular, when $c_1=a$ and $c_2=0$, S is a sphere.

(b) For $K = -1/b^2 < 0$,

$$u = c_1 e^{u_1/b} + c_2 e^{-u_1/b},$$

$$x_3 = \int \left[1 - \frac{1}{b^2}\left(c_1 e^{u_1/b} - c_2 e^{-u_1/b}\right)^2 \right]^{1/2} du_1.$$

In particular, when $c_1 = b$ and $c_2 = 0$, S is a pseudosphere (see Exercise 8 of Section 4, Chapter 3).

***2.** A surface is called a *Weingarten surface* if its principal curvatures κ_1 and κ_2 satisfy a linear relation

$$\lambda_1 \, d\kappa_1 + \lambda_2 \, d\kappa_2 = 0. \tag{2.24}$$

A Weingarten surface with relation (2.24) is called a *special Weingarten* surface if both λ_1 and λ_2 are positive. Show that a compact convex (Definition 5.1) special Weingarten surface S is a sphere [Chern, 10], and deduce that a compact convex surface with its Gaussian curvature K and mean curvature H satisfying

$$aK + 2bH + c = 0, \tag{2.25}$$

where a, b, c are constant such that $b^2 - ac > 0$, is a sphere.

3. THE GAUSS–BONNET FORMULA

The Gauss-Bonnet formula yields one of the most important results in the differential geometry of surfaces. The formula for a compact surface S expresses the integral of the Gaussian curvature of S as 2π times the Euler characteristic of S. Since the Euler characteristic is a topological invariant, the formula gives a relation between the differential invariant ds^2 and a topological property of the surface. This section presents the formula in its most general form, that is, for a general region on a surface. The formula is also valid on an abstract surface with a Riemann metric, that is, with an element of arc, but we shall prove it only for surfaces in Euclidean space E^3.

3.1. Definition. Let α: $[0, l] \to S$ be an oriented simple closed piecewise regular parametrized curve on an oriented surface S; let $\alpha(t_i)$, $i = 0, 1, \cdots, k$ and $0 = t_0 < t_1 < \cdots < t_k < t_{k+1} = l$, be the vertices of α, and let the image sets $\alpha([t_i, t_{i+1}])$ be the arcs of α (see Definition 1.2.2 of Chapter 2). Let $|\theta_i|$, $0 < |\theta_i| \leqslant \pi$, be the smallest angle from $\alpha'(t_i - 0)$ to $\alpha'(t_i + 0)$. Then the angle θ_i, $-\pi < \theta_i < \pi$, with a sign to be chosen as follows is called the *exterior angle*, and $\pi - \theta_i$ the *interior angle*, of α at the vertex $\alpha(t_i)$.

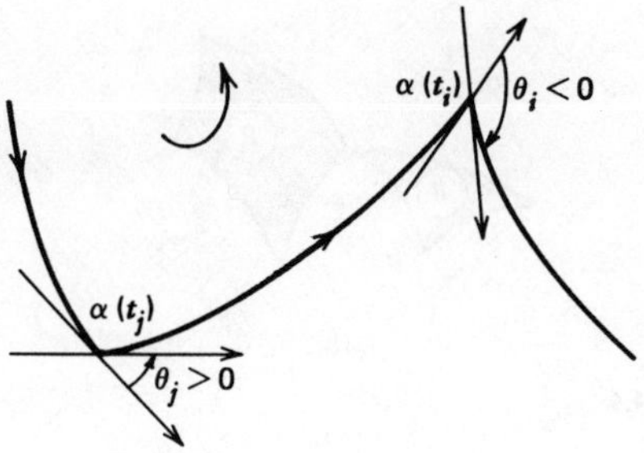

Figure 4.2

If $|\theta_i| \neq \pi$, we define the sign of θ_i to be that of the determinant $|\alpha'(t_i - 0), \alpha'(t_i + 0), \mathbf{e}_3|$, where $\mathbf{e}_3$ is the oriented unit normal vector of S. This means that if the vertex $\alpha(t_i)$ is not a "cusp" (Fig. 4.2), the sign of θ_i is given by the orientation of S.

If $|\theta_i| = \pi$, that is, if the vertex $\alpha(t_i)$ is a cusp, we choose the sign of θ_i to be that of the determinant $|\alpha'(t_i - \varepsilon), \alpha'(t_i + \varepsilon), \mathbf{e}_3|$ since by the regularity condition of α there exists a number $\varepsilon' > 0$ such that the determinant does not change sign for $0 < \varepsilon < \varepsilon'$ (Fig. 4.3).

Before we derive the Gauss-Bonnet formula, we need to review a few notions from topology.

3.2. Definition. Let S be a surface. A region $R \subset S$, the union of a connected open subset of S with its boundary, is said to be *regular* if it is compact and its boundary is a finite union of nonintersecting (simple), closed piecewise regular curves.

The region in Fig. 4.4*a* is regular, but that in Fig. 4.4*b* is not. For convenience, we shall consider a compact surface as a regular region with empty boundary.

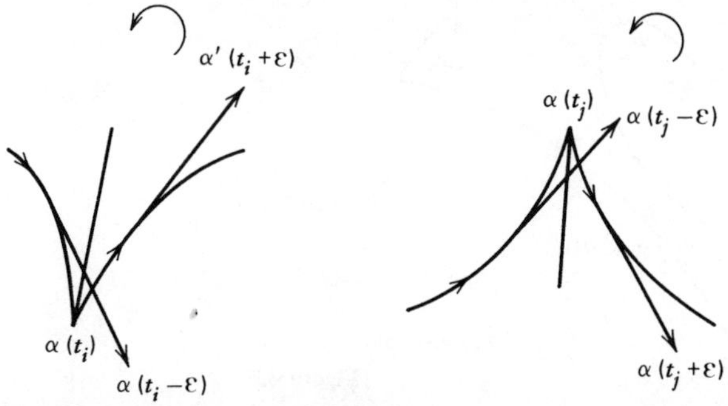

Figure 4.3

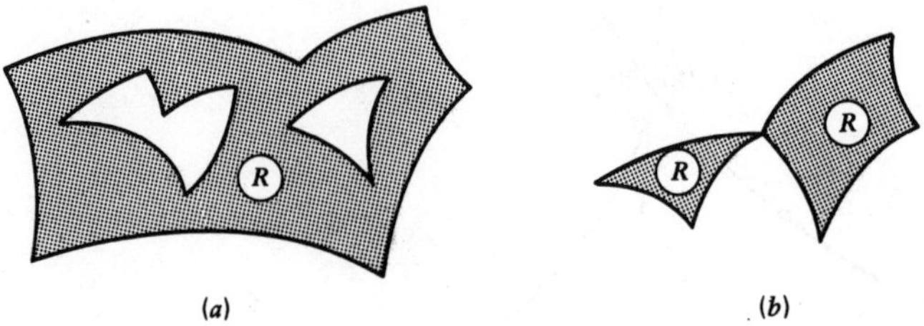

Figure 4.4

3.3. Definition. Let S be an oriented surface. A region $R \subset S$ is called a *simple region* if R is homeomorphic to a disk, and the boundary ∂R of R is the image set of a simple closed piecewise regular parametrized curve α: $I \to S$. α is said to be *positively oriented* if at each point $\alpha(t)$ the determinant $|\alpha'(t), \mathbf{h}(t), \mathbf{e}_3| > 0$, where $\mathbf{h}(t)$ is the inward pointing unit normal vector of α, and $\mathbf{e}_3$ is the oriented unit normal vector of S.

Intuitively, α having a positive orientation means that if one goes along α in the positive direction with one's head pointing to $\mathbf{e}_3$, the region R remains to one's left-hand side (Fig. 4.5).

3.4. Definition. A simple region that has only three vertices is a *triangle*. A *triangulation* of a regular region $R \subset S$ is a finite family $\mathfrak{T}$ of triangles $\{T_1, \cdots, T_n\}$ such that

(a) $\cup_{i=1}^{n} T_i = R$.
(b) For each pair of distinct i and j, $T_i \cap T_j$ is either empty or a common side of T_i and T_j, or a common vertex of T_i and T_j.

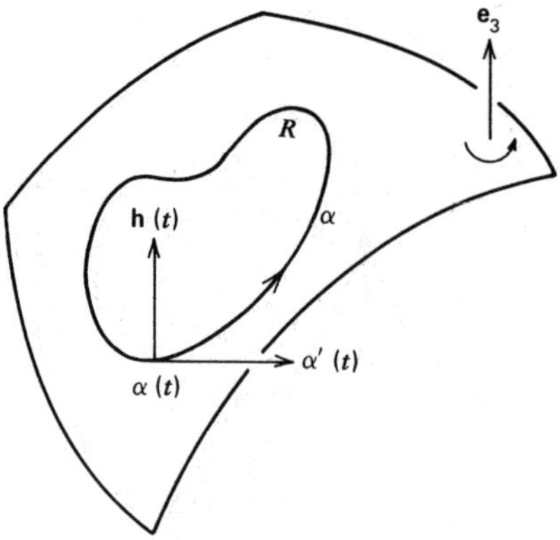

Figure 4.5

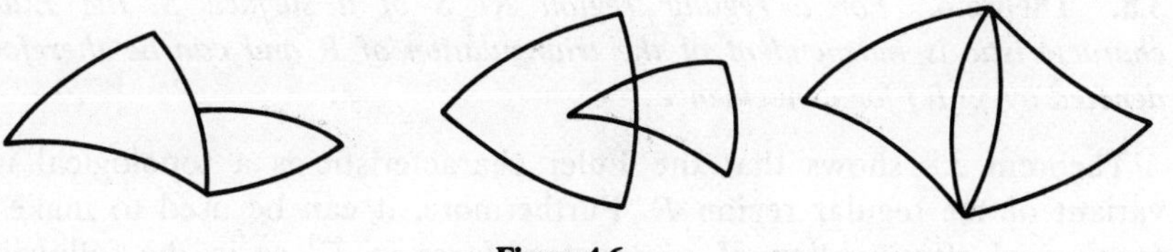

Figure 4.6

By way of clarifying condition (b), Fig. 4.6 shows three *unallowable* types of intersection of triangles.

3.5. Definition. For a given triangulation $\mathfrak{T}$ of a regular region $R \subset S$ of a surface S, let f be the number of triangles (faces), e the number of sides (edges), and v the number of vertices. Then the number

$$v - e + f = \chi \tag{3.1}$$

is the *Euler characteristic* of the triangulation.

The following theorems are given here without proofs; for more details see, for instance, an exposition in L. V. Ahlfors and L. Sario [1, Chapter I].

3.6. Theorem. *Every regular region of a surface admits a triangulation.*

3.7. Theorem. *Let S be an oriented surface, and $\{\mathbf{x}_\alpha\}, \mathbf{x}_\alpha : U \subset E^2 \to S$, a family of parametrizations compatible with the orientation of S, that is, for each α, $|\mathbf{x}_{\alpha u}, \mathbf{x}_{\alpha v}, \mathbf{e}_3| > 0$, where $(u, v) \in E^2$, $\mathbf{x}_{\alpha u}$ (respectively, $\mathbf{x}_{\alpha v}$) is the partial derivative of $\mathbf{x}_\alpha$ with respect to u (respectively, v), and $\mathbf{e}_3$ is the oriented unit normal vector of S. Let $R \subset S$ be a regular region of S. Then there is a triangulation $\mathfrak{T}$ of R such that every triangle $T \in \mathfrak{T}$ is contained in some neighborhood of the family $\{\mathbf{x}_\alpha\}$. Furthermore, if the boundary of every triangle of $\mathfrak{T}$ is positively oriented, opposite orientations are induced on the common side of every pair of adjacent triangles (Fig. 4.7).*

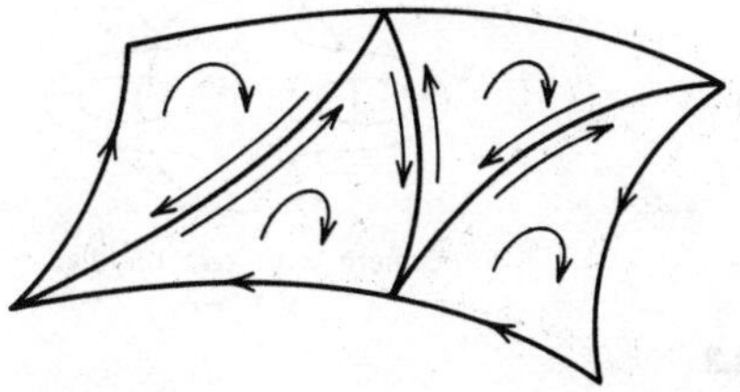

Figure 4.7

3.8. Theorem. *For a regular region $R \subset S$ of a surface S, the Euler characteristic is independent of the triangulation of R and can be therefore denoted by $\chi(R)$ for convenience.*

Theorem 3.8 shows that the Euler characteristic is a topological invariant of the regular region R. Furthermore, it can be used to make a topological classification of compact surfaces in E^3 as in the following theorem.

3.9. Theorem. *Let $S \subset E^3$ be a compact connected surface. Then the Euler characteristic $\chi(S)$ of S takes one of the values $2, 0, -2, \cdots, -2g, \cdots$, where g is a positive integer. Furthermore, if $S^* \subset E^3$ is another compact surface and $\chi(S) = \chi(S^*)$, then S is homeomorphic to S^*.*

By a direct calculation, we obtain that the Euler characteristic of the sphere is 2, that of the torus (sphere with one handle; see Fig. 4.8) is zero, that of the double torus (sphere with two handles) is -2, and, in general, that of the g-torus (sphere with g handles) is $-2(g-1)$. Thus from Theorem 3.9 it follows that every compact connected surface $S \subset E^3$ is homeomorphic to a sphere with g handles. The number

$$g = \frac{1}{2}(2 - \chi(S)) \tag{3.2}$$

is called the *genus* of S.

Now we can state and prove our main theorem.

3.10. Theorem (K. F. Gauss-O. Bonnet). *Let $R \subset S$ be a regular region of an oriented surface S with a boundary ∂R formed by n closed, simple, piecewise regular curves $C_1, \cdots, C_n$. Suppose that each C_i is positively oriented, and let $\phi_1, \cdots, \phi_p$ be the set of all interior angles of the curves*

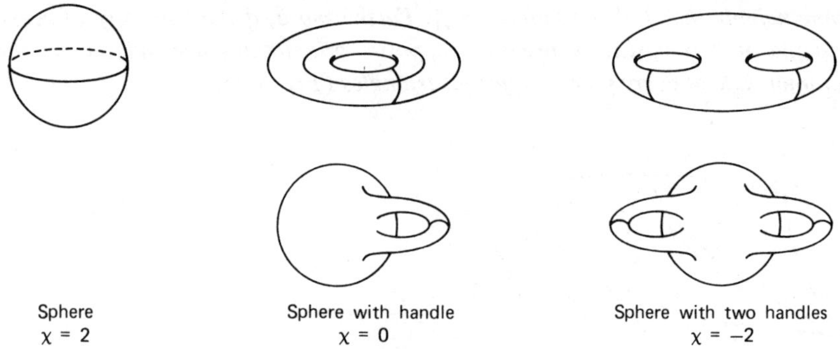

Figure 4.8

$C_1, \cdots, C_n$. *Then*

$$\sum_{i=1}^{n} \int_{C_i} \kappa_g(s)\,ds + \int\!\!\int_R K\,dA = 2\pi\chi(R) - \sum_{i=1}^{p} (\pi - \phi_i), \tag{3.3}$$

where s and $\kappa_g(s)$ are, respectively, the arc length and the geodesic curvature of C_i, and K and dA are, respectively, the Gaussian curvature and the element of area [*see* (*8.13*) *of Chapter 3*] *of R.*

Proof. Let C be a curve on S, and denote, as usual, the oriented unit normal vector of S by $\mathbf{e}_3$. Suppose that along C there is a unit tangent vector field $\mathbf{z}_1(s)$ of S such that the components of $\mathbf{z}_1$ are functions of class $\geqslant 1$ of the arc length s of C. Let $\mathbf{y}_1(s)$ be a unit parallel tangent vector field of S along C, and τ the angle between $\mathbf{y}_1$ and $\mathbf{z}_1$. As in the case of geodesic curvature (see part 1 of Remark 10.2, Chapter 3), $d\tau/ds$ is then independent of the choice of the vectors $\mathbf{y}_1(s)$ and is called the *variation of the vector field* $\mathbf{z}_1(s)$ along C. When $\mathbf{z}_1$ is the field of the unit tangent vectors of C, its variation is the geodesic curvature of C.

At first, we restrict our discussion to a coordinate neighborhood of an orthogonal parametrization $\mathbf{x}$: $U \subset E^2 \to S$ of S, which is compatible with the orientation of S. Let $(u,v) \in U$, and let $\mathbf{e}_1$ be the unit tangent vectors of the u-curves in the sense of increasing u. Denote the angle between $\mathbf{e}_1$ and $\mathbf{z}_1$ by θ, and introduce the unit tangent vectors $\mathbf{y}_2, \mathbf{z}_2, \mathbf{e}_2$ of S orthogonal to the tangent vectors $\mathbf{y}_1, \mathbf{z}_1, \mathbf{e}_1$, respectively, such that

$$|\mathbf{y}_1, \mathbf{y}_2, \mathbf{e}_3| = |\mathbf{z}_1, \mathbf{z}_2, \mathbf{e}_3| = |\mathbf{e}_1, \mathbf{e}_2, \mathbf{e}_3| = 1. \tag{3.4}$$

Then between the vectors $\mathbf{y}_1, \mathbf{y}_2$ and $\mathbf{z}_1, \mathbf{z}_2$ we have the relations

$$\begin{aligned} \mathbf{y}_1 &= \mathbf{e}_1 \cos(\theta - \tau) + \mathbf{e}_2 \sin(\theta - \tau), \\ \mathbf{y}_2 &= -\mathbf{e}_1 \sin(\theta - \tau) + \mathbf{e}_2 \cos(\theta - \tau). \end{aligned} \tag{3.5}$$

By (3.5) and $\mathbf{e}_1 \cdot \mathbf{e}_2 = 0$, in the same way as we derive (10.5) of Chapter 3, we find that

$$\frac{d\mathbf{y}_1}{ds} \cdot \mathbf{y}_2 = \mathbf{e}_2 \cdot \frac{d\mathbf{e}_1}{ds} + \frac{d(\theta - \tau)}{ds}.$$

Since the vectors $\mathbf{y}_1$ are parallel along C, from (9.9) of Chapter 3 it follows that $(d\mathbf{y}_1/ds) \cdot \mathbf{f}_i = 0$ for any two orthogonal tangent vectors $\mathbf{f}_i (i = 1,2)$. Therefore $(d\mathbf{y}_1/ds) \cdot \mathbf{y}_1 = 0$, $(d\mathbf{y}_1/ds) \cdot \mathbf{y}_2 = 0$, or $\mathbf{e}_2 \cdot (d\mathbf{e}_1/ds) + d\theta/ds - d\tau/ds = 0$. Hence we obtain the following formula for the variation ν of the vector field $\mathbf{z}_1(s)$:

$$\nu = \frac{d\tau}{ds} = \frac{d\theta}{ds} + \frac{d\mathbf{e}_1}{ds} \cdot \mathbf{e}_2, \tag{3.6}$$

or

$$\nu\, ds = d\theta + \mathbf{e}_2 \cdot d\mathbf{e}_1. \tag{3.7}$$

Now we want to evaluate the total variation of a vector field along a closed curve C:

$$\int_C \nu\, ds = \int_C d\theta + \int_C \mathbf{e}_2 \cdot d\mathbf{e}_1. \tag{3.8}$$

For this purpose we need the following Green's theorem in the uv-plane: *Let R be a region of the uv-plane, bounded by a closed simple piecewise smooth curve C. If $P(u,v)$ and $Q(u,v)$ are C^1 functions in R, then*

$$\int_C P\,du + Q\,dv = \int_R\int \left(\frac{\partial Q}{\partial u} - \frac{\partial P}{\partial v} \right) du\,dv. \tag{3.9}$$

Since $F=0$ and $\mathbf{e}_1 = \mathbf{x}_u/\sqrt{E}$, $\mathbf{e}_2 = \mathbf{x}_v/\sqrt{G}$, we have

$$\mathbf{e}_2 \cdot d\mathbf{e}_1 = \frac{\mathbf{x}_v}{\sqrt{G}} \cdot \left[\left(\frac{\mathbf{x}_{uu}}{\sqrt{E}} - \frac{\mathbf{x}_u E_u}{2E^{3/2}} \right) du + \left(\frac{\mathbf{x}_{uv}}{\sqrt{E}} - \frac{\mathbf{x}_u E_v}{2E^{3/2}} \right) dv \right], \tag{3.10}$$

and, in consequence of (7.20) of Chapter 3,

$$\begin{gathered} {\Gamma_1}^1{}_1 = \frac{E_u}{2E}, \qquad {\Gamma_1}^2{}_1 = -\frac{E_v}{2G}, \qquad {\Gamma_1}^1{}_2 = \frac{E_v}{2E}, \\ {\Gamma_1}^2{}_2 = \frac{G_u}{2G}, \qquad {\Gamma_2}^2{}_2 = \frac{G_v}{2G}. \end{gathered} \tag{3.11}$$

On the other hand, from $E = \mathbf{x}_u \cdot \mathbf{x}_u$, $F = \mathbf{x}_u \cdot \mathbf{x}_v = 0$, $G = \mathbf{x}_v \cdot \mathbf{x}_v$, it follows that $E_v = 2\mathbf{x}_u \cdot \mathbf{x}_{uv} = -2\mathbf{x}_{uu} \cdot \mathbf{x}_v$, $G_u = 2\mathbf{x}_{uv} \cdot \mathbf{x}_v$, so that (3.10) is reduced to

$$\mathbf{e}_2 \cdot d\mathbf{e}_1 = \sqrt{\frac{G}{E}} \left({\Gamma_1}^2{}_1\, du + {\Gamma_1}^2{}_2\, dv \right). \tag{3.12}$$

If the closed curve C bounds a region R, applying (3.9) to (3.12) gives

$$\begin{aligned} &\int_C \sqrt{\frac{G}{E}} \left({\Gamma_1}^2{}_1\, du + {\Gamma_1}^2{}_2\, dv \right) \\ &= \int_R\int \left[\left(\sqrt{\frac{G}{E}}\, {\Gamma_1}^2{}_2 \right)_u - \left(\sqrt{\frac{G}{E}}\, {\Gamma_1}^2{}_1 \right)_v \right] du\,dv. \end{aligned} \tag{3.13}$$

But by using (7.22) of Chapter 3 and (3.11), we can easily show

$$\left(\sqrt{\frac{G}{E}}\, {\Gamma_1}^2{}_2 \right)_u - \left(\sqrt{\frac{G}{E}}\, {\Gamma_1}^2{}_1 \right)_v = -K\sqrt{EG}. \tag{3.14}$$

Substituting (3.12) through (3.14) in (3.8) we thus obtain

$$\int_C \nu\, ds = \int_C d\theta - \int_R\!\!\int K\, dA, \tag{3.15}$$

where

$$dA = \sqrt{EG}\, du\, dv$$

is the element of area of S.

Now let $R \subset \mathbf{x}(U)$ be a simple region of S such that its boundary C is a positively oriented curve α parametrized by the arc length s, and let $\alpha(s_0), \cdots, \alpha(s_k)$ and $\phi_0, \cdots, \phi_k$ be, respectively, the vertices and the interior angles of α. Let $\mathbf{z}_1(s)$ be the unit tangent vector of the boundary α of R. Then the left-hand side of (3.15) becomes the integral of the geodesic curvature $\kappa_g(s)$ of α.

Let θ_i: $[s_i, s_{i+1}] \to E^1$ be the differentiable functions that measure at each $s \in [s_i, s_{i+1}]$ the positive angle from $\mathbf{x}_u$ to $\alpha'(s)$. Then by using (2.67) of Chapter 2 and assuming $s_{k+1} = s_0$, we obtain

$$\int_C d\theta = \sum_{i=0}^{k} [\theta(s_{i+1}) - \theta(s_i)] = 2\pi - \sum_{i=0}^{k} (\pi - \phi_i), \tag{3.16}$$

since the curve α is positively oriented so that its rotation index is $+1$. Hence (3.15) becomes

$$\sum_{i=0}^{k} \int_{s_i}^{s_{i+1}} \kappa_g\, ds + \int_R\!\!\int K\, dA = 2\pi - \sum_{i=0}^{k} (\pi - \phi_i), \tag{3.17}$$

which is the *Gauss-Bonnet formula* for a region R lying in a coordinate neighborhood of a surface.

Finally, let $R \subset S$ be a regular region of an oriented surface S. Then by Theorem 3.7 there exists a triangulation $\mathfrak{T}$ of R such that every triangle $T_j \in \mathfrak{T}, j = 1, \cdots, k$, is contained in a coordinate neighborhood $\mathbf{x}_j(U_j)$ of a family of orthogonal parametrizations $\{\mathbf{x}_\alpha\}$, compatible with the orientation of S. Furthermore, if the boundary of every triangle of $\mathfrak{T}$ is positively oriented, oppositive orientations are induced on the common side of every pair of adjacent triangles (Fig. 4.9).

By applying the local Gauss-Bonnet formula (3.17) to every triangle of $\mathfrak{T}$, adding up the resulting equations, and using the fact that the integrals of geodesic curvature along each interior side cancel each other, we obtain

$$\sum_{i=1}^{n} \int_{C_i} \kappa_g(s)\, ds + \int_R\!\!\int K\, dA = 2\pi f - \sum_{j} (\pi - \phi_j') - \sum_{i=0}^{p} (\pi - \phi_i), \tag{3.18}$$

where f is the number of the triangles T_j of $\mathfrak{T}$, ϕ_j' are the interior angles of the triangles at the interior vertices, (i.e., at the vertices, not on the

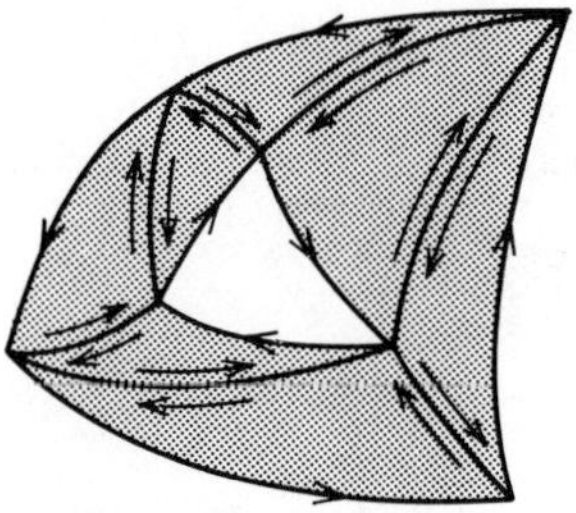

Figure 4.9

boundary α of R), and ϕ_i are the interior angles of the boundary curves $C_1, \cdots, C_n$.

In the first sum $\Sigma_j(\pi - \phi_j')$ on the right-hand side of (3.18), each term $\pi - \phi_j$ is an exterior angle at an interior vertex of a triangle. Since each such exterior angle (therefore each π) is associated with an interior side, that is, with a side that is not on the boundary ∂R, and each interior side is on exactly two triangles, the first part of the sum is equal to $2\pi e$, where e is the number of the interior sides. Since the sum of all the interior angles at an interior vertex is 2π, the second part of the sum is equal to $-2\pi v$, where v is the number of the interior vertices. Since on the boundary ∂R the number of vertices is equal to that of sides, $\chi(R) = v - e + f$, and by substituting all the quantities above in (3.18) we hence arrive at the required formula (3.3). Q.E.D.

If R is a simple region, its Euler characteristic is 1; therefore Theorem 3.10 becomes Corollary 3.11.

3.11. Corollary. *If R is a simple region of S, then* (3.3) *becomes* (3.17).

Similarly from (3.3) we obtain Corollary 3.12.

3.12. Corollary. *If the boundary ∂R of a region R has no vertex, then*

$$\sum_{i=0}^{k} \int_{s_i}^{s_{i+1}} \kappa_g \, ds + \int_R \int K \, dA = 2\pi\chi(R). \tag{3.19}$$

Since a compact surface may be considered as a regular region with empty boundary, Corollary 3.13 follows immediately from Corollary 3.12.

3.13. Corollary. *For a compact surface,*

$$\int_S \int K \, dA = 2\pi\chi(S). \tag{3.20}$$

Corollary 3.13 is the result mentioned at the beginning of this section. It shows that on the one hand $\chi(S)$ is independent of the triangulation of S and, on the other hand, $\int_S\int K\,dA$ is a topological invariant.

3.14. Applications. Several applications of the Gauss-Bonnet theorem (Theorem 3.10) include the following:

(a) *A compact surface S of positive Gaussian curvature K is homeomorphic to a sphere.*

Since $K>0$ by (3.20) the Euler characteristic is positive, and by Theorem 3.9, S is homeomorphic to a sphere.

(b) *Let S be an orientable surface of negative or zero Gaussian curvature. Then two geodesics γ_1 and γ_2 starting from a point $\mathbf{p}\in S$ cannot meet again at a point $\mathbf{q}\in S$ to form the boundary of a simple region R of S.*

Suppose the contrary. Then Corollary 3.11 and (3.17) give

$$\int_R\int K\,dA=\phi_1+\phi_2,$$

where ϕ_1 and ϕ_2 are the interior angles of R. Thus we have a contradiction, since $K\leqslant 0$ and $\phi_i>0$ for $i=1,2$.

When $\phi_1=\phi_2=\pi$, γ_1 and γ_2 form a simple closed geodesic of S. Thus *on a surface of zero or negative Gaussian curvature there exists no simple closed geodesic bounding a simple region of S.*

(c) *A surface S with negative Gaussian curvature and homeomorphic to a cylinder has at most one simple closed geodesic.*

At first we notice that there is a homeomorphism h of a cylinder $C=S^1\times L$, where $S^1=\{(x,y)\in E^2\,|\,x^2+y^2=1\}$ and $L=(0,\infty)$, to a plane π minus a point $\mathbf{p}\in\pi$. In fact, by taking $\mathbf{p}$ to be (0,0), we can express

$$h\colon E^2-(0,0)\to C$$

by

$$h(x,y)=\left(\left(\frac{x}{l},\frac{y}{l}\right),l\right),\qquad l=\sqrt{x^2+y^2}\,.$$

To prove Application c we suppose that the surface S has a simple closed geodesic C_1. Since there is a homeomorphism f of S to a plane π minus a point $\mathbf{p}\in\pi$, by Application b, $f(C_1)$ is the boundary of a simple region of π containing $\mathbf{p}$.

If S has another simple closed geodesic C_2, then C_1 does not intersect C_2: if this were the case, the arcs of $f(C_1)$ and $f(C_2)$ between two "consecutive" points $\mathbf{r}_1$ and $\mathbf{r}_2$ of intersection would be the boundary of a simple region, contradicting Application b (see Fig. 4.10). By the argument

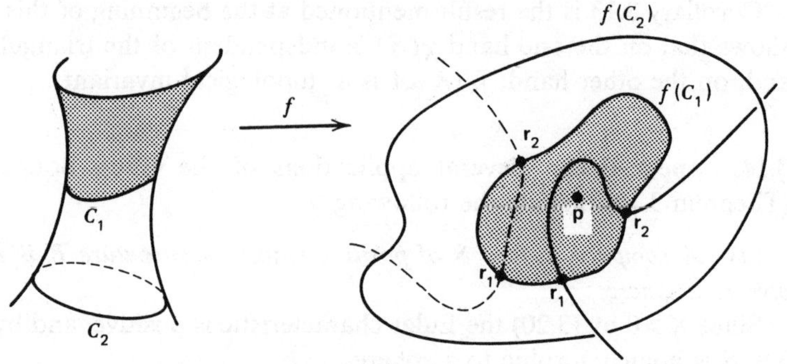

Figure 4.10

above, $f(C_2)$ is the boundary of a simple region R of π containing **p**, the interior of which is homeomorphic to a cylinder. Thus $\chi(R)=0$, since the Gaussian curvature of a cylinder vanishes identically. On the other hand, by the Gauss-Bonnet formula

$$\int\!\!\int_{f^{-1}(R)} K\,dA = 2\pi\chi(R) = 0,$$

which does not hold because $K<0$.

(d) *On a compact surface S of positive Gaussian curvature any two simple closed geodesics C_1 and C_2 intersect.*

By Application a, S is homeomorphic to a sphere. If C_1 and C_2 do not intersect, then C_1 and C_2 together form the boundary of a region R. Since a sphere minus a point is homeomorphic to a plane (see Exercise 11b of Section 1, Chapter 3), R is so and therefore $\chi(R)=0$. By (3.19) we have

$$\int\!\!\int_R K\,dA = 0,$$

which does not hold because $K>0$.

(e) Let C be a closed space curve of length L given by $\mathbf{x}(s)$ with arc length s, and S^2 the unit sphere with center at the origin $\mathbf{0}$ in the space E^3. Then the mapping f: $C \to S^2$ given by

$$f(\mathbf{x}(s)) = \mathbf{n}(s), \qquad s \in [0, L],$$

where $\mathbf{n}(s)$ is the unit principal normal vector of C at the point $\mathbf{x}(s)$, is called the *principal normal mapping* of C, and the image Γ of C under f is called the *principal normal indicatrix* of C. By the Frenet formula, (1.3.11) of Chapter 2, we have

$$\frac{d\mathbf{t}}{ds}=\kappa\mathbf{n},\quad \frac{d\mathbf{n}}{ds}=-\kappa\mathbf{t}+\tau\mathbf{b},\quad \frac{d\mathbf{b}}{ds}=-\tau\mathbf{n}, \tag{3.21}$$

where κ and τ are the curvature and torsion, and $\mathbf{t}$ and $\mathbf{b}$ the unit tangent and binormal vectors of C at $\mathbf{x}(s)$, respectively, such that the determinant $|\mathbf{t},\mathbf{n},\mathbf{b}|=1$. Thus Γ has a nonzero tangent whenever

$$\kappa^2+\tau^2\neq 0. \tag{3.22}$$

As another application of the Gauss-Bonnet theorem we have Theorem 3.14.1.

3.14.1. Theorem (C. G. J. Jacobi). *If the principal normal indicatrix Γ of a closed space curve C is regular and simple, then Γ divides the unit sphere S^2 into two regions of equal areas.*

Proof. Let σ be the arc length of Γ. Then from (3.21) we have

$$\frac{d\sigma}{ds}=\sqrt{\left(\frac{d\mathbf{n}}{ds}\right)^2}=\sqrt{\kappa^2+\tau^2}, \tag{3.23}$$

$$\frac{d\mathbf{n}}{d\sigma}=\frac{d\mathbf{n}}{ds}\frac{ds}{d\sigma}=-\mathbf{t}\cos\phi+\mathbf{b}\sin\phi, \tag{3.24}$$

where

$$\kappa=\sqrt{\kappa^2+\tau^2}\,\cos\phi,\qquad \tau=\sqrt{\kappa^2+\tau^2}\,\sin\phi. \tag{3.25}$$

Furthermore, differentiation of (3.24) and use of (3.25) and (3.23) give

$$\frac{d^2\mathbf{n}}{d\sigma^2}=(\mathbf{t}\sin\phi+\mathbf{b}\cos\phi)\frac{d\phi}{d\sigma}-\mathbf{n}. \tag{3.26}$$

Hence by (10.6) of Chapter 3 and (3.24) and (3.26), we can easily obtain the geodesic curvature of Γ:

$$\kappa_g=\left|\mathbf{n},\frac{d\mathbf{n}}{d\sigma},\frac{d^2\mathbf{n}}{d\sigma^2}\right|=\frac{d\phi}{d\sigma}. \tag{3.27}$$

Let R be one of the regions bounded by Γ on S^2, and A its area. Since the Gauss curvature of S^2 is 1, and $\chi(R)=1$, by applying Corollary 3.11 to R we have

$$2\pi=\int_R dA+\int_\Gamma d\phi=A,$$

which is one-half of the area of S^2.

(f) Let T be a geodesic triangle (i.e., each side of T is a geodesic) with interior angles ϕ_1, ϕ_2, ϕ_3 on an oriented surface S. Then $\chi(T)=1$; therefore (3.3) gives

$$\int_T\int K\,dA=\sum_{i=1}^{3}\phi_i-\pi. \tag{3.28}$$

Thus *the sum of the interior angles*, $\Sigma_{i=1}^{3}\phi_i$, *of a geodesic triangle T on an oriented surface S is greater than, equal to, or less than* π *depending on whether* $K>0$, $=0$, *or* <0. Furthermore, from Exercise 1 of Section 8, Chapter 3, and (3.28) it follows that $\Sigma_{i=1}^{3}\phi_i-\pi$, called the *excess* of T, is equal to the area of the spherical image of T under the Gauss mapping (Definition 2.12 of Chapter 3).

(g) A vector field $\mathbf{v}$ on a surface S is a function that assigns a vector $\mathbf{v}(\mathbf{p})\in E^3$ to each point $\mathbf{p}\in S$. Let $\mathbf{p}=\mathbf{x}(u,v)$. Then $\mathbf{v}(\mathbf{p})$ can be written as $\mathbf{v}(\mathbf{p})=(v_1(u,v), v_2(u,v), v_3(u,v))$. The vector field $\mathbf{v}$ is said to be of class C^r, $r>0$, if the functions v_1, v_2, v_3 are C^r functions for all u, v. $\mathbf{p}\in S$ is called a *singular point* of a C^r vector field $\mathbf{v}$ of S if $\mathbf{v}(\mathbf{p})=0$. The singular point $\mathbf{p}$ is said to be *isolated* if there exists a neighborhood V of $\mathbf{p}$ on S such that $\mathbf{v}$ has no singular points in V other than $\mathbf{p}$.

With each isolated singular point $\mathbf{p}$ of a C^r, $r\geqslant 1$, vector field $\mathbf{v}$ on an oriented surface S we associate an integer, called the *index* of v, as follows.

Let $\mathbf{x}$: $U\subset E^2\to S$ be an orthogonal parametrization of S at $\mathbf{p}=\mathbf{x}(0,0)$ compatible with the orientation of S, and let α: $[0,l]\to S$ be a simple closed piecewise regular parametrized curve C on S such that $\alpha((0,l))\subset\mathbf{x}(U)$ is the boundary of a simple region R containing $\mathbf{p}$ as its only singular point. Let $\mathbf{v}=\mathbf{v}(t)$, $t\in[0,l]$, be the restriction of $\mathbf{v}$ along α, and let $\theta=\theta(t)$ be some differentiable determination of the angle from $\mathbf{x}_u$ to $\mathbf{v}(t)$. Since the curve C is closed, the vector $\mathbf{v}$ returns to the same position after a circuit around C; hence I defined by

$$\int_C d\theta=\int_0^l\frac{d\theta}{dt}dt=\theta(l)-\theta(0)=2\pi I \tag{3.29}$$

is an integer, which is called the *index* of the vector field $\mathbf{v}$ at the singular point $\mathbf{p}$.

To show that the index I is well defined, we first notice that it is independent of the choice of the parametrization $\mathbf{x}$, since from (3.15) and (3.29) we have

$$2\pi I=\int_C \nu\,ds+\int_R\int K\,dA, \tag{3.30}$$

where the right-hand side is independent of the choice of $\mathbf{x}$.

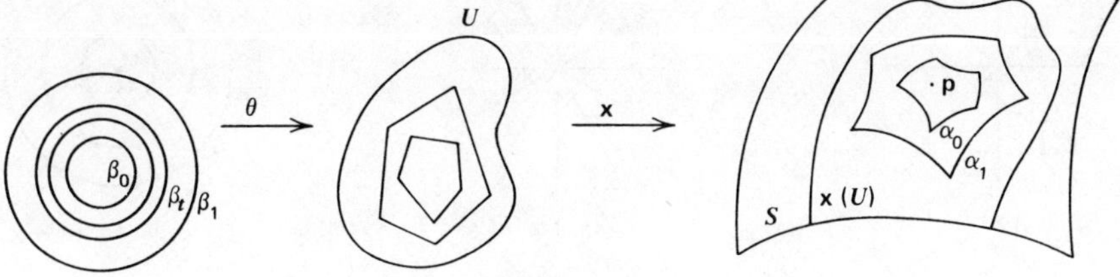

Figure 4.11

Next, we shall show that the index I does not depend on the choice of the curve α. Let α_0 and α_1 be the two curves as in the definition of I, and first suppose that the image sets of α_0 and α_1 do not intersect. Then there is a homeomorphism of the region bounded by the image sets of α_0 and α_1 onto a region of the plane bounded by two concentric circles β_0 and β_1. Since we can obtain a family of concentric circles β_t that depend continuously on t and deform β_0 into β_1, there is a family of curves α_t that depend continuously on t and deform α_0 into α_1 (Fig. 4.11). Denote by I_t the index of $\mathbf{v}$ with respect to the curve α_t. Since the index is an integral, I_t depends continuously on t, $t \in [0, l]$. Being an integer, I_t is constant under this deformation; hence $I_0 = I_1$.

If the image sets of α_0 and α_1 intersect, we choose a curve so small that it has no intersection with both α_0 and α_1, and then apply the previous result.

It should be remarked that when $\mathbf{p}$ is not a singular point of $\mathbf{v}$. We still can define the index of $\mathbf{v}$ at $\mathbf{p}$, but then it is zero. This follows because, since I is independent of the choice of $\mathbf{x}_u$, we can choose $\mathbf{x}_u$ to be $\mathbf{v}$ so that $\theta(t) \equiv 0$.

In Fig. 4.12 we give some examples of indices of vector fields in a plane at an isolated singular point. The curves are the trajectories or the integral curves of the vector fields and the singularities are called, respectively: (a) a source or maximum, (b) a sink or minimum, (c) a center, (d) a simple saddle point, (e) a monkey saddle, and (f) a dipole.

Concerning the indices of vector fields, we have the following important theorem.

3.14.2. Theorem (H. Poincaré). *The sum of the indices of a C^1 vector field* $\mathbf{v}$ *with only isolated singular points on a compact surface S is equal to the Euler characteristic of S.*

Proof. At first we notice that the number of the singular points is finite. Otherwise, by the Bolzano-Weierstrass theorem† the set of the singular

†Every infinite subset of a compact set A in E^2 has a least one limit point in A.

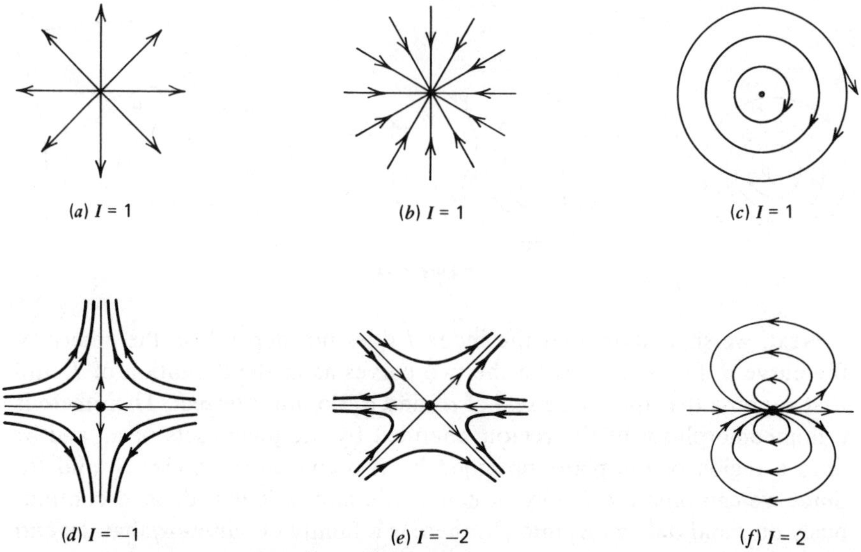

Figure 4.12

points has a limit point that is a nonisolated singular point, a contradiction to the hypothesis.

Now let $\{\mathbf{x}_u\}$ be a family of orthogonal parametrizations of S, compatible with the orientation of S. Let $\mathfrak{T}$ be a triangulation of S such that the following conditions hold:

1. Each triangle $T \in \mathfrak{T}$ is contained in a coordinate neighborhood of the family $\{\mathbf{x}_u\}$.
2. Each $T \in \mathfrak{T}$ contains at most one singular point.
3. The boundary of each $T \in \mathfrak{T}$ contains no singular points and is positively oriented.

Applying (3.17) to each triangle $T \in \mathfrak{T}$, summing up the results, using the fact that the integrals of geodesic curvature along each interior side cancelled with each other, and using (3.30) and Corollary 3.13 we finally arrive at

$$\sum_{i=1}^{k} I_i = \frac{1}{2\pi}\int\!\!\int_S K\,dA = \chi(S), \qquad (3.31)$$

where I_i is the index of the singular point $\mathbf{p}_i$, $i = 1, \cdots, k$. Q.E.D.

It should be remarked that Theorem 3.14.2 implies that $\Sigma_{i=1}^{k} I_i$ does not depend on $\mathbf{v}$, but depends only on the topology of the surface S. For instance, on any surface homeomorphic to a sphere the sum of the indices of any C^1 vector field with isolated singular points must be equal to 2. Therefore any such surface cannot have a C^1 vector field without singular points.

Exercises

1. Prove (3.14).

*2. Show that an orientable compact surface with positive Gaussian curvature is simply connected (J. L. Synge).

3. Let S be a regular compact orientable surface that is not homeomorphic to a sphere. Prove that there are points on S at which the Gaussian curvature is positive, negative, and zero.

*4. Use the Gauss-Bonnet formula to find the Euler characteristic of the torus with parametrization (1.18) of Chapter 3.

*5. Find the Euler characteristic of the surface $x_1^2+x_2^4+x_3^6=1$.

6. Let T be the torus with parametrization (1.18) of Chapter 3. Prove that the vector field on T obtained by parametrizing all its meridians (v=const) by arc length and taking their tangent vectors is nonzero everywhere and is differentiable.

7. Show that (0,0) is an isolated singular point, and compute the index at (0,0) of the following vector fields $\mathbf{v}=(a(t), b(t))$ in the x_1x_2-plane satisfying the differential equations $dx_1/dt=a$, $dx_2/dt=b$:

   *(a) $\mathbf{v}=(x_1, x_2)$.
   (b) $\mathbf{v}=(x_2, -x_1)$.
   (c) $\mathbf{v}=(x_1, -x_2)$.
   *(d) $\mathbf{v}=(x_1^2-x_2^2, -2x_1x_2)$.

8. Prove that an orientable compact surface S in E^3 has a differentiable vector field without singular points if and only if S is homeomorphic to a torus.

4. EXTERIOR DIFFERENTIAL FORMS AND A UNIQUENESS THEOREM FOR SURFACES

Section 6 of Chapter 1 introduced the exterior multiplication $\wedge$, the exterior differentiation d, and the mixed operation $\hat{\mathbf{x}}$ on differential forms in a space E^3, and obtained the structural equations for a positively oriented frame at a point (Definition 5.4.1 of Chapter 1). In this section we first express all these on a surface, then in Sections 5 through 7 we use them to prove a uniqueness theorem and some global theorems for surfaces.

Let S be a surface in E^3 so that on S the natural coordinate functions x_1, x_2, x_3 for points in E^3 are functions of two variables u, v. Then

$$dx_i(u,v)=\frac{\partial x_i}{\partial u}du+\frac{\partial x_i}{\partial v}dv, \qquad i=1,2,3. \tag{4.1}$$

By substituting (4.1) in (6.1.10), (6.2.15), (6.2.14), and (6.2.16) of Chapter 1, and using $du\wedge dv=-dv\wedge du$, we obtain

$$\begin{aligned} \alpha&=a(u,v)\,du+b(u,v)\,dv,\\ \beta&=c(u,v)\,du+f(u,v)\,dv, \end{aligned} \tag{4.2}$$

$$\alpha\wedge\beta=(af-bc)\,du\wedge dv, \tag{4.3}$$

$$\alpha\wedge\beta=-\beta\wedge\alpha, \tag{4.4}$$

$$d\alpha=(b_u-a_v)\,du\wedge dv, \tag{4.5}$$

where α, β are two 1-forms on S, and the subscripts u, v denote the partial derivatives. From (4.5) it follows that Green's formula (3.9) becomes

$$\int_C \alpha=\int_R\int d\alpha, \tag{4.6}$$

where R is a region of the uv-plane, bounded by a closed simple piecewise smooth curve C.

Let $\mathbf{x}\mathbf{e}_1\mathbf{e}_2\mathbf{e}_3$ be a positively oriented frame at a point $\mathbf{x}\in E^3$. The space $\mathfrak{F}$ of all such frames has dimension 6, since it takes three coordinates to specify the positive vector $\mathbf{x}$, two more coordinates to specify the unit vector $\mathbf{e}_1$, and one further coordinate to fix $\mathbf{e}_2$, and after $\mathbf{e}_1, \mathbf{e}_2$ the unit vector $\mathbf{e}_3$ is completely determined.

Now on a surface S in E^3 we consider a two-parameter subfamily of the family $\mathfrak{F}$ such that the vectors $\mathbf{x}, \mathbf{e}_1, \mathbf{e}_2$, and $\mathbf{e}_3$ have components whose Euclidean natural coordinates are C^2 functions of two variables u, v in an open set $U\subset E^2$. Then from Section 6.3 of Chapter 1 we have

$$\begin{aligned} d\mathbf{x}&=\sum_{i=1}^{3}\omega_i\mathbf{e}_i,\\ d\mathbf{e}_i&=\sum_{j=1}^{3}\omega_{ij}\mathbf{e}_j, \qquad i=1,2,3, \end{aligned} \tag{4.7}$$

and therefore the structural equations

$$d\omega_i = \sum_{j=1}^{3} \omega_j \wedge \omega_{ji}, \qquad i=1,2,3, \tag{4.8}$$

$$d\omega_{ij} = \sum_{k=1}^{3} \omega_{ik} \wedge \omega_{kj}, \qquad i,j=1,2,3, \tag{4.9}$$

where ω_i, ω_{ij} are 1-forms on the surface S, and ω_{ij} satisfy

$$\omega_{ij} + \omega_{ji} = 0, \qquad i,j=1,2,3, \tag{4.10}$$

which imply

$$\omega_{ii} = 0, \qquad i=1,2,3. \tag{4.11}$$

These structural equations are equivalent to the integrability conditions (7.21), (7.22) and its similar equation, and (7.24) of Chapter 3.

Now let $\mathbf{x}$: $U \subset E^2 \to S$ be a parametrization of a surface S in E^3. To apply the theory of frame fields to the theory of surfaces S, we choose the frames $\mathbf{x}\mathbf{e}_1\mathbf{e}_2\mathbf{e}_3(u,v)$ such that $\mathbf{x}(u,v) \in S$, and $\mathbf{e}_3(u,v)$ is the unit normal vector of the surface S. The latter condition involves a choice of one of the two normal vectors, and means geometrically that the discussion applies to an oriented surface. Quantities invariant under the change $\mathbf{e}_3 \to -\mathbf{e}_3$ are the invariants of an unoriented surface.

By this choice, the vectors $\mathbf{e}_1$ and $\mathbf{e}_2$ are tangent vectors of S at $\mathbf{x}$. Since $d\mathbf{x} = \mathbf{x}_u\,du + \mathbf{x}_v\,dv$ is a linear combination of two tangent vectors, from (4.7) we have $\omega_3 = 0$. Moreover, ω_1 and ω_2 are linearly independent 1-forms (Lemma 6.3.1 of Chapter 1). Put

$$\omega_{13} = a\omega_1 + b\omega_2, \qquad \omega_{23} = b'\omega_1 + c\omega_2. \tag{4.12}$$

Then (4.8) for $i=3$ gives

$$0 = \omega_1 \wedge \omega_{13} + \omega_2 \wedge \omega_{23}. \tag{4.13}$$

Substituting (4.12) in (4.13) and observing that $\omega_1 \wedge \omega_2 \neq 0$, we obtain

$$b = b'. \tag{4.14}$$

The first and second fundamental forms of the surface S have the expressions

$$\begin{aligned} \mathrm{I} &= d\mathbf{x}^2 = \omega_1^2 + \omega_2^2, \\ \mathrm{II} &= -d\mathbf{x} \cdot d\mathbf{e}_3 = \omega_1\omega_{13} + \omega_2\omega_{23} \\ &= a\omega_1^2 + 2b\omega_1\omega_2 + c\omega_2^2. \end{aligned} \tag{4.15}$$

Since the mean curvature H and the Gaussian curvature K of the surface S

at the point $\mathbf{x}(u, v)$ are the two elementary symmetric functions of the roots of the equation:

$$\begin{vmatrix} a-\lambda & b \\ b & c-\lambda \end{vmatrix} = 0$$

[cf. (4.6) of Chapter 3], we have

$$H = \tfrac{1}{2}(a+c), \qquad K = ac - b^2. \tag{4.16}$$

It should be noticed that if we change the direction of $\mathbf{e}_3$, then a, b, and c change sign. Hence when the orientation of the surface is reversed, H changes its sign, while K is an invariant of the unoriented surface.

4.1. Lemma. *H and K are in a sense the only invariants of the surface S obtained algebraically from the two fundamental forms.*

Proof. To make the statement of Lemma 4.1 precise, we study the effect of a rotation of the tangent vectors $\mathbf{e}_1, \mathbf{e}_2$ yielding the new vectors:

$$\begin{aligned} \mathbf{e}_1^* &= \mathbf{e}_1 \cos\Delta + \mathbf{e}_2 \sin\Delta, \\ \mathbf{e}_2^* &= -\mathbf{e}_1 \sin\Delta + \mathbf{e}_2 \cos\Delta, \end{aligned} \tag{4.17}$$

where the angle $\Delta = \Delta(u, v)$. Then with respect to the new frame $\mathbf{x}\mathbf{e}_1^*\mathbf{e}_2^*\mathbf{e}_3$ we have equations of the same form as (4.7) through (4.12) and (4.14), expressed by the same letters, but each of which, except $d, \mathbf{x}, \mathbf{e}_3$, has an asterisk. We also denote each new equation by the same numbers as those of the corresponding old equation but with an asterisk. From (4.7)*, (4.7), and (4.17) it follows that

$$\begin{aligned} \omega_1^* &= d\mathbf{x} \cdot \mathbf{e}_1^* = \omega_1 \cos\Delta + \omega_2 \sin\Delta, \\ \omega_2^* &= d\mathbf{x} \cdot \mathbf{e}_2^* = -\omega_1 \sin\Delta + \omega_2 \cos\Delta. \end{aligned} \tag{4.18}$$

By introducing matrices

$$T = \begin{pmatrix} \cos\Delta & \sin\Delta \\ -\sin\Delta & \cos\Delta \end{pmatrix}, \qquad \omega^* = \begin{pmatrix} \omega_1^* \\ \omega_2^* \end{pmatrix}, \qquad \omega = \begin{pmatrix} \omega_1 \\ \omega_2 \end{pmatrix}, \tag{4.19}$$

(4.18) are thus reduced to the simplified form

$$\omega^* = T\omega. \tag{4.20}$$

Moreover, from (4.7)*, (4.17), and (4.10)* we have the differential forms

$$\omega_{13}^* = -d\mathbf{e}_3 \cdot \mathbf{e}_1^* = -\omega_{31}^*, \qquad \omega_{23}^* = -d\mathbf{e}_3 \cdot \mathbf{e}_2^* = -\omega_{32}^*. \tag{4.21}$$

By introducing matrices

$$\theta = \begin{pmatrix} \omega_{13} \\ \omega_{23} \end{pmatrix}, \qquad \theta^* = \begin{pmatrix} \omega_{13}^* \\ \omega_{23}^* \end{pmatrix}, \tag{4.22}$$

and using (4.7) for de_3, (4.17), (4.21), and (4.19) we obtain

$$\theta^* = T\theta. \tag{4.23}$$

Suppose that

$$A = \begin{pmatrix} a & b \\ b & c \end{pmatrix}, \qquad A^* = \begin{pmatrix} a^* & b^* \\ b^* & c^* \end{pmatrix}. \tag{4.24}$$

Then use of (4.22), (4.24), (4.19), (4.12), (4.14), (4.12)*, and (4.14)* yields

$$\theta = A\omega, \qquad \theta^* = A^*\omega^*. \tag{4.25}$$

Equations 4.23, 4.25, and 4.20 imply that $T\theta = TA\omega = A^*T\omega$ or

$$A^* = TAT^{-1}, \tag{4.26}$$

which gives the change of the coefficients in II arising from a change (4.17) of the frames.

With an indeterminate λ we have, in consequence of (4.26),

$$\det(A^* - \lambda I) = \det(T(A - \lambda I)T^{-1}) = \det(A - \lambda I).$$

Thus under a change of frames the equation $\det(A - \lambda I) = 0$ is invariant, so that the set of the two roots λ_1, λ_2 of the equation is unchanged. Hence the two elementary symmetric functions $\frac{1}{2}(\lambda_1 + \lambda_2)$ and $\lambda_1\lambda_2$ of λ_1 and λ_2 are the unique independent invariants of the surface S obtained algebraically from the two fundamental forms I and II. (In general, a function of two variables x_1, x_2 is a symmetric function of x_1, x_2 if it is unaltered when x_1, x_2 are interchanged, and every symmetric function can be expressed in terms of the elementary symmetric functions.) Since $\det(A - \lambda I) = \lambda^2 - (a+c)\lambda + (ac - b^2)$, the two elementary symmetric functions of the roots of the equation $\det(A - \lambda I) = 0$ are just H and K, and Lemma 4.1 is proved.

4.2. Lemma. *$\omega_1 \wedge \omega_2$ is the element of area dA of the surface.*

Proof. From (4.18) it follows readily that

$$\omega_1^* \wedge \omega_2^* = \omega_1 \wedge \omega_2,$$

so that $\omega_1 \wedge \omega_2$ is an invariant under the rotation (4.17).

By means of (4.7), the identity of Lagrange (3.3.8) of Chapter 1, and (2.13) and (8.13) of Chapter 3, we obtain

$$\begin{aligned}
\omega_1 \wedge \omega_2 &= (d\mathbf{x} \cdot \mathbf{e}_1) \wedge (d\mathbf{x} \cdot \mathbf{e}_2) \\
&= [(\mathbf{x}_u\, du + \mathbf{x}_v\, dv) \cdot \mathbf{e}_1] \wedge [(\mathbf{x}_u\, du + \mathbf{x}_v\, dv) \cdot \mathbf{e}_2] \\
&= [(\mathbf{x}_u \cdot \mathbf{e}_1)(\mathbf{x}_v \cdot \mathbf{e}_2) - (\mathbf{x}_v \cdot \mathbf{e}_1)(\mathbf{x}_u \cdot \mathbf{e}_2)]\, du \wedge dv \qquad (4.27) \\
&= [(\mathbf{x}_u \times \mathbf{x}_v) \cdot (\mathbf{e}_1 \times \mathbf{e}_2)]\, du \wedge dv = [(\mathbf{x}_u \times \mathbf{x}_v) \cdot \mathbf{e}_3]\, du \wedge dv \\
&= \sqrt{EG - F^2}\, du \wedge dv = dA.
\end{aligned}$$

4.3. Another Proof of Theorem 7.2 of Chapter 3. We have the first fundamental form $I=\omega_1^2+\omega_2^2$ and the 1-form ω_{12} satisfying (4.8):

$$\begin{aligned} d\omega_1 &= \omega_2\wedge\omega_{21} = -\omega_2\wedge\omega_{12}, \\ d\omega_2 &= \omega_1\wedge\omega_{12}. \end{aligned} \tag{4.28}$$

For given ω_1 and ω_2, ω_{12} is uniquely determined by (4.28); if there is another solution of (4.28) denoted by ω'_{12}, then

$$d\omega_1 = -\omega_2\wedge\omega'_{12}, 2q \quad d\omega_2 = \omega_1\wedge\omega'_{12}. \tag{4.29}$$

From (4.28) and (4.29) it follows that $\omega_2\wedge(\omega'_{12}-\omega_{12})=0$, so that

$$\omega'_{12}-\omega_{12}=f\omega_1, \tag{4.30}$$

f being a function of u, v, since any 1-form on the surface S is a linear combination of ω_1 and ω_2 (see Exercise 2 of Section 4). Similarly,

$$\omega'_{12}-\omega_{12}=g\omega_2, \tag{4.31}$$

where g is a function of u, v. Since ω_1 and ω_2 are linearly independent, both (4.30) and (4.31) can hold when $f=g=0$, so that $\omega_{12}-\omega'_{12}=0$. Next from (4.9), (4.12), (4.14), and (4.16) we have

$$\begin{aligned} d\omega_{12} &= \omega_{13}\wedge\omega_{32} = -(a\omega_1+b\omega_2)\wedge(b\omega_1+c\omega_2) \\ &= -(ac-b^2)\omega_1\wedge\omega_2 = -K\omega_1\wedge\omega_2, \end{aligned} \tag{4.32}$$

which asserts that K depends only on I. Q.E.D.

With the method of frame fields so far developed we are in a position to prove the following uniqueness theorem for surfaces, which is the second part of Theorem 7.3 of Chapter 3.

4.4. Theorem. *Let S and S^* be two surfaces, f: $S\to S^*$ a diffeomorphism,* I, I* *and* II, II*, *respectively, the corresponding first and second fundamental forms of S, S^* under f. Then f is an isometry (orientation preserving or reversing) of E^3 if and only if* $\mathrm{I}=\mathrm{I}^*$, $\mathrm{II}=\pm\mathrm{II}^*$.

The "if" part of the theorem is immediate, since I is invariant under an isometry, and II up to a sign.

The "only if" part of the theorem can be proved in the same way as the corresponding theorem for curves (Theorem 1.5.3 of Chapter 2), with the one new feature that there will be integrability conditions or, equivalently, structural equations. We first prove Lemma 4.5.

4.5. Lemma. *Let* $\mathbf{x}\mathbf{e}_1\mathbf{e}_2\mathbf{e}_3(u,v)$ *and* $\mathbf{x}^*\mathbf{e}_1^*\mathbf{e}_2^*\mathbf{e}_3^*(u,v)$ *be two families of frames depending on two parameters u, v. Let ω_i, ω_{ij}, $i, j=1,2,3$, be defined* *by (4.7), and ω_i^*, ω_{ij}^* the corresponding forms defined by similar equations for the second family of frames. If*

$$\omega_i = \omega_i^*, \qquad \omega_{ij} = \omega_{ij}^*, \tag{4.33}$$

the two families of frames differ by an orientation-preserving isometry of E^3.

Proof of Lemma 4.5. By applying an orientation-preserving isometry we may assume that for a fixed set of values (u_0, v_0) we have

$$\mathbf{x}\mathbf{e}_1\mathbf{e}_2\mathbf{e}_3(u_0, v_0) = \mathbf{x}^*\mathbf{e}_1^*\mathbf{e}_2^*\mathbf{e}_3^*(u_0, v_0). \tag{4.34}$$

Let (e_{i1}, e_{i2}, e_{i3}) and $(e_{i1}^*, e_{i2}^*, e_{i3}^*)$, $i=1,2,3$, be, respectively, the components of the vectors $\mathbf{e}_i$ and $\mathbf{e}_i^*$ with respect to a given positively oriented frame in the space. Then the second equation of (4.7) and its corresponding equation for the second family of frames $\mathbf{x}^*\mathbf{e}_1^*\mathbf{e}_2^*\mathbf{e}_3^*$ can be written as

$$de_{ij}(u,v) = \sum_{k=1}^{3} \omega_{ik} e_{kj}, \qquad de_{ij}^*(u,v) = \sum_{k=1}^{3} \omega_{ik} e_{kj}^*.$$

Since $\omega_{ik} = -\omega_{ki}$, we have

$$\begin{aligned} d\left(\sum_{i=1}^{3} e_{ij} e_{ik}^*\right) &= \sum_{i,l=1}^{3} \omega_{il} e_{lj} e_{ik}^* + \sum_{i,l=1}^{3} e_{ij} \omega_{il} e_{lk}^* \\ &= \sum_{i,l=1}^{3} (\omega_{il} + \omega_{li}) e_{lj} e_{ik}^* = 0, \end{aligned} \tag{4.35}$$

showing that $\sum_{i=1}^{3} e_{ij} e_{ik}^*$ is independent of u and v. Since $\sum_{k=1}^{3} e_{ik} e_{jk} = \delta_{ij}$, by interchanging rows and columns of the matrix (e_{ij}) we have

$$\sum_{i=1}^{3} e_{ij} e_{ik} = \delta_{jk}. \tag{4.36}$$

From (4.34) through (4.36) it thus follows that

$$\begin{aligned} \sum_{i=1}^{3} e_{ij} e_{ik}^* &= \sum_{i=1}^{3} e_{ij}(u_0, v_0) e_{ik}^*(u_0, v_0) \\ &= \sum_{i=1}^{3} e_{ij}(u_0, v_0) e_{ik}(u_0, v_0) = \delta_{jk}. \end{aligned} \tag{4.37}$$

Subtracting (4.37) from (4.36) we obtain, for a fixed k,

$$\sum_{i=1}^{3} e_{ij}(e_{ik} - e_{ik}^*) = 0, \qquad j = 1,2,3, \tag{4.38}$$

which implies that $e_{ik} = e_{ik}^*$ or $\mathbf{e}_i = \mathbf{e}_i^*$, since $\det(e_{ij}) \neq 0$. Furthermore, from the first equation of (4.7) and its corresponding equation we have

$$d(\mathbf{x}-\mathbf{x}^*)=\sum_{i=1}^{3}\omega_i(\mathbf{e}_i-\mathbf{e}_i^*)=0.$$

Thus $\mathbf{x}-\mathbf{x}^*$ is independent of u and v, and

$$\mathbf{x}-\mathbf{x}^*=\mathbf{x}(u_0,v_0)-\mathbf{x}^*(u_0,v_0)=0.$$

Hence the lemma is proved.

Proof of Theorem 4.4. Suppose the two surfaces S and S^* to have the same parameters (u,v) so that each pair of corresponding points under the diffeomorphism f has the same parameters. Without loss of generality we can further suppose that the u-curves and v-curves are orthogonal to each other on one surface; the same is then true on the other surface because it is assumed that $\mathrm{I}=\mathrm{I}^*$. To the surface S attach the family of frames $\mathbf{x}\mathbf{e}_1\mathbf{e}_2\mathbf{e}_3(u,v)$, such that $\mathbf{x}(u,v)$ is the position vector of the point with the parameters $u,v,\mathbf{e}_1(u,v)$ and $\mathbf{e}_2(u,v)$ are the unit tangent vectors of the u-, v-curves, respectively, and $\mathbf{e}_3(u,v)$ is the unit normal vector of the surface S satisfying $|\mathbf{e}_1,\mathbf{e}_2,\mathbf{e}_3|=\pm 1$. Similarly, we have a family of frames $\mathbf{x}^*\mathbf{e}_1\mathbf{e}_2^*\mathbf{e}_3^*(u,v)$ attached to the surface S^*. From (4.7) we see that ω_1,ω_1^* are multiples of du, and ω_2,ω_2^* are multiples of dv. But we have also

$$\omega_1^2+\omega_2^2=\omega_1^{*2}+\omega_2^{*2},$$

which is possible only when ω_1,ω_2 differ from ω_1^*,ω_2^*, respectively, by at most a sign. If ω_i and ω_i^* differ by a sign, we change the corresponding tangent vector to its negative. It may then be necessary to change the sign of $\mathbf{e}_3^*$ to keep the frame $\mathbf{e}_1^*\mathbf{e}_2^*\mathbf{e}_3^*$ positively oriented. Therefore in all cases it is possible to choose the frames $\mathbf{x}^*\mathbf{e}_1^*\mathbf{e}_2^*\mathbf{e}_3^*(u,v)$ such that

$$\omega_1^*=\omega_1,\qquad \omega_2^*=\omega_2. \tag{4.39}$$

To prove the theorem it suffices to show that by a reflection of a surface if necessary, the conditions (4.33) of Lemma 4.5 are fulfilled for our two families of frames. Since $\mathbf{e}_3,\mathbf{e}_3^*$ are normal vectors of the surfaces S,S^*, respectively, we have $\omega_3^*=\omega_3=0$. Exterior differentiation of (4.39) gives

$$d\omega_1^*=d\omega_1,\qquad d\omega_2^*=d\omega_2,$$

which become, in consequence of (4.8) and its corresponding equation for the second family of frames $\mathbf{x}^*\mathbf{e}_1^*\mathbf{e}_2^*\mathbf{e}_3^*$,

$$(\omega_{12}^*-\omega_{12})\wedge\omega_2=0,\qquad (\omega_{12}^*-\omega_{12})\wedge\omega_1=0.$$

It follows that $\omega_{12}^*-\omega_{12}$ is a multiple of both ω_1 and ω_2, and therefore that $\omega_{12}^*-\omega_{12}=0$, since ω_1 and ω_2 are linearly independent.

If we have $\mathrm{II}^* = -\mathrm{II}$, we apply a reflection in a plane (part 3 of Examples 5.4.7, Chapter 1) to the surface S^* to get a surface S^{**}. All the frames attached to S^* will be mapped into negatively oriented frames, and we make them positively oriented by taking the negative of the unit normal vector. The second fundamental form II^{**} of the surface S^{**} with orientation reversed will then be equal to II. This show that by a reflection if necessary we can suppose $\mathrm{II}^* = \mathrm{II}$, which can be written explicitly as

$$a^*\omega_1^{*2} + 2b^*\omega_1^*\omega_2^* + c^*\omega_2^{*2} = a\omega_1^2 + 2b\omega_2\omega_2 + c\omega_2^2. \tag{4.40}$$

Equation 4.40 is possible only when $a^* = a$, $b^* = b$, $c^* = c$ or

$$\omega_{13}^* = \omega_{13}, \qquad \omega_{23}^* = \omega_{23}.$$

Thus the conditions (4.33) of Lemma 4.5 are all fulfilled, and the proof of the theorem is complete.

Exercises

1. Prove (4.2), (4.3), and (4.5).

2. Prove (4.12); that is, prove that any 1-form on the surface S is a linear combination of ω_1 and ω_2.

***3.** Let S be a surface given by (4.7) and $\omega_3 = 0$. Show that for any smooth closed curve C on S, $\dfrac{1}{2\pi}\displaystyle\int_C \omega_{12}$, mod 1, is invariant under a rotation field given by (4.17) along C.

5. RIGIDITY OF CONVEX SURFACES AND MINKOWSKI'S FORMULAS

5.1. Definition. A surface in a Euclidean space E^3 is *convex* if its Gaussian curvature is positive everywhere.

5.2. Theorem (S. Cohn-Vossen). *An isometry f between two compact convex surfaces in E^3 is either an isometry of E^3 or an isometry of E^3 and a reflection in a plane. In other words, such an isometry f is always trivial.*

Proof. Suppose that there is an isometry f between two compact convex surfaces S, S^*. Then by Theorem 4.4 it is sufficient to show that under the isometry f the second fundamental forms $\mathrm{II}, \mathrm{II}^*$ of S, S^* are equal. To this end we use (4.7) through (4.9), (4.12), and (4.14) for S; under f the corresponding quantities and equations for S^* are denoted, respectively, by the same notation and numbers with an asterisk. Without loss of generality we may assume that at corresponding points under f we have (4.39) and therefore

$$\omega_{12}^* = \omega_{12}. \tag{5.1}$$

By means of (6.2.19) of Chapter 1 we can have the exterior differential of the determinant $|\mathbf{x}, \omega_{13}^*\mathbf{e}_1 + \omega_{23}^*\mathbf{e}_2, \mathbf{e}_3|$ in the usual way:

$$\begin{aligned} d|\mathbf{x}, \omega_{13}^*\mathbf{e}_1 + \omega_{23}^*\mathbf{e}_2, \mathbf{e}_3| = &|d\mathbf{x}, \omega_{13}^*\mathbf{e}_1 + \omega_{23}^*\mathbf{e}_2, \mathbf{e}_3| \\ &+ |\mathbf{x}, d(\omega_{13}^*\mathbf{e}_1 + \omega_{23}^*\mathbf{e}_2), \mathbf{e}_3| \\ &+ |\mathbf{x}, \omega_{13}^*\mathbf{e}_1 + \omega_{23}^*\mathbf{e}_2, -d\mathbf{e}_3|. \end{aligned} \tag{5.2}$$

To compute the right-hand side of (5.2) we should first notice that each determinant is the same as an ordinary one except that the product of any two 1-forms that are scalar factors of the determinant is not ordinary but exterior.

By using (4.7), (4.12)*, (4.14)*, (4.16)*, Lemma 4.2, and $|\mathbf{e}_1, \mathbf{e}_2, \mathbf{e}_3| = 1$, we obtain

$$\begin{aligned} |d\mathbf{x}, \omega_{13}^*\mathbf{e}_1 + \omega_{22}^*\mathbf{e}_2, \mathbf{e}_3| &= |\omega_1\mathbf{e}_1 + \omega_2\mathbf{e}_2, \omega_{13}^*\mathbf{e}_1 + \omega_{23}^*\mathbf{e}_2, \mathbf{e}_3| \\ &= |\omega_1\mathbf{e}_1, \omega_{23}^*\mathbf{e}_2, \mathbf{e}_3| + |\omega_2\mathbf{e}_2, \omega_{13}^*\mathbf{e}_1, \mathbf{e}_3| \\ &= \omega_1 \wedge \omega_{23}^* - \omega_2 \wedge \omega_{13}^* \\ &= (a^* + c^*)\omega_1 \wedge \omega_2 = 2H^*\, dA. \end{aligned} \tag{5.3}$$

Similarly, from (4.7), (4.9)*, and (5.1) it follows that

$$|\mathbf{x}, d(\omega_{13}^*\mathbf{e}_1 + \omega_{23}^*\mathbf{e}_2), \mathbf{e}_3| = 0, \tag{5.4}$$

since

$$d(\omega_{13}^*\mathbf{e}_1 + \omega_{23}^*\mathbf{e}_2) = (d\omega_{13}^*)\mathbf{e}_1 - \omega_{13}^* \wedge d\mathbf{e}_1 + (d\omega_{23}^*)\mathbf{e}_2 - \omega_{23}^* \wedge d\mathbf{e}_2.$$

Also, by (4.7), (4.12), (4.12)*, (4.14), (4.14)*, (4.16), (4.16)*, and $K = K^*$, we have

$$|\mathbf{x}, \omega_{13}^*\mathbf{e}_1 + \omega_{23}^*\mathbf{e}_2, -d\mathbf{e}_3| = p(\omega_{13}^* \wedge \omega_{23} - \omega_{23}^* \wedge \omega_{13}) = pJ\, dA, \tag{5.5}$$

where

$$p = \mathbf{x} \cdot \mathbf{e}_3, \tag{5.6}$$

$$J = ac^* + a^*c - 2bb^* = 2K - \begin{vmatrix} a^* - a & b^* - b \\ b^* - b & c^* - c \end{vmatrix}. \tag{5.7}$$

Here p is called the *support function* of the oriented surface S with parametrization $\mathbf{x}(u, v)$ and is geometrically the oriented distance from the origin $\mathbf{0}$ of E^3 to the tangent plane of S at the point $\mathbf{x}(u, v)$.

Substitution of (5.3) through (5.5) in (5.2) gives

$$d|\mathbf{x}, \omega_{13}^*\mathbf{e}_1 + \omega_{23}^*\mathbf{e}_2, \mathbf{e}_3| = 2H^*\, dA + pJ\, dA. \tag{5.8}$$

Integrating (5.8) over the surface S and applying Green's formula (4.6) we thus obtain the integral formula

$$2\int_S\int H^*\,dA+\int_S\int pJ\,dA=0. \tag{5.9}$$

In particular, when S and S^* are identical, $J=2K$ and $H^*=H$, and (5.9) therefore takes the form

$$2\int_S\int H\,dA+2\int_S\int pK\,dA=0. \tag{5.10}$$

Subtracting (5.10) from (5.9) we obtain

$$\begin{aligned}2\int_S\int H^*\,dA-2\int_S\int H\,dA&=\int_S\int(2K-J)p\,dA\\&=\int_S\int\begin{vmatrix}a^*-a & b^*-b\\ b^*-b & c^*-c\end{vmatrix}p\,dA.\end{aligned} \tag{5.11}$$

To complete the proof we need the following lemma.

5.3. Lemma. *Let*

$$\lambda x^2+2\mu xy+\nu y^2, \qquad \lambda^* x^2+2\mu^* xy+\nu^* y^2 \tag{5.12}$$

be two positive definite quadratic forms with

$$\lambda^*\nu^*-\mu^{*2}=\lambda\nu-\mu^2. \tag{5.13}$$

Then

$$\begin{vmatrix}\lambda^*-\lambda & \mu^*-\mu\\ \mu^*-\mu & \nu^*-\nu\end{vmatrix}\leq 0, \tag{5.14}$$

where the equality holds when and only when the two forms are identical.

Proof of Lemma 5.3. At first we observe that $\lambda x^2+2\mu xy+\nu y^2$ is positive definite if and only if $\lambda>0$, $\lambda\nu-\mu^2>0$ (Lemma 2.4.3 of Chapter 1), which imply that $\nu>0$. Then the statement of the lemma remains unchanged under a nonsingular linear transformation of the variables x, y. In fact, if we consider a nonsingular linear transformation

$$x=a_1\bar{x}+b_1\bar{y}, \qquad y=a_2\bar{x}+b_2\bar{y}, \qquad (a_1b_2-a_2b_1\neq 0),$$

under which the two forms of (5.12) become

$$\bar{\lambda}\bar{x}^2+2\bar{\mu}\bar{x}\bar{y}+\bar{\nu}\bar{y}^2, \qquad \bar{\lambda}^*\bar{x}^2+2\bar{\mu}^*\bar{x}\bar{y}+\nu^*\bar{y}^2, \tag{5.15}$$

then

$$\begin{vmatrix} \bar{\lambda} & \bar{\mu} \\ \bar{\mu} & \bar{\nu} \end{vmatrix} = \begin{vmatrix} a_1 & a_2 \\ b_1 & b_2 \end{vmatrix}^2 \begin{vmatrix} \lambda & \mu \\ \mu & \nu \end{vmatrix}, \qquad \begin{vmatrix} \bar{\lambda}^* & \bar{\mu}^* \\ \bar{\mu}^* & \bar{\nu}^* \end{vmatrix} = \begin{vmatrix} a_1 & a_2 \\ b_1 & b_2 \end{vmatrix}^2 \begin{vmatrix} \lambda^* & \mu^* \\ \mu^* & \nu^* \end{vmatrix},$$

$$\begin{vmatrix} \bar{\lambda}^* - \bar{\lambda} & \bar{\mu}^* - \bar{\mu} \\ \bar{\mu}^* - \bar{\mu} & \bar{\nu}^* - \bar{\nu} \end{vmatrix} = \begin{vmatrix} a_1 & a_2 \\ b_1 & b_2 \end{vmatrix}^2 \begin{vmatrix} \lambda^* - \lambda & \mu^* - \mu \\ \mu^* - \mu & \nu^* - \nu \end{vmatrix}.$$

Thus the statement of the lemma remains unchanged under any nonsingular linear transformation. Applying such a linear transformation when necessary, we can therefore assume that $\mu^* = \mu$. Then the left-hand side of (5.14) is reduced to

$$(\lambda^* - \lambda)(\nu^* - \nu) = -\frac{\nu}{\lambda^*}(\lambda^* - \lambda)^2 \leqslant 0,$$

since (5.13) becomes $\nu^* = \lambda\nu/\lambda^*$, and ν, λ^* are positive. Moreover, $(\lambda^* - \lambda)(\nu^* - \nu) = 0$ when and only when $\lambda^* = \lambda$ and $\nu^* = \nu$. Hence the lemma is proved.

Now we continue to prove Theorem 5.2. Let us choose the origin $\mathbf{0}$ to be inside S, so that we have $p > 0$. Then by Lemma 5.3, the integrand on the right-hand side of (5.11) is nonpositive, and it follows that

$$\int_S\int H\,dA \geqslant \int_S\int H^*\,dA.$$

Since the relation between S and S^* is symmetric, we must also have

$$\int_S\int H^*\,dA \leqslant \int_S\int H\,dA.$$

Hence

$$\int_S\int H\,dA = \int_S\int H^*\,dA,$$

from which it follows that the determinant on the right-hand side of (5.11) vanishes. Therefore by Lemma 5.3 again, we obtain

$$a^* = a, \qquad b^* = b, \qquad c^* = c,$$

which complete the proof of Theorem 5.2.

5.4. Remarks. 1. The proof of Theorem 5.2, is essentially due to G. Herglotz [24].

2. Theorem 5.2 is certainly not true in general if we remove the restriction that the surfaces be convex. Fig. 4.13 gives two C^∞ surfaces of revolution, which are obviously isometric but not congruent (under an isometry of E^3).

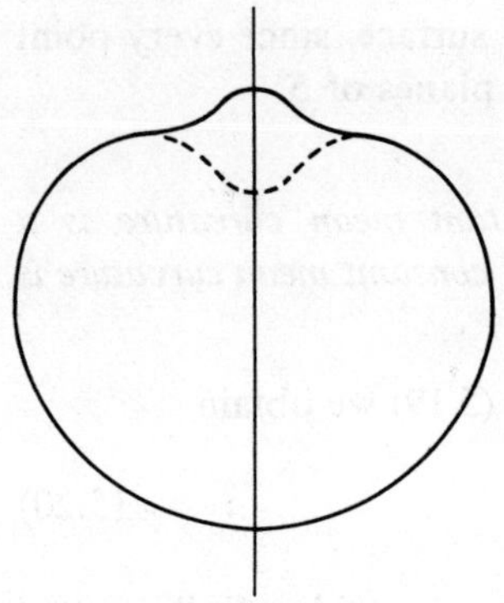

Figure 4.13

Formula 5.10, which is now written as

$$\iint_S H\,dA + \iint_S pK\,dA = 0, \tag{5.16}$$

was first proved by Minkowski for convex surfaces. According to our proof here the formula is also valid for any compact surface and can be derived directly from one surface, since in consequence of (4.7), for identical S and S^*, (5.8) is reduced to

$$d\,|\mathbf{x}, \mathbf{e}_3, d\mathbf{e}_3| = 2H\omega_1\wedge\omega_2 + 2pK\omega_1\wedge\omega_2. \tag{5.17}$$

By the same method we can obtain a generalization of another formula of Minkowski for a compact convex surface. Let S be a compact surface. Then in the same way used in computing (5.2), we can easily have the differential form

$$\begin{aligned} d\,|\mathbf{e}_3, \mathbf{x}, d\mathbf{x}| &= |d\mathbf{e}_3, \mathbf{x}, d\mathbf{x}| + |\mathbf{e}_3, d\mathbf{x}, d\mathbf{x}| \\ &= p\,|\omega_{32}\wedge\omega_1 - \omega_{31}\wedge\omega_2| + 2\omega_1\wedge\omega_2 \\ &= 2pH\omega_1\wedge\omega_2 + 2\omega_1\wedge\omega_2. \end{aligned} \tag{5.18}$$

Integration of (5.18) over the surface S and application of Green's theorem thus yield the integral formula

$$\iint_S pH\,dA + A = 0, \tag{5.19}$$

which was also first proved by Minkowski for convex surfaces.

5.5. Definition. A compact surface S in E^3 is a *star-shaped surface* if there exists a point in E^3 lying on the same side of all the tangent planes of S.

A compact convex surface S is a star-shaped surface, since every point inside S lies on the same side of all the tangent planes of S.

5.6. Theorem. *A star-shaped surface of constant mean curvature is a sphere. In particular, a compact convex surface of constant mean curvature is a sphere.*

Proof. When H is constant, from (5.16) and (5.19) we obtain

$$\int_S\int (H^2-K)p\,dA=0. \tag{5.20}$$

Take the origin $\mathbf{0}$ to be the point lying on the same side of all tangent planes of S so that we can assume that $p>0$. Since

$$H^2-K\geqslant 0 \tag{5.21}$$

by (4.12) of Chapter 3, (5.20) holds when and only when $H^2-K=0$, which implies that every point of S is an umbilical point by Corollary 4.4 of Chapter 3. Hence from Theorem 4.5 of Chapter 3 it follows that S is a sphere.

Exercises

1. If a line l intersects a closed convex surface S, show that l either is tangent to S or intersects S at exactly two points.
2. Prove (5.4) and (5.5).

*3. Show that for the mean curvature H of the torus T with the parametrization (1.18) of Chapter 3

$$\int_T\int H\,dA>0.$$

*4. Prove that a star-shaped surface S of constant Gaussian curvature is a sphere (this is a special case of Theorem 2.4). In particular, a compact convex surface of constant Gaussian curvature is a sphere.

6. SOME TRANSLATION AND SYMMETRY THEOREMS

This section gives some translation and symmetry theorems on surfaces that were obtained in a joint paper by H. Hopf and K. Voss [29].

6.1. Theorem. *Let S, S^* be two compact oriented surfaces of class C^2 in E^3. Suppose that there is an orientation-preserving diffeomorphism $f\colon S\to S^*$*

such that the lines joining $\mathbf{p}$ *and* $\mathbf{p}^*=f(\mathbf{p})$, *for all* $p\in S$ *are parallel to a fixed direction* $\mathbf{r}$ *in* E^3, *and such that* S, S^* *have the same mean curvature at each pair of the corresponding points* $\mathbf{p},\mathbf{p}^*$ *but have no cylindrical elements whose generators are parallel to* $\mathbf{r}$. *Then* f *is a translation in* E^3.

Proof. At first, let us suppose that S is an oriented surface of class C^2 with a boundary C in E^3, and $\mathbf{x}$: $U\subset E^2\to S$ a parametrization of S. Let $\mathbf{i}$ be the unit vector in the fixed direction $\mathbf{r}$, and w a function of class C^2 over S. Then all the equations of Section 4 can be applied to S, and we shall use the same symbols with an asterisk for the corresponding quantities for the surface S^* defined by the vector equation

$$\mathbf{x}^*=\mathbf{x}+\mathbf{w}, \tag{6.1}$$

where

$$\mathbf{w}=w\mathbf{i}. \tag{6.2}$$

By (6.2.25) and (6.2.26) of Chapter 1, (7.1) and (4.10) of Chapter 3, and (4.27) we can easily obtain

$$d\mathbf{x}\hat{\times}d\mathbf{x}=2(\mathbf{x}_u\times\mathbf{x}_v)\,du\wedge dv=2\mathbf{e}_3\,dA, \tag{6.3}$$

$$d\mathbf{x}\hat{\times}d\mathbf{e}_3=(\mathbf{x}_u\times\mathbf{e}_{3v}-\mathbf{x}_v\times\mathbf{e}_{3u})\,du\wedge dv=-2\mathbf{e}_3H\,dA, \tag{6.4}$$

where dA is the element of area of S. Then from (6.1) through (6.4) it follows that

$$\mathbf{w}\times d\mathbf{w}=w\mathbf{i}\times\mathbf{i}\,dw=(\mathbf{i}\times\mathbf{i})w\,dw=0, \tag{6.5}$$

$$d\mathbf{w}\hat{\times}d\mathbf{w}=d(w\mathbf{i})\hat{\times}d(w\mathbf{i})=(\mathbf{i}\times\mathbf{i})\,dw\wedge dw=0, \tag{6.6}$$

$$\begin{aligned}2(e_3^*dA^*-\mathbf{e}_3dA)&=d\mathbf{x}^*\hat{\times}d\mathbf{x}^*-d\mathbf{x}\hat{\times}d\mathbf{x}\\&=(d\mathbf{x}+d\mathbf{w})\hat{\times}(d\mathbf{x}+d\mathbf{w})-d\mathbf{x}\hat{\times}d\mathbf{x}\\&=d\mathbf{x}\hat{\times}d\mathbf{x}+2d\mathbf{x}\hat{\times}d\mathbf{w}+d\mathbf{w}\hat{\times}d\mathbf{w}-d\mathbf{x}\hat{\times}d\mathbf{x}\\&=2d\mathbf{x}\hat{\times}d\mathbf{w},\end{aligned}$$

$$\begin{aligned}2(\mathbf{e}_3^*\,dA^*-\mathbf{e}_3\,dA)&=d\mathbf{x}^*\hat{\times}d\mathbf{x}^*-d\mathbf{x}\hat{\times}d\mathbf{x}\\&=d\mathbf{x}^*\hat{\times}d\mathbf{x}^*-(d\mathbf{x}^*-d\mathbf{w})\hat{\times}(d\mathbf{x}^*-d\mathbf{w})\\&=d\mathbf{x}^*\hat{\times}d\mathbf{x}^*-d\mathbf{x}^*\hat{\times}d\mathbf{x}^*+2d\mathbf{x}^*\hat{\times}d\mathbf{w}-d\mathbf{w}\hat{\times}d\mathbf{w}\\&=2d\mathbf{x}^*\hat{\times}d\mathbf{w}.\end{aligned}$$

Comparison of the last two equations gives immediately

$$\mathbf{e}_3^*\,dA^*-\mathbf{e}_3\,dA=d\mathbf{x}\hat{\times}d\mathbf{w}=d\mathbf{x}^*\hat{\times}d\mathbf{w}. \tag{6.7}$$

By taking the inner products of the vector $\mathbf{w}$ with both sides of (6.7) we obtain, by virtue of (6.5),

$$\mathbf{w}\cdot(\mathbf{e}_3^*\,dA^*-\mathbf{e}_3\,dA)=\mathbf{w}\cdot(d\mathbf{x}\hat{\times}d\mathbf{w})=|\mathbf{w},d\mathbf{x},d\mathbf{w}|=-d\mathbf{x}\cdot(\mathbf{w}\times d\mathbf{w})=0,$$

or

$$\mathbf{w}\cdot\mathbf{e}_3^*\,dA^* = \mathbf{w}\cdot\mathbf{e}_3\,dA. \tag{6.8}$$

From (6.1) and (6.5) it follows that

$$|\mathbf{e}_3^*, \mathbf{w}, d\mathbf{w}| = \mathbf{e}_3^*\cdot(\mathbf{w}\times d\mathbf{w}) = 0, \tag{6.9}$$

$$\begin{aligned} |\mathbf{e}_3^*, \mathbf{w}, d\mathbf{x}| &= |\mathbf{e}_3^*, \mathbf{w}, d\mathbf{x}^* - d\mathbf{w}| \\ &= |\mathbf{e}_3^*, \mathbf{w}, d\mathbf{x}^*| - |\mathbf{e}_3^*, \mathbf{w}, d\mathbf{w}| = |e_3^*, \mathbf{w}, d\mathbf{x}^*|. \end{aligned} \tag{6.10}$$

By means of the relation $d^2\mathbf{x}=0$ and equations (6.4) and (6.7), we have

$$\begin{aligned} d\,|\mathbf{e}_3, \mathbf{w}, d\mathbf{x}| &= -\mathbf{w}\cdot(d\mathbf{e}_3 \hat{\times} d\mathbf{x}) + \mathbf{e}_3\cdot(d\mathbf{w}\hat{\times} d\mathbf{x}) \\ &= 2H(\mathbf{w}\cdot\mathbf{e}_3)\,dA + \mathbf{e}_3\cdot(\mathbf{e}_3^*\,dA^* - \mathbf{e}_3\,dA) \\ &= 2H(\mathbf{w}\cdot\mathbf{e}_3)\,dA + \mathbf{e}_3\cdot\mathbf{e}_3^*\,dA^* - dA. \end{aligned} \tag{6.11}$$

Similarly, in consequence of (6.10), (6.7), (6.8), and the equation analogous to (6.4) for the surface S^*, we obtain

$$\begin{aligned} d\,|\mathbf{e}_3^*, \mathbf{w}, d\mathbf{x}| &= d\,|\mathbf{e}_3^*, \mathbf{w}, d\mathbf{x}^*| \\ &= -\mathbf{w}\cdot(d\mathbf{e}_3^* \hat{\times} d\mathbf{x}^*) + \mathbf{e}_3^*\cdot(d\mathbf{w}\hat{\times} d\mathbf{x}^*) \\ &= 2H^*(\mathbf{w}\cdot\mathbf{e}_3^*)\,dA^* + \mathbf{e}_3^*\cdot(\mathbf{e}_3^*\,dA^* - \mathbf{e}_3\,dA) \\ &= 2H^*(\mathbf{w}\cdot\mathbf{e}_3)\,dA + dA^* - \mathbf{e}_3\cdot\mathbf{e}_3^*\,dA. \end{aligned} \tag{6.12}$$

Subtraction of (6.11) from (6.12) gives

$$\begin{aligned} &d\,|\mathbf{e}_3^* - \mathbf{e}_3, \mathbf{w}, d\mathbf{x}| \\ &\quad = 2(H^* - H)(\mathbf{w}\cdot\mathbf{e}_3)\,dA + (1 - \mathbf{e}_3\cdot\mathbf{e}_3^*)(dA + dA^*). \end{aligned} \tag{6.13}$$

Integrating (6.13) over the surface S and applying Green's formula (4.6) to the left-hand side of the equation, we arrive at the integral formula

$$\begin{aligned} &\int_C |\mathbf{e}_3^* - \mathbf{e}_3, \mathbf{w}, d\mathbf{x}| \\ &\quad = \iint_S \left[2(H^* - H)(\mathbf{w}\cdot\mathbf{e}_3)\,dA + (1 - \mathbf{e}_3\cdot\mathbf{e}_3^*)(dA + dA^*)\right]. \end{aligned} \tag{6.14}$$

In particular, when the surface S has empty boundary C, the integral on the left-hand side of (6.14) vanishes; hence

$$2\iint_S (H^* - H)(\mathbf{w}\cdot\mathbf{e}_3)\,dA + \iint_S (1 - \mathbf{e}_3\cdot\mathbf{e}_3^*)(dA + dA^*) = 0. \tag{6.15}$$

Since from the assumption of the theorem $H^* = H$ over S, the formula (6.15) becomes

$$\iint_S (1 - \mathbf{e}_3 \cdot \mathbf{e}_3^*)(dA + dA^*) = 0. \tag{6.16}$$

Since f preserves the orientation, $dA + dA^* > 0$. Furthermore, $1 - \mathbf{e}_3 \cdot \mathbf{e}_3^* \geqslant 0$ because $\mathbf{e}_3$ and $\mathbf{e}_3^*$ are unit vectors. Thus the integrand of (6.16) is nonnegative, and therefore (6.16) holds when and only when $1 - \mathbf{e}_3 \cdot \mathbf{e}_3^* = 0$, which implies that

$$\mathbf{e}_3^* = \mathbf{e}_3. \tag{6.17}$$

Now in the space E^3 we choose the rectangular frame $\mathbf{0x}_1\mathbf{x}_2\mathbf{x}_3$ in such a way that the $\mathbf{x}_3$-axis is along the fixed unit vector $\mathbf{i}$. Since the surface S has no cylindrical elements whose generators are parallel to $\mathbf{i}$, the closed set M of all points of the surface S, at each of which the $\mathbf{x}_3$-component of the unit normal vector $\mathbf{e}_3$ of the surface S is zero, has no inner points; therefore the open set $S - M$ is everywhere dense over S. Thus in neighborhoods of any point of the set $S - M$ and its corresponding point on the surface S^*, x_1, x_2, x_3 are regular parameters of the two surfaces S, S^* so that S, S^* can be represented, respectively, by the equations

$$\begin{aligned} x_3 &= x_3(x_1, x_2), \\ x_3 &= x_3^*(x_1, x_2) = x_3(x_1, x_2) + w(x_1, x_2). \end{aligned} \tag{6.18}$$

Then the unit normal vectors $\mathbf{e}_3, \mathbf{e}_3^*$ of S, S^* are

$$\begin{aligned} \mathbf{e}_3 &= \frac{1}{\sqrt{1 + (\partial x_3/\partial x_1)^2 + (\partial x_3/\partial x_2)^2}} \left(\frac{\partial x_3}{\partial x_1}, \frac{\partial x_3}{\partial x_2}, -1 \right), \\ \mathbf{e}_3^* &= \frac{1}{\sqrt{1 + (\partial x_3^*/\partial x_1)^2 + (\partial x_3^*/\partial x_2)^2}} \left(\frac{\partial x_3^*}{\partial x_1}, \frac{\partial x_3^*}{\partial x_2}, -1 \right). \end{aligned} \tag{6.19}$$

From (6.17) through (6.19) it follows immediately that in a neighborhood of any point of the set $S - M$,

$$\frac{\partial x_3^*}{\partial x_1} = \frac{\partial x_3}{\partial x_1}, \qquad \frac{\partial x_3^*}{\partial x_2} = \frac{\partial x_3}{\partial x_2},$$

and the function w is constant. Thus $\partial w/\partial x_1$ and $\partial w/\partial x_2$ are zero in the everywhere dense set $S - M$ and therefore on the whole surface S by continuity. Hence the function w is constant on the whole surface S, and the proof of the theorem is complete.

6.2. Remarks. 1. In Theorem 6.1 we must require the condition that neither the surface S nor the surface S^* contains pieces of a cylinder whose generators are parallel to the fixed direction $\mathbf{r}$. This is because if the surface S is a cylinder capped smoothly by almost-hemispheres at both ends, while the surface S^* is obtained from the surface S by elongating the cylindrical portion of the surface S, the surface S^* is not obtainable from the surface S by a translation.

2. Theorem 6.1 is still true if the mean curvature is replaced by the Gaussian curvature.

3. Theorem 6.1 was extended to hypersurfaces in a Euclidean n-space E^n for $n>3$ independently by C. C. Hsiung [31] and K. Voss [59].

Theorem 6.1 can be extended to surfaces with boundaries as follows.

6.3. Corollary. *Let S, S^* be compact oriented surfaces of class C^2 with boundaries C and C^*, respectively, in space E^3. Suppose that there is a diffeomorphism f: $S \to S^*$ with the same properties as those of the diffeomorphism f in Theorem* 6.1.

(a) *If the two boundaries C, C^* are coincident, the two surfaces S, S^* are coincident.*

(b) *If the normals of the surfaces S, S^* at every pair of corresponding points, under f, on the boundaries C, C^* are parallel, then f is a translation in E^3.*

Proof. In both parts of this corollary, the integral over the boundary C on the left-hand side of (6.14) also vanishes, since over the boundary $C, \mathbf{w}=0$ and $\mathbf{e}_3^*=\mathbf{e}_3$ in parts a and b, respectively. By the same argument as that in the proof of Theorem 6.1, we see that f is a translation, which in particular reduces to an identity in part a. Hence Corollary 6.3 is proved.

6.4. Definition. A compact surface S in E^3 is *convex in a given direction* if no line in this direction intersects S at more than two points.

It is obvious that a compact surface S is convex in the usual sense if it is convex in every direction in E^3.

An application of Theorem 6.1 gives the following symmetry theorem.

6.5. Theorem. *Let a compact oriented surface S of class C^2 in E^3 be convex in a given direction* $\mathbf{r}$. *If S has the same mean curvature at its points of intersection with any line in the direction* $\mathbf{r}$, *then the surface S has a plane of symmetry orthogonal to the direction* $\mathbf{r}$.

Proof. Let f be a mapping of the surface S onto itself such that the two points of intersection of S with any line in the direction $\mathbf{r}$ are mapped into each other. In particular, if a line in the direction $\mathbf{r}$ is tangent to S at a point $\mathbf{p}$, then $f(\mathbf{p})=\mathbf{p}$. Let r be the reflection with respect to an arbitrary plane orthogonal to the direction $\mathbf{r}$, and $\mathbf{p}$ any point on S. Then the mapping $rf(\mathbf{p})=\mathbf{p}^*$ maps the surface S onto the surface $S^*=rf(S)$ generated by the point $\mathbf{p}^*$, and the two surfaces S, S^* satisfy the conditions of Theorem 6.1 so that $rf=t$ is a translation. Therefore $f=rt$ is a reflection with respect to a plane orthogonal to the direction $\mathbf{r}$, and hence Theorem 6.5 follows. Q.E.D.

By noting that a compact surface S in E^3 must be a sphere if it has a plane of symmetry orthogonal to every direction in E^3, we arrive readily from THeorem 6.5 at the following known result (the particular case of Theorem 5.6).

6.6. Corollary. *A compact convex surface S of class C^2 with constant mean curvature in E^3 is a sphere.*

7. UNIQUENESS THEOREMS FOR MINKOWSKI'S AND CHRISTOFFEL'S PROBLEMS

In global differential geometry few problems have attracted as much attention as those relating to the existence and uniqueness under certain conditions on compact convex surfaces in space E^3. Here uniqueness is to be understood as uniqueness within isometries of E^3, possibly combined with a reflection in a plane. The following are three of those classical problems.

WEYL'S PROBLEM [60]. It is the problem of the realization by a convex surface in E^3 of a differential geometric metric and positive Gaussian curvature given on the unit sphere Σ. In more detail, given a positive definite quadratic form

$$ds^2 = E(u,v)\,du^2 + 2F(u,v)\,du\,dv + G(u,v)\,dv^2$$

defined at every point (u,v) on Σ such that the Gaussian curvature K of the form ds^2 is positive every where, does there exist in E^3 a compact convex surface that may be diffeomorphic to Σ so that its first fundamental form, in terms of the parameters u, v on Σ, is ds^2?

MINKOWSKI'S PROBLEM [40]. Given a positive point function $K(\mathbf{e}_3)$ defined on the unit sphere Σ with unit inner normal vector $\mathbf{e}_3$, does there exist a compact convex surface having $K(\mathbf{e}_3)$ as its Gaussian curvature at the point where the unit inner vector is $\mathbf{e}_3$?

CHRISTOFFEL'S PROBLEM [15]. The problem is the same as Minkowski's except that the positive point function $K(\mathbf{e}_3)$ is replaced by the sum of the principal radii of curvature (or equivalently the ratio of the mean curvature to the Gaussian curvature).

The literature of these three problems is very long, so it is not given here. This section gives uniqueness theorems for both Minkowski's and Christoffel's problems; the uniqueness theorem for Weyl's problem appears in Section 5 as Theorem 5.2 of Cohn-Vossen. For the existence theorems for Weyl's and Minkowski's problems, see Nirenberg [43].

First we have the following uniqueness theorem for Minkowski's problem.

7.1. Theorem. *Let S and S^* be two compact oriented convex surfaces of class C^2 in E^3. Suppose that there is a diffeomorphism $f: S \to S^*$ such that at each pair of corresponding points, S and S^* have the same unit inner normal vector $\mathbf{e}_3$ and equal Gaussian curvatures K and K^*, respectively. Then f is a translation in E^3.*

Proof. As in the proof of Theorem 6.1, we first suppose that two surfaces S and S^* with boundaries C and C^*, respectively, satisfy all the other conditions of Theorem 7.1 so that all equations of Section 4 can be applied to S, and for the corresponding quantities and equations for S^*, respectively, we shall use the same symbols and numbers with an asterisk. Since under the given diffeomorphism f the two surfaces S and S^* have the same unit normal vectors at corresponding points, $\mathbf{e}_3^* = \mathbf{e}_3$ and without loss of generality we may assume that at corresponding points

$$\mathbf{e}_\alpha^* = \mathbf{e}_\alpha, \qquad \alpha = 1,2. \tag{7.1}$$

Then equations (4.7)* become

$$d\mathbf{x}^* = \sum_{\alpha=1}^{2} \omega_\alpha^* \mathbf{e}_\alpha, \tag{7.2}$$

$$d\mathbf{e}_3 = \sum_{\alpha=1}^{2} \omega_{3\alpha}^* \mathbf{e}_\alpha, \tag{7.3}$$

so that by comparing (7.3) with the second equation of (4.7) for $i=3$, we have

$$\omega_{3\alpha}^* = \omega_{3\alpha}. \tag{7.4}$$

Since $K>0$, from (4.16) the matrix

$$A=\begin{pmatrix} a & b \\ b & c \end{pmatrix}$$

is nonsingular. Therefore we can write, by using (4.12) and (4.14),

$$\omega_\alpha = \sum_{\beta=1}^{2} \lambda_{\alpha\beta}\omega_{\beta 3}, \qquad \alpha=1,2, \tag{7.5}$$

where $(\lambda_{\alpha\beta})$ is the inverse matrix of A and

$$\lambda_{\alpha\beta}=\lambda_{\beta\alpha}, \qquad \alpha,\beta=1,2. \tag{7.6}$$

From (4.16) and the definition of $\lambda_{\alpha\beta}$ it follows that

$$\lambda_{11}\lambda_{22}-\lambda_{12}\lambda_{21}=(ac-b^2)^{-1}=\frac{1}{K}. \tag{7.7}$$

Since $K^*=K>0$, we have (7.5)*, (7.6)*, (7.7)* and, in consequence of (7.4) and (7.5)*,

$$\omega_\alpha^* = \sum_{\beta=1}^{2} \lambda_{\alpha\beta}^*\omega_{\beta 3}, \tag{7.8}$$

$$\lambda_{11}^*\lambda_{22}^*-\lambda_{12}^*\lambda_{21}^*=\lambda_{11}\lambda_{22}-\lambda_{12}\lambda_{21}=\frac{1}{K}. \tag{7.9}$$

From (4.27) and (4.32) it follows that

$$K\,dA=\omega_{13}\wedge\omega_{23}, \tag{7.10}$$

which together with (7.5), (7.8), and (7.9) implies

$$\begin{aligned} &\omega_1^*\wedge\omega_2-\omega_2^*\wedge\omega_1 \\ &\quad=(\lambda_{11}^*\lambda_{22}-\lambda_{12}^*\lambda_{21})\omega_{13}\wedge\omega_{23}-(\lambda_{21}^*\lambda_{12}-\lambda_{22}^*\lambda_{11})\omega_{13}\wedge\omega_{23} \\ &\quad=(\lambda_{11}^*\lambda_{22}-\lambda_{12}^*\lambda_{21}-\lambda_{21}^*\lambda_{12}+\lambda_{22}^*\lambda_{11})K\,dA \\ &\quad=[-\Lambda+2(\lambda_{11}\lambda_{22}-\lambda_{12}\lambda_{21})]K\,dA=-\Lambda K\,dA+2dA, \end{aligned} \tag{7.11}$$

where

$$\Lambda=\begin{vmatrix} \lambda_{11}^*-\lambda_{11} & \lambda_{12}^*-\lambda_{12} \\ \lambda_{21}^*-\lambda_{21} & \lambda_{22}^*-\lambda_{22} \end{vmatrix}.$$

Using (6.3), (7.2), (4.7), and (7.11), we can easily obtain

$$\begin{aligned} d|\mathbf{x}^*,\mathbf{x},d\mathbf{x}| &= \mathbf{x}^*\cdot(d\mathbf{x}\,\hat{\times}\,d\mathbf{x})-\mathbf{x}\cdot(d\mathbf{x}^*\,\hat{\times}\,d\mathbf{x}) \\ &=2\mathbf{x}^*\cdot\mathbf{e}_3\,dA-\mathbf{x}\cdot\left(\sum_{\alpha=1}^{2}\omega_\alpha^*\mathbf{e}_\alpha\,\hat{\times}\,\sum_{\alpha=1}^{2}\omega_\alpha\mathbf{e}_\alpha\right) \\ &=2\mathbf{x}^*\cdot\mathbf{e}_3\,dA-\mathbf{x}\cdot\mathbf{e}_3(\omega_1^*\wedge\omega_2-\omega_2^*\wedge\omega_1) \\ &=2p^*\,dA+\Lambda pK\,dA-2p\,dA, \end{aligned} \tag{7.12}$$

where we have placed

$$p = \mathbf{x} \cdot \mathbf{e}_3, \qquad p^* = \mathbf{x}^* \cdot \mathbf{e}_3. \tag{7.13}$$

From (7.10), (7.10)*, and (7.4) it follows that

$$K^* \, dA^* = K \, dA, \tag{7.14}$$

which together with the assumption that $K^* = K$ implies that $dA^* = dA$. Interchanging the roles of the two surfaces S and S^* in (7.12), we thus obtain

$$d|\mathbf{x}, \mathbf{x}^*, d\mathbf{x}^*| = 2p \, dA + \Lambda p^* K \, dA - 2p^* \, dA. \tag{7.15}$$

Addition of (7.12) and (7.15) gives immediately

$$d|\mathbf{x}^*, \mathbf{x}, d\mathbf{x}| + d|\mathbf{x}, \mathbf{x}^*, d\mathbf{x}^*| = \Lambda(p + p^*) K \, dA. \tag{7.16}$$

Integrating (7.16) over the surface S and applying Green's formula to the left-hand side of the equation, we then arrive at the integral formula

$$\int_C (|\mathbf{x}^*, \mathbf{x}, d\mathbf{x}| + |\mathbf{x}, \mathbf{x}^*, d\mathbf{x}^*|) = \int_S \int \Lambda(p + p^*) K \, dA. \tag{7.17}$$

In particular, when the boundary C is empty, the integral on the left-hand side of (7.17) vanishes; hence

$$\int_S \int \Lambda(p + p^*) K \, dA = 0. \tag{7.18}$$

When necessary applying a translation in E^3 along the line joining any two corresponding points $\mathbf{x}$ and $\mathbf{x}^*$ of the two surfaces S and S^*, we may assume without loss of generality that the intersection D of the two surfaces S and S^* is nonempty. Then we can choose the origin $\mathbf{0}$ of the fixed rectangular frame $\mathbf{0}\mathbf{x}_1\mathbf{x}_2\mathbf{x}_3$ in E^3 to be in the region D, so that $p > 0$, $p^* > 0$. By using Lemma 5.3 and (7.9) we thus see that the integrand of (7.18) is nonpositive; therefore (7.18) holds when and only when

$$\Lambda = 0, \tag{7.19}$$

which implies, by Lemma 5.3 again,

$$\lambda^*_{\alpha\beta} = \lambda_{\alpha\beta}, \qquad \alpha, \beta = 1, 2. \tag{7.20}$$

Using (7.20), (7.5), and (7.8) we obtain

$$\omega^*_\alpha = \omega_\alpha, \qquad \alpha = 1, 2. \tag{7.21}$$

From (7.21), (4.7), and (7.2) it follows that over the whole surface S

$$d\mathbf{x}^* = d\mathbf{x},$$

which implies that

$$\mathbf{x}^* = \mathbf{x} + \mathbf{c}, \tag{7.22}$$

where $\mathbf{c}$ is a constant vector. Hence the proof of the theorem is complete.

This proof is due to S. S. Chern [13], and is a modification of Herglotz's proof of Theorem 5.2.

By the proof of Theorem 7.1 we can readily extend Theorem 7.1 to surfaces with boundaries as follows.

7.2. Corollary [C. C. Hsiung, 32]. *Let S and S^* be two compact oriented convex surfaces of class C^2 with boundaries C and C^*, respectively, in space E^3. Suppose that there is a diffeomorphism f: $S \to B^*$ with the same properties as those of the diffeomorphism f in Theorem 7.1. If f restricted to C is a translation in E^3 carrying C onto C^*, then f is a translation for the whole surface S.*

Proof. From the assumption of f restricted to C it follows that along C, $d\mathbf{x}^* = d\mathbf{x}$ and therefore $|\mathbf{x}^*, \mathbf{x}, d\mathbf{x}| + |\mathbf{x}, \mathbf{x}^*, d\mathbf{x}^*| = 0$. So (7.18) still holds, and Theorem 7.2 is proved by the arguments in the remainder of the proof of Theorem 7.1, except that D now is the intersection of the convex hulls of S and S^*. Q.E.D.

The uniqueness theorem for Christoffel's problem can be given as follows.

7.3. Theorem. *Let S and S^* be two compact oriented convex surfaces of class C^2 in space E^3. Suppose that there is a diffeomorphism f: $S \to S^*$ such that at each pair of corresponding points, S and S^* have the same unit inner normal vector $\mathbf{e}_3$ and equal sums of the principal radii of curvature. Then f is a translation in E^3.*

Proof. As in the proof of Theorem 7.1, we first suppose that two surfaces S and S^* with boundaries C and C^*, respectively, satisfy all the other conditions of Theorem 7.3 so that all equations of Section 4 can be applied to S, and for the corresponding quantities and equations for S^* we shall use the same symbols and numbers with an asterisk, respectively. Furthermore, we may assume (7.1) at corresponding points under f so that we have (7.2) through (7.7).

From (4.7) it follows that

$$d\mathbf{x} \hat{\times} d\mathbf{e}_3 = \mathbf{e}_3(\omega_1 \wedge \omega_{32} - \omega_2 \wedge \omega_{31}), \tag{7.23}$$

which compared with (6.4) implies

$$2H\,dA = \omega_{32} \wedge \omega_1 - \omega_{31} \wedge \omega_2. \tag{7.24}$$

Using (7.24), (7.5), and $\omega_{ij} = -\omega_{ji}$, we thus obtain

$$2H\,dA = (\lambda_{11} + \lambda_{22})\omega_{13} \wedge \omega_{23}. \tag{7.25}$$

From (7.25) and (7.10) it follows immediately that the sum of the principal radii of curvature of the surface S at the point $\mathbf{x}$ is equal to

$$2\frac{H}{K}=\lambda_{11}+\lambda_{22}. \tag{7.26}$$

By means of (7.23), (7.24), (7.2), (4.7), (7.25), and (7.11), we have the differential form

$$\begin{aligned} d|\mathbf{x}^*,\mathbf{e}_3,d\mathbf{x}| &= \mathbf{x}^*\cdot(d\mathbf{e}_3 \hat{\times} d\mathbf{x})-\mathbf{e}_3\cdot(d\mathbf{x}^*\hat{\times} d\mathbf{x}) \\ &= -2(\mathbf{x}^*\cdot\mathbf{e}_3)H\,dA-(\omega_1^*\wedge\omega_2-\omega_2^*\wedge\omega_1) \\ &= -p^*(\lambda_{11}+\lambda_{22})\omega_{13}\wedge\omega_{23} \\ &\quad -(\lambda_{11}^*\lambda_{22}-2\lambda_{12}^*\lambda_{21}+\lambda_{22}^*\lambda_{11})\omega_{13}\wedge\omega_{23}, \end{aligned} \tag{7.27}$$

where p^* is given by (7.13). Integrating (7.27) over the surface S and applying Green's formula to the left-hand side of the equation, we arrive at the integral formula

$$\begin{aligned} \int_C|\mathbf{x}^*,\mathbf{e}_3,d\mathbf{x}| &= -\int_S\!\!\int p^*(\lambda_{11}+\lambda_{22})\omega_{13}\wedge\omega_{23} \\ &\quad -\int\!\!\int(\lambda_{11}^*\lambda_{22}-2\lambda_{12}^*\lambda_{21}+\lambda_{22}^*\lambda_{11})\omega_{13}\wedge\omega_{23}. \end{aligned} \tag{7.28}$$

From (7.26) and (7.26)* and the assumption that the sums of the principal radii of curvature of S and S^* at corresponding points under f are equal, it follows readily that

$$\lambda_{11}^*+\lambda_{22}^*=\lambda_{11}+\lambda_{22}. \tag{7.29}$$

Replacing S by S^* in (7.28) and using (7.29) and (7.4), we obtain

$$\begin{aligned} \int_C|\mathbf{x}^*,\mathbf{e}_3,d\mathbf{x}^*| &= -\int_S\!\!\int p^*(\lambda_{11}+\lambda_{22})\omega_{13}\wedge\omega_{23} \\ &\quad -\int_S\!\!\int 2(\lambda_{11}^*\lambda_{22}^*-\lambda_{12}^*\lambda_{21}^*)\omega_{13}\wedge\omega_{23}. \end{aligned} \tag{7.30}$$

Subtracting (7.30) from (7.29) gives

$$\int_C|\mathbf{x}^*,\mathbf{e}_3,d\mathbf{x}-d\mathbf{x}^*|$$

$$=\int_S\!\!\int[2(\lambda_{11}^*\lambda_{22}^*-\lambda_{12}^*\lambda_{21}^*)-(\lambda_{11}^*\lambda_{22}-2\lambda_{12}^* <_{21}+\lambda_{22}^*\lambda_{11})]\omega_{13}\wedge\omega_{23}. \tag{7.31}$$

In particular, when the boundary C is empty, the integral on the left-hand side of (7.31) vanishes, hence

$$\int_S\int [2(\lambda_{11}^*\lambda_{22}^* - \lambda_{12}^*\lambda_{21}^*) \\ -(\lambda_{11}^*\lambda_{22} - 2\lambda_{12}^*\lambda_{21} + \lambda_{22}^*\lambda_{11})]\,\omega_{13}\wedge\omega_{23} = 0. \tag{7.32}$$

Subtracting (7.32) from the equation obtained by interchanging S and S^* in (7.32), we have

$$\int_S\int (\lambda_{11}\lambda_{22} - \lambda_{12}\lambda_{21})\,\omega_{13}\wedge\omega_{23} \\ = \int_{S^*}\int (\lambda_{11}^*\lambda_{22}^* - \lambda_{12}^*\lambda_{21}^*)\,\omega_{13}\wedge\omega_{23}. \tag{7.33}$$

Thus the addition of (7.32) to (7.33) gives

$$\int_S\int \left[(\lambda_{11}^* - \lambda_{11})(\lambda_{22}^* - \lambda_{22}) - (\lambda_{12}^* - \lambda_{12})^2\right]\omega_{13}\wedge\omega_{23} = 0. \tag{7.34}$$

On the other hand,

$$(\lambda_{11}^* - \lambda_{11})(\lambda_{22}^* - \lambda_{22}) = \frac{1}{2}[(\lambda_{11}^* - \lambda_{11}) + (\lambda_{22}^* - \lambda_{22})]^2 \\ - \frac{1}{2}\left[(\lambda_{11}^* - \lambda_{11})^2 + (\lambda_{22}^* - \lambda_{22})^2\right],$$

which is reduced, by means of (7.29), to

$$(\lambda_{11}^* - \lambda_{11})(\lambda_{22}^* - \lambda_{22}) = -\frac{1}{2}\left[(\lambda_{11}^* - \lambda_{11} - \lambda_{11})^2 + (\lambda_{22}^* - \lambda_{22})^2\right] \leqslant 0. \tag{7.35}$$

Moreover, from (7.10) and the assumption $K > 0$ we have

$$\omega_{13}\wedge\omega_{23} > 0.$$

Thus the integrand of (7.34) is nonpositive, and therefore (7.34) holds when and only when (7.20) does. By the same argument as that in the remainder of the proof of Theorem 7.1, f is a translation in E^3 and Theorem 7.3 is completely proved.

As we did for Theorem 7.1, we can extend Theorem 7.3 to surfaces with boundaries as follows.

7.4. Corollary [C. C. Hsiung, 30]. *Let S, S^* be two compact oriented convex surfaces of class C^2 with boundaries C, C^*, respectively, in E^3. Suppose that there is a diffeomorphism f: $S \to S^*$ with the same properties as those of the diffeomorphism f in Theorem 7.3. If f restricted to C is a translation in E^3 carrying C onto C^*, then f is a translation for the whole surface S.*

Proof. From the assumption of f restricted to C it follows that along C, $d\mathbf{x}^* = d\mathbf{x}$ and therefore $|\mathbf{x}^*, \mathbf{e}_3, d\mathbf{x} - d\mathbf{x}^*| = 0$. So (7.32) still holds, and Theorem 7.4 is proved by the arguments in the remainder of the proof of Theorem 7.3.

Exercises

1. Prove (7.10) by computing $d\mathbf{e}_3 \hat{\times} d\mathbf{e}_3$ in two different ways.

8. COMPLETE SURFACES

In this section we study complete surfaces in space E^3, which form a class larger than that of compact surfaces in E^3. We begin by continuing our discussion of the intrinsic distance between two points on a surface S in E^3 (cf. Definition 9.6 of Chapter 3).

8.1. Definition. A continuous mapping α: $[a, b] \to S$ of a closed interval $[a, b] \subset E^1$ of the line E^1 into a surface S in E^3 is said to be a *parametrized piecewise* $C^k (k \geqslant 1)$ *curve* on S joining $\alpha(a)$ to $\alpha(b)$ if there exists a partition of $[a, b]$ by points $a = t_0 < t_1 < \cdots < t_{n-1} < t_n = b$ such that α is C^k in interval $[t_{i-1}, t_i]$, $i = 1, \cdots, n$. The *length* $L(\alpha)$ of α is defined to be

$$L(\alpha) = \sum_{i=1}^{n} \int_{t_{i-1}}^{t_i} \|\alpha'(t)\| \, dt, \tag{8.1}$$

where the prime denotes the derivative with respect to t (cf. Definition 1.2.2 of Chapter 2 for a parametrized piecewise curve in E^3).

8.2. Lemma. *Given two points* $\mathbf{p}, \mathbf{q} \in S$ *of a regular connected surface* S, *there exists a parametrized piecewise* C^k *curve joining* $\mathbf{p}$ *and* $\mathbf{q}$.

Proof. Since S is connected, any two points of S can be joined by arcwise connected paths, so that there exists a continuous curve α: $[a, b] \to S$ with $\alpha(a) = \mathbf{p}$, $\alpha(b) = \mathbf{q}$. Assume that $t \in [a, b]$, and let I_t be an open interval in $[a, b]$ containing t such that $\alpha(I_t)$ is contained in a coordinate neighborhood of $\alpha(t)$. Then the union $\cup I_t$, $t \in [a, b]$, covers $[a, b]$, and by compactness a finite number of intervals $I_1, \cdots, I_n$ still covers $[a, b]$. Thus it is possible to divide I by points $a = t_0 < t_1 < \cdots < t_{m-1} < t_m = b$ such that each subinterval $[t_{i-1}, t_i]$ is contained in some I_j, $j = 1, \cdots, n$. Hence $\alpha(t_{i-1}, t_i)$ is contained in a coordinate neighborhood.

Since $\mathbf{p} = \alpha(t_0)$ and $\alpha(t_1)$ lie in a same coordinate neighborhood $\mathbf{x}(U) \subset S$, it is possible to join them by a C^k curve, namely, the image by $\mathbf{x}$ of a C^k

curve in $U \subset E^2$ joining $\mathbf{x}^{-1}(\alpha(t_0))$ and $\mathbf{x}^{-1}(\alpha(t_1))$. By this process we joint $\alpha(t_{i-1})$ and $\alpha(t_i)$, $i=1,\cdots,m$, by a C^k curve, and thus obtain a piecewise C^k curve joining $\mathbf{p}=\alpha(t_0)$ and $\mathbf{q}=\alpha(t_m)$. Q.E.D.

Now let $\mathbf{p},\mathbf{q} \in S$ be two points of a (regular) surface S. Denote by $\alpha_{p,q}$ a parametrized piecewise C^k curve on S joining $\mathbf{p}$ and $\mathbf{q}$, and by $L(\alpha_{p,q})$ the length of $\alpha_{p,q}$. Then Lemma 8.2 shows that the set of all such curves $\alpha_{p,q}$ is not empty. Thus we can have the following definition.

8.3. Definition. The (intrinsic) distance $d(\mathbf{p},\mathbf{q})$ between two points $\mathbf{p},\mathbf{q} \in S$ is

$$d(\mathbf{p},\mathbf{q}) = \inf L(\alpha_{p,q}), \tag{8.2}$$

where the infimum is taken over all piecewise C^k curve joining $\mathbf{p}$ and $\mathbf{q}$.

8.4. Lemma. *The distance* $d(\mathbf{p},\mathbf{q})$ *given by* (8.2) *defines a metric on a surface S and the same topology as that of S.*

Proof. d defines a metric if it has the following properties:

$$d(\mathbf{p},\mathbf{q}) \geqslant 0, \tag{8.3}$$

$$d(\mathbf{p},\mathbf{q}) = 0 \quad \text{if and only if} \quad \mathbf{p}=\mathbf{q}, \tag{8.4}$$

$$d(\mathbf{p},\mathbf{q}) = d(\mathbf{q},\mathbf{p}), \tag{8.5}$$

$$d(\mathbf{p},\mathbf{q}) + d(\mathbf{q},\mathbf{r}) \geqslant d(\mathbf{p},\mathbf{r}), \tag{8.6}$$

where $\mathbf{p},\mathbf{q},\mathbf{r}$ are arbitrary points of S. Now we verify these properties as follows.

Equation 8.3 follows because the infimum of positive numbers is positive or zero.

To prove (8.4) we first see that $d(\mathbf{p},\mathbf{p})=0$ is obviously true, since a point as a degenerate curve has zero length.

Now suppose that $d(\mathbf{p},\mathbf{q}) = \inf L(\alpha_{p,q}) = 0$ and $\mathbf{p} \neq \mathbf{q}$. Let V be a neighborhood of $\mathbf{p}$ on S with $\mathbf{q} \notin V$, such that every point of V may be joined to $\mathbf{p}$ by a unique geodesic of V (cf. Lemma 10.7 of Chapter 3). Let $B_r(\mathbf{p}) \subset V$ be the region, called a *geodesic disk*, bounded by a geodesic circle of radius r, centered at $\mathbf{p}$, and contained in V. By the definition of infimum, given $0 < \varepsilon < r$, there exists a parametrized piecewise C^k curve α: $[a,b] \to S$ joining $\mathbf{p}$ to $\mathbf{q}$ and with $L(\alpha) < \varepsilon$. Since $\alpha([a,b])$ is connected and $\mathbf{q} \notin B_r$, there exists a point $t_0 \in [a,b]$ such that $\alpha(t_0)$ belongs to the boundary of $B_r(\mathbf{p})$. Thus $L(\alpha) \geqslant r > \varepsilon$, which contradicts the assumption $L(\alpha) < \varepsilon$. Hence $\mathbf{p}=\mathbf{q}$, and property (8.4) is completely verified.

Equation 8.5 is immediate because $\alpha_{p,q}$ and $\alpha_{q,p}$ are the same curve.

Equation 8.6 follows because the length of any piecewise curve joining $\mathbf{p}$ and $\mathbf{q}$ is greater than or equal to $d(\mathbf{p},\mathbf{q})$.

Hence d defines a metric in S.

A basis for the open sets in the metric topology consists of the geodesic disks $B_r(\mathbf{p})$, $r>0, \mathbf{p}\in S$, which are open sets in the usual topology. Conversely, given a neighborhood $U(\mathbf{p})$ of $\mathbf{p}$ on S, there exists an $r>0$ with $B_r(\mathbf{p})\subset U(\mathbf{p})$.

8.5. Corollary. $|d(\mathbf{p},\mathbf{r})-d(\mathbf{r},\mathbf{q})|\leqslant d(\mathbf{p},\mathbf{q})$.

Proof. From

$$d(\mathbf{p},\mathbf{r})\leqslant d(\mathbf{p},\mathbf{q})+d(\mathbf{q},\mathbf{r}),\qquad d(\mathbf{r},\mathbf{q})\leqslant d(\mathbf{r},\mathbf{p})+d(\mathbf{p},\mathbf{q})$$

it follows that

$$-d(\mathbf{p},\mathbf{q})\leqslant d(\mathbf{p},\mathbf{r})-d(\mathbf{r},\mathbf{q})\leqslant d(\mathbf{p},\mathbf{q}).$$

8.6. Definition. A surface S is called a *complete surface* if every Cauchy sequence of points of S (see Definition 1.4.7 of Chapter 1) converges on S.

8.7. Examples. 1. From Theorem 1.4.6 of Chapter 1 it follows that the space E^3 is complete. Moreover, any closed subset S of E^3 is complete. In fact, any Cauchy sequence $\{\mathbf{p}_i\}$ of points in S is also a Cauchy sequence of points in E^3, which has a limit point $\mathbf{p}$ because of the completeness of E^3. Since S is closed, $\mathbf{p}\in S$; hence S is complete.

2. The Cartesian plane of pairs of real numbers (x, y) with the origin $(0,0)$ removed is not complete. The distance function $d(\mathbf{p},\mathbf{q})$ is the Euclidean distance function defined by

$$d(\mathbf{p},\mathbf{q})=\left[(x_p-x_q)^2+(y_p-y_q)^2\right]^{1/2},$$

where (x_p, y_p), (x_q, y_q) are the rectangular coordinates of the points $\mathbf{p}$ and $\mathbf{q}$, respectively. The sequence of points $\{(1/n,0)\}$ is easily seen to be a Cauchy sequence that does not converge on the plane, so this plane is not complete.

3. Consider the paraboloid S given by

$$\{(x_1, x_2, x_2)\in E^3 | x_3=x_1^2+x_2^2\}.$$

Since every point not on S is exterior to S, S is closed. (In general, the set $\{(x_1, x_2, x_3)\in E^3 | f(x_1, x_2, x_3)=0$ and f is a continuous function$\}$ is closed, since it is the inverse image $f^{-1}(0)$ of zero under the continuous mapping $f\colon E^3\to E^1$.) Thus by the above part 1, S is complete.

8.8. Theorem [H. Hopf and W. Rinow, 28]. *For a connected surface S, the following three properties are equivalent.*

(a) *Every Cauchy sequence of points of S is convergent.*
(b) *Every geodesic can be extended indefinitely in either direction, or else it forms a closed curve.*
(c) *Every bounded set of points of S is relatively compact.*

Proof. Condition c implies a, since every infinite subset of a compact set has a limit point (see Exercise 1 of Section 1.5, Chapter 1).

Now we prove that condition a implies b. Let α be a geodesic of S, which cannot be extended indefinitely. If α is a closed curve, then condition b is satisfied. If α is not a closed curve and if $\mathbf{p}(x)$ is some point on α, there is some number L such that α can be extended for distances (measured along α) less the L but cannot be extended for distances greater than L. Consider now the sequence of points $\{\mathbf{x}_n\}$ lying on α at distances from $\mathbf{p}$ along α given by $(1-1/n)L$. Since

$$\lim_{\substack{n\to\infty\\ m\to\infty}} d(\mathbf{x}_n,\mathbf{x}_m)=\lim_{\substack{n\to\infty\\ m\to\infty}}\left|\left(1-\frac{1}{n}\right)L-\left(1-\frac{1}{m}\right)L\right|=\lim_{\substack{n\to\infty\\ m\to\infty}}\left|\frac{1}{m}-\frac{1}{n}\right|L=0,$$

$\{\mathbf{x}_n\}$ is a Cauchy sequence, which by condition a converges to some point $\mathbf{q}$ on α whose distance from $\mathbf{p}$ is precisely L. If $\{\mathbf{x}'_n\}$ is another Cauchy sequence such that $d(\mathbf{x}_1,\mathbf{x}'_n)\to L$, then $\{\mathbf{x}'_n\}$ tends to some limit point $\mathbf{q}'$. Now the sequence $\mathbf{x}_1,\mathbf{x}'_1,\mathbf{x}_2,\mathbf{x}'_2,\mathbf{x}_3,\mathbf{x}'_3,\cdots$, is also a Cauchy sequence tending to both $\mathbf{q}$ and $\mathbf{q}'$. Thus $\mathbf{q}=\mathbf{q}'$, and there exists a unique end point $\mathbf{q}$ distant L from $\mathbf{p}$ along α. Consider now a coordinate neighborhood of S that contains $\mathbf{q}$. At $\mathbf{q}$ there is a uniquely determined direction $\mathbf{t}$ that is the direction of the geodesic $-\alpha$ that starts at $\mathbf{q}$. In this coordinate neighborhood there is a unique geodesic at $\mathbf{q}$ that has the direction $(-\mathbf{t})$ and this gives a continuation of α beyond $\mathbf{q}$, contrary to our assumption. It follows that α must satisfy condition b, so we have proved that a implies b.

Since c implies a, we conclude that c implies b. Thus to complete the proof of Theorem 8.8 it remains only to prove that b implies c.

Suppose now that the surface S has property b. Consider a point $\mathbf{a}$ of S, and geodesic arcs that start at $\mathbf{a}$. We define the *initial vector* of a geodesic arc starting at $\mathbf{a}$ to be the tangent vector of this arc at $\mathbf{a}$, which has the same sense and length as those of the geodesic arc. Since S has property b, it follows that every tangent vector of S at $\mathbf{a}$, whatever its length, is the initial vector of a unique geodesic arc starting at $\mathbf{a}$. This arc may eventually cut itself or, if it forms part of a closed geodesic, may even cover part of itself.

Let $S(r)$ be the set of points $\mathbf{x}$ of S whose distance from $\mathbf{a}$ does not exceed r [i.e., $d(\mathbf{x},\mathbf{a})\leqslant r$], and let $E(r)$ be the set of points $\mathbf{x}$ of $S(r)$ that can be joined to **a** by a geodesic arc whose length is actually equal to $d(\mathbf{x},\mathbf{a})$.

We first prove that the set $E(r)$ is compact. Let $\{\mathbf{x}_h\}$, $h=1,2,\cdots$, be a sequence of points of $E(r)$, and let $\mathbf{t}_h$ be the initial vector of a geodesic arc of length $d(\mathbf{a},\mathbf{x}_h)$ joining **a** to $\mathbf{x}_h$. Then by Bolzano-Weierstrass theorem the sequence of vectors $\{\mathbf{t}_h\}$, regarded as a sequence of points in a Euclidean plane, admits at least one limit vector **t**. Moreover, this vector **t** is the initial vector of a geodesic arc whose extremity belongs to $E(r)$ and is a limit point of the sequence $\{\mathbf{x}_h\}$. Thus $E(r)$ is compact.

We next show

$$E(r)=S(r). \tag{8.7}$$

It is obvious that (8.7) is true when $r=0$. Furthermore, if it is true for $r=R>0$, it is surely true for $r<R$. Now we shall show the converse: that is, if (8.7) is true for $r<R$, it is also true for $r=R$. Now every point of $S(R)$ is the limit of a sequence of points whose distance from **a** is less than R. By assumption these points belong to $E(r)$, and since $E(r)$ is closed, it follows that their limit belongs to $E(r)$. Thus (8.7) is true for $r=R$. To establish (8.7) completely, it is necessary only to show that if (8.7) holds for $r=R$, it also holds for $r=R+s$ with arbitrary positive s.

We next show that to any point **y** of S such that $d(\mathbf{a},\mathbf{y})>R$, there is a point **x** such that

$$d(\mathbf{a},\mathbf{x})=R, \tag{8.8}$$

$$d(\mathbf{a},\mathbf{y})=R+d(\mathbf{y},\mathbf{x}). \tag{8.9}$$

From (8.2) it follows that we can join **a** to **y** by a curve α whose length is less than $d(\mathbf{a},\mathbf{y})+h^{-1}$ for any integer . Let $\mathbf{x}_h$ be the last point of this curve α belonging to $E(R)=S(R)$ (see Fig. 4.14). Since $d(\mathbf{a},\mathbf{x}_h)=R$, by (8.6) we

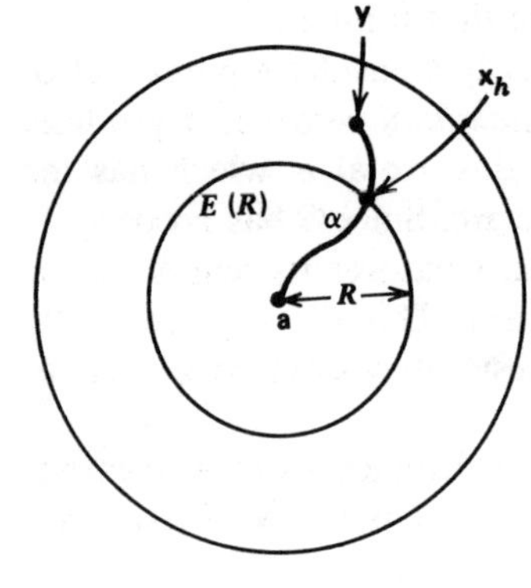

Figure 4.14

obtain

$$d(\mathbf{a},\mathbf{y}) \leqslant d(\mathbf{a},\mathbf{x}_h) + d(\mathbf{x}_h,\mathbf{y}) = R + d(\mathbf{x}_h,\mathbf{y}), \tag{8.10}$$

$$d(\mathbf{x}_h,\mathbf{y}) \geqslant d(\mathbf{a},\mathbf{y}) - R. \tag{8.11}$$

Since the arc length of α from $\mathbf{a}$ to $\mathbf{y}$ is the sum of the arc lengths from $\mathbf{a}$ to $\mathbf{x}_h$ and from $\mathbf{x}_h$ to $\mathbf{y}$, we have

$$\begin{aligned} d(\mathbf{x}_h,\mathbf{y}) &\leqslant \operatorname{arc}(\mathbf{x}_h,\mathbf{y}) = \operatorname{arc}(\mathbf{a},\mathbf{y}) - \operatorname{arc}(\mathbf{a},\mathbf{x}_h) \\ &\leqslant d(\mathbf{a},\mathbf{y}) + h^{-1} - \operatorname{arc}(\mathbf{a},\mathbf{x}_h) \\ &\leqslant d(\mathbf{a},\mathbf{y}) + h^{-1} - R. \end{aligned}$$

Now let $h \to \infty$. Then $\{\mathbf{x}_h\}$ will have at least one limit point $\mathbf{x}$ with the property

$$d(\mathbf{x},\mathbf{y}) \leqslant d(\mathbf{a},\mathbf{y}) - R. \tag{8.12}$$

Comparison of (8.11) with (8.12) gives (8.9) at this point $\mathbf{x}$. Thus we have proved the existence of a point $\mathbf{x}$ satisfying (8.8) and (8.9).

We have already seen in Chapter 3 that if two points $\mathbf{x},\mathbf{y}$ on the surface S are not far apart, the point $\mathbf{y}$ is the extremity of a unique geodesic arc with origin $\mathbf{x}$ and length $d(\mathbf{x},\mathbf{y})$. More precisely, we can have Lemma 8.9.

8.9. Lemma. *On a surface S in space E^3, for every point $\mathbf{x} \in S$ there exists a positive continuous function $s(\mathbf{x})$ such that any two points $\mathbf{y},\mathbf{z}$ of $S_x(s(\mathbf{x})) = \{\mathbf{y} \in S \mid d(\mathbf{x},\mathbf{y}) \leqslant s(\mathbf{x})\}$ can be joined by a unique geodesic arc with length $d(\mathbf{y},\mathbf{z})$.*

Proof. For each $\mathbf{x} \in S$, let $s(\mathbf{x})$ be the supremum of $s > 0$ such that any two points $\mathbf{y}$ and $\mathbf{z}$ with $d(\mathbf{x},\mathbf{y}) \leqslant s$ and $d(\mathbf{x},\mathbf{z}) \leqslant s$ can be joined by a geodesic arc with length $d(\mathbf{y},\mathbf{z})$. From the foregoing remark we know that $s(\mathbf{x}) > 0$. If $s(\mathbf{x}) = \infty$ for some point $\mathbf{x}$, then $s(\mathbf{y}) = \infty$ for every point $\mathbf{y}$ of S, and any positive continuous function on S satisfies the condition of the lemma. Now assume that $s(\mathbf{x}) < \infty$ for every $\mathbf{x} \in S$. We shall prove the continuity of $s(\mathbf{x})$ by showing that $|s(\mathbf{x}) - s(\mathbf{y})| \leqslant d(\mathbf{x},\mathbf{y})$. Without loss of generality, we may assume that $s(\mathbf{x}) > s(\mathbf{y})$. If $d(\mathbf{x},\mathbf{y}) \geqslant s(\mathbf{x})$, obviously $|s(\mathbf{x}) - s(\mathbf{y})| < d(\mathbf{x},\mathbf{y})$. If $d(\mathbf{x},\mathbf{y}) < s(\mathbf{x})$, then $S_y(s') = \{\mathbf{z} \in S \mid d(\mathbf{y},\mathbf{z}) \leqslant s'\}$ is contained in $S_x(s(\mathbf{x}))$, where $s' = s(\mathbf{x}) - d(\mathbf{x},\mathbf{y})$. Hence $s(\mathbf{y}) \geqslant s(\mathbf{x}) - d(\mathbf{x},\mathbf{y})$, that is, $|s(\mathbf{x}) - s(\mathbf{y})| \leqslant d(\mathbf{x},\mathbf{y})$, completing the proof of the lemma.

Now we come back to prove that condition b implies c for Theorem 8.8. From Lemma 8.9 it follows that for each $\mathbf{x} \in S$, there exists a continuous function $s(\mathbf{x}) > 0$ such that if $d(\mathbf{x},\mathbf{y}) < s(\mathbf{x})$, the point $\mathbf{y}$ is the extremity of the unique geodesic arc of length $d(\mathbf{x},\mathbf{y})$ joining $\mathbf{x}$ to $\mathbf{y}$. Moreover, the continuous function $s(x)$ attains a positive minimum value on the compact set $E(R)$, and we take s to be this minimum.

If (8.7) is true for $r=R$, and if $R<d(\mathbf{a},\mathbf{y})\leqslant R+s$, there exists an $\mathbf{x}\in E(R)$ such that $d(\mathbf{a},\mathbf{x})=R$ and $d(\mathbf{x},\mathbf{y})=d(\mathbf{a},\mathbf{y})-R\leqslant s$. Thus there exist a geodesic arc β of length $d(\mathbf{a},\mathbf{x})$ joining $\mathbf{a}$ to $\mathbf{x}$, and a geodesic γ of length $d(\mathbf{x},\mathbf{y})$ joining $\mathbf{x}$ to $\mathbf{y}$. The composite arc ρ formed by β and γ joins $\mathbf{a}$ to $\mathbf{y}$, and has $d(\mathbf{a},\mathbf{y})$ as its length. Thus ρ is a geodesic arc, and $\mathbf{y}$ is joined to $\mathbf{a}$ by a geodesic arc whose length is equal to $d(\mathbf{a},\mathbf{y})$. Hence $\mathbf{y}\in E(R+s)$, and the range of validity of (8.7) is extended from $E(R)$ to $E(R+s)$. We have incidentally proved that condition c implies that any two points of S can be joined by a geodesic arc whose length is equal to the distance between them.

Now it is evident that for a given bounded set M of points on S there is some R such that M is contained in $S(R)$ and therefore that M is relatively compact since $S(R)(=E(R))$ is compact. Thus condition b implies c, and the equivalence of the three conditions is established. Q.E.D.

We have also proved Theorem 8.10.

This proof is due to G. de Rham [46].

8.10. Theorem. *On a complete surface S any two points $\mathbf{p},\mathbf{q}$ can be joined by a geodesic arc with length $d(\mathbf{p},\mathbf{q})$ that is shortest among the lengths of all piecewise arcs between $\mathbf{p}$ and $\mathbf{q}$.*

Theorem 8.10 can be easily verified for the following three simple complete surfaces (cf. part 3 of Example 8.7).

8.11. Examples. 1. In a plane, the geodesics are straight lines, and any two points $\mathbf{p}$ and $\mathbf{q}$ can be joined by a unique line segment with length $d(\mathbf{p},\mathbf{q})$.

2. On a sphere, the geodesics are the great circles (Exercise 6 of Section 10, Chapter 3), and any two points $\mathbf{p}$ and $\mathbf{q}$ that are not antipodal points (i.e., the end points of a diameter), can be joined by two great circular arcs, of which only one has length $d(\mathbf{p},\mathbf{q})$. Moreover, between two antipodal points $\mathbf{p}$ and $\mathbf{q}$ there are infinitely many great circular arcs with the same length $d(\mathbf{p},\mathbf{q})$.

3. On a circular cylinder, any two points $\mathbf{p}$ and $\mathbf{q}$ on the same generator can be joined not only by means of the generator [with length $d(\mathbf{p},\mathbf{q})$] but also by infinitely many circular helices of varying pitch, which wind around the cylinder and all are geodesics (Exercise 9 of Section 10, Chapter 3).

Since a compact surface evidently possesses property c of Theorem 8.8, we have Theorem 8.12.

8.12. Theorem. *All compact surfaces are complete.*

Concerning the shortest arcs between two points on a surface, we showed first in Lemma 10.7 of Chapter 3 that on a surface two points **p** and **q** not too far apart (more precisely, lying in a geodesic field) can be joined by a unique geodesic arc with length $d(\mathbf{p},\mathbf{q})$, and then in Theorem 8.10 that on a complete surface any two points **p** and **q** can be joined by a geodesic arc with length $d(\mathbf{p},\mathbf{q})$; however the arc may not be unique, as we have seen in part 3 of Example 8.11. For a general case we have Theorem 8.13.

8.13. Theorem. *If there exists a shortest C^2 piecewise arc α between two points* **p** *and* **q** *on a surface S, the arc α is geodesic.*

Proof. Let α be given by $[a,b]\to S$ with $\alpha(a)=\mathbf{p}$ and $\alpha(b)=\mathbf{q}$, and denote the segment of α from $\alpha(c)$ to $\alpha(d)$, $a<c<d<b$, by α_{cd}. Let $\alpha_{t_1t_2}$ with $a<t_1<t_2<b$ be any segment of α in a coordinate neighborhood of S. Then $\alpha_{t_1t_2}$ is the shortest arc between $\alpha(t_1)$ and $\alpha(t_2)$, since otherwise there is an arc γ between $\alpha(t_1)$ and $\alpha(t_2)$, which is shorter than $\alpha_{t_1t_2}$, so that the composite arc formed by the arcs α_{at_1}, γ, and α_{t_2b} is shorter than α, contradicting the assumption of the theorem. Thus by applying the Heine-Borel theorem to a covering of α, we need only to show that any segment of α in a coordinate neighborhood of S is geodesic.

Let the segment of α in the neighborhood of an orthogonal parametrization $\mathbf{x}(u,v)$ of S be taken as the curve $u=0$ from $v=v_0$ to v_1. With respect to $\mathbf{x}(u,v)$, the first fundamental form of S is given by

$$ds^2=E\,du^2+G\,dv^2. \tag{8.13}$$

Let β be any arc between $\alpha(v_0)$ and $\alpha(v_1)$ close to α and determined by a small variation $\delta n=\sqrt{E}\;\delta u$ along the curves $v=$const measured from α (see Fig. 4.15). If β is of equation $u=u(v)$, we take u so small that it can be identified with δu. Since β passes through $\alpha(v_0)$ and $\alpha(v_1)$, $u(v_0)=u(v_1)=0$ and the lengths of $\alpha_{v_0v_1}$ and β are, respectively,

$$L=\int_{v_0}^{v_1}\sqrt{g}\;dv,\qquad L+\delta L=\int_{v_0}^{v_1}\sqrt{Eu'^2+G_1}\;dv,$$

where $G=G(0,v)$, $G_1=G(\delta u,v)$, $E=E(\delta u,v)$, and $u'=du/dv$.

Since

$$G(\delta u,v)=G(0,v)+\delta u G_u+\cdots,\qquad G_u=\left(\frac{\partial G(u,v)}{\partial u}\right)_{u=0},$$

by neglecting all terms of order higher than δu (including $(u')^2$) we obtain

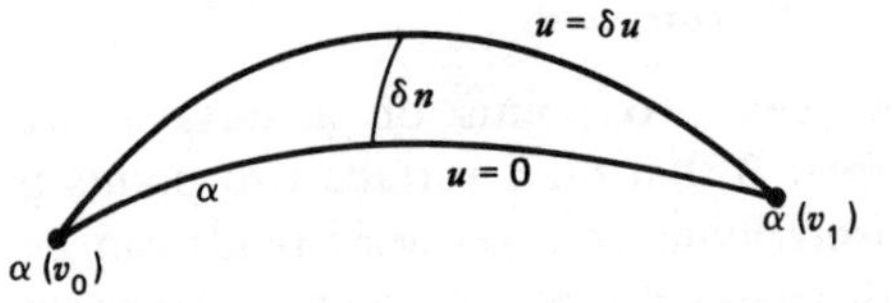

Figure 4.15

the *first variation* δL of the length L:

$$\delta L = \int_{v_0}^{v_1}\left[\sqrt{G}\left(1+\delta u\frac{G_u}{G}\right)^{1/2} - \sqrt{G}\right]dv$$
$$= \int_{v_0}^{v_1}\left(\sqrt{G} + \frac{1}{2}\delta u\frac{G_u}{\sqrt{G}} - \sqrt{G}\right)dv = \int_{v_0}^{v_1}\frac{1}{2}\delta u\frac{G_u}{\sqrt{G}}dv \tag{8.14}$$
$$= \int_{v_0}^{v_1}\frac{G_u}{2G\sqrt{E}}\delta n\, ds, \qquad ds = \sqrt{G}\,dv \text{ measured along } \alpha.$$

Thus substituting in (8.14) the equation obtained by interchanging u and v in the equation just before (10.10) of Chapter 3 we have

$$\delta L = -\int_{v_0}^{v_1}\kappa_g \delta n\, ds, \tag{8.15}$$

where κ_g is the geodesic curvature along $\alpha_{v_0v_1}(u=0)$.

A necessary condition that L be a minimum is that $\delta L = 0$ for all δn. By Lemma 7.1 of Chapter 1 this condition is fulfilled if and only if the geodesic curvature κ_g vanishes along $\alpha_{v_0v_1}$. Hence by Definition 10.1 of Chapter 3, $\alpha_{v_0v_1}$ is a geodesic. Q.E.D.

Next we use the second variation of the length of an arc to find a necessary condition for a shortest geodesic arc α between two points $\mathbf{p}$ and $\mathbf{q}$ in a certain coordinate neighborhood of a surface S. To this end we embed the arc α in a parametrization $\mathbf{x}(u,v)$ of the surface S. In Chapter 3 it was shown that local coordinates u, v could be introduced in a neighborhood of α such that $u=0$ represents α and such that the curves $v=\text{const}$ are geodesics orthogonal to α. If v is chosen as the arc length along α, so that v ranges over the interval $0 \leqslant v \leqslant L$, with L the length of α, the first fundamental form of S is given by

$$ds^2 = du^2 + G(u,v)\,dv^2, \tag{8.16}$$

which is (10.10) of Chapter 3. Since $u=0$ represents α, we have

$$G(0,v) = 1. \tag{8.17}$$

Since α is a geodesic, the geodesic curvature κ_g vanishes on α, so from the equation obtained by interchanging u and v in the equation just before (10.10) of Chapter 3, it follows that

$$G_u(0, v)=0. \tag{8.18}$$

Among all piecewise C^1 arcs through $\mathbf{p}$ and $\mathbf{q}$ on the surface S, our discussion is concerned only with those, called *admissible arcs* of the geodesic arc α, lying in a neighborhood of α within which the geodesic coordinate system for the form (8.16) is valid. For what follows it is useful to consider a special class of such arcs, namely, arcs defined by $u=\varepsilon\eta(v)$ in the geodesic coordinate system, where $\eta(v)$ is a piecewise C^2 function defined for $0 \leqslant v \leqslant L$ such that $\eta(0)=\eta(L)=0$, and ε is so small that the arcs will lie close to α. The lengths $L(\varepsilon)$ of such arcs in the neighborhood of α are given by

$$L(\varepsilon)=\int_0^L \sqrt{\varepsilon^2\eta'^2+G(\varepsilon\eta(v), v)}\, dv, \tag{8.19}$$

where $\eta'=d\eta/dv$. Since α is assumed to be a shortest arc,

$$\left.\frac{dL(\varepsilon)}{d\varepsilon}\right|_{\varepsilon=0}=0,$$

which can be easily verified by means of (8.18) and

$$\left.\frac{dL(\varepsilon)}{d\varepsilon}\right|_{\varepsilon=0}=\int_0^L \left.\frac{\varepsilon\eta'^2+\frac{\eta}{2}G_u(\varepsilon\eta(v), v)}{\sqrt{\varepsilon^2\eta'^2+G(\varepsilon\eta(v), v)}}\right|_{\varepsilon=0} dv. \tag{8.20}$$

Here we obtain (8.20) by using $(d/d\varepsilon)\int=\int(d/d\varepsilon)$ for the integral in (8.19), since its integrand has a continuous derivative. It is well known that a further necessary condition $L(0)$ to furnish a minimum is

$$\left.\frac{d^2L(\varepsilon)}{d\varepsilon^2}\right|_{\varepsilon=0} \geqslant 0.$$

Now let the integrand in (8.19) be simply written as $F(\varepsilon)=\sqrt{f(\varepsilon)}$. Since

$$f'(0)=0, \qquad \left.\frac{dL(\varepsilon)}{d\varepsilon}\right|_{\varepsilon=0}=0,$$

we have

$$F''(0)=\frac{1}{2}\frac{f''(0)}{\sqrt{f(0)}}. \tag{8.21}$$

Also $f(0)=G(0, v)=1$, and Maclaurin's expansion of $L(\varepsilon)$,

$$L(\varepsilon)=L+\left.\frac{dL(\varepsilon)}{d\varepsilon}\right|_{\varepsilon=0}\varepsilon+\frac{1}{2}\left.\frac{d^2L(\varepsilon)}{d\varepsilon^2}\right|_{\varepsilon=0}\varepsilon^2+\cdots,$$

then takes the form

$$L(\varepsilon)=L+\int_0^L \frac{1}{2}\left[\eta'^2+\frac{\eta^2}{2}G_{uu}(0,v)\right]\varepsilon^2\,dv+\cdots, \tag{8.22}$$

where the unwritten terms are of degree $\geqslant 3$ in ε. From (10.26) of Chapter 3 and (8.17) and (8.18), it follows that

$$G_{uu}=-2K \qquad \text{for} \quad u=0, \tag{8.23}$$

where K is the Gaussian curvature of the surface S. Thus (8.22) can be written as

$$L(\varepsilon)-L=\frac{\varepsilon^2}{2}\int_0^L\left[\eta'^2-K(0,v)\eta^2\right]dv+\cdots. \tag{8.24}$$

Clearly, if the length L of α is to furnish a minimum, the *second variation* $\delta^2 L$ of the length L,

$$\delta^2 L=\int_0^L\left[\eta'^2-K(0,v)\eta^2\right]dv, \tag{8.25}$$

must be $\geqslant 0$ for all admissible arcs of α. This is therefore another necessary condition that α should satisfy.

Since the minimizing arc α is assumed given in the present case by $\eta=0$, from (8.25) it follows readily that $\delta^2 L=0$ with respect to α. Thus the new necessary condition on α will be satisfied if the integral $\delta^2 L$ in (8.25) has $\delta^2 L=0$ as its minimum for $\eta\equiv 0$ within the class of admissible functions introduced above. The idea of introducing an auxiliary minimum problem based on the second variation is due to Jacobi. A necessary condition for such a minimum is that a function $\eta(v)$ that vanishes at the two end points should satisfy the corresponding Euler equation

$$\frac{d^2\eta}{dv^2}+K(0,v)\eta=0, \tag{8.26}$$

which can be easily obtained by means of Theorem 7.2 of Chapter 1, and is called the *Jacobi equation*.

8.14. Definition. A *conjugate point* p^* on a geodesic arc α is an interior point of the arc where a solution $\eta(v)$ of (8.26), which vanishes at $p(v=0)$ but not identically in v, also vanishes at $p^*(v=v^*)$.

8.15. Example. Let α be a geodesic starting at any point $\mathbf{p}$ of a sphere S of radius r. Then the Jacobi equation (8.26) becomes

$$\frac{d^2\eta}{ds^2}+\frac{1}{r^2}\eta=0,$$

where s is the arc length of α. The general solution of this equation is given by

$$\eta(s)=a\sin\frac{s}{r}+b\cos\frac{s}{r},$$

where a, b are arbitrary constants. The initial conditions $\eta(0)=0$, $\eta'(0)=1$ then give $\eta(s)=r\sin(s/r)$ whose first positive zero occurs at $s=\pi r$. Thus *the first conjugate point of* $\alpha(0)=\mathbf{p}$ *on* α *is at the antipodal point* (*i.e.*, $-\mathbf{p}$) *of* $\mathbf{p}$.

Concerning the relationship between conjugate points and shortest geodesic arcs we have

8.16. Theorem. *For a geodesic arc* α *between any two points* $\mathbf{p},\mathbf{q}$ *on a surface* S *to be shortest among all its admissible arcs, it is necessary and sufficient that no conjugate point* $\mathbf{p}^*\not\equiv\mathbf{q}$ *exist between* $\mathbf{p}$ *and* $\mathbf{q}$ *on* α.

Proof. Here we prove only the "necessity part" of the theorem.

For this purpose we assume that on a geodesic arc joining $\mathbf{p}$ and $\mathbf{q}$ on S, which is shortest among all its admissible arcs, there is a conjugate point $\mathbf{p}^*$ lying between $\mathbf{p}$ and $\mathbf{q}$. By Definition 8.14, we have a solution $\eta=\phi(v)$ of the Jacobi equation (8.26) such that $\phi(\mathbf{p})=0$, $\phi(\mathbf{p}^*)=0$. Of course, $\eta=\varepsilon\phi(v)$ for an arbitrary constant ε is also such a solution.

Now define a function $\tilde{\eta}$ by

$$\tilde{\eta}=\begin{cases}\phi & \text{from} \quad \mathbf{p} \text{ to } \mathbf{p}^*,\\ 0 & \text{from} \quad \mathbf{p}^* \text{ to } \mathbf{q}.\end{cases}$$

The next step in the argument is to show that $\tilde{\eta}$ is a "corner" solution (cf. Theorem 7.3 of Chapter 1) of the problem of giving δ^2L an extremal value.

Since

$$\int_{\mathbf{p}}^{\mathbf{p}^*}\eta\eta''\,dv=[\eta\eta']_{\mathbf{p}}^{\mathbf{p}^*}-\int_{\mathbf{p}}^{\mathbf{p}^*}\eta'^2\,dv=-\int_{\mathbf{p}}^{\mathbf{p}^*}\eta'^2\,dv,$$

where $\eta=\phi(v)$, it follows that

$$\begin{aligned}\int_{\mathbf{p}}^{\mathbf{q}}\left[\tilde{\eta}'^2-K(0,v)\tilde{\eta}^2\right]dv&=\int_{\mathbf{p}}^{\mathbf{p}^*}\left[\eta'^2-K(0,v)\eta^2\right]dv\\&=-\int_{p}^{p^*}\eta\left[\eta''+K(0,v)\eta\right]dv=0,\end{aligned}$$

because of (8.26). Thus $\tilde{\eta}$ satisfies the condition $\delta^2L=0$ and can be chosen as near to the curve $\eta=0$ as we like, since ε is arbitrary. Hence $\eta=0$ gives δ^2L its minimal value, and η must be a "corner" solution of the problem of finding a minimum of δ^2L. By Theorem 7.3 of Chapter 1, we have $\eta'_+=\eta'_-=0$. But this is impossible, since there is no nontrivial solution of (8.26), which vanishes simultaneously with its derivative. Hence the "necessity part" of the theorem is proved.

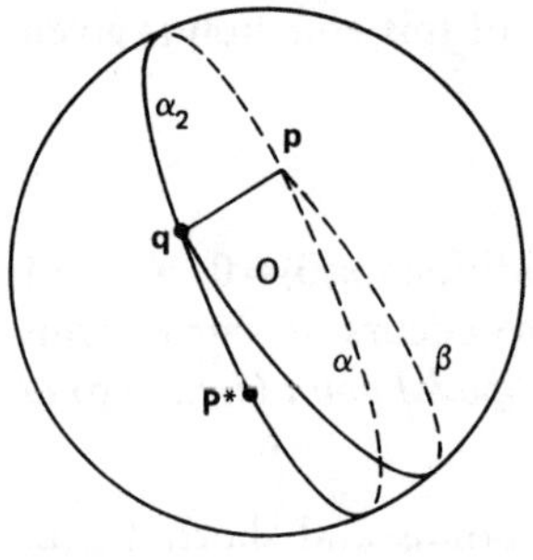

Figure 4.16

8.17. Remark. It is easy to verify Theorem 8.17 for a sphere S. Consider a great circle α on S through any two points **p** and **q**. Let α_1 be the arc **pp*q** on α, where **p*** is the antipodal point of **p**, and let α_2 be the other arc between **p*** and **q** on α (see Fig. 4.16). Point **q** lies between **p** and its first conjugate point **p*** on α, and **p*** lies between **p** and **q** on α. It is true that α_2 is shortest among all its admissible arcs between **p** and **q** (actually, in this case, among all arcs between **p** and **q** on S), but α_1 is not. Actually, if the plane of α is rotated slightly about an axis through **p** and **q**, it will slice from the sphere S an arc β, which can be verified analytically to be strictly shorter than α_1.

As applications of Theorem 8.16 to special surfaces, we first have Theorem 8.18.

8.18. Theorem. *If the Gaussian curvature K of a connected surface S is nonpositive, then there are no conjugate points on any geodesic on S, and a geodesic arc between any two points on S is always shortest among all its admissible arcs.*

Proof. Let α be a geodesic arc of length L between any two points **p** and **q** on S, and use the same local coordinates u, v as before, so that α is given by $u=0$, **p** by $u=0$, $v=0$ and **q** by $u=0$, $v=L$. Since $K(0,v) \leqslant 0$, from the Euler equation (8.26) it follows that

$$\eta \frac{d^2\eta}{dv^2} = -K(0,v)\eta^2 \geqslant 0,$$

so that

$$\frac{d}{dv}\left(\eta \frac{d\eta}{dv}\right) \geqslant 0.$$

Thus the function $\eta\, d\eta/dv$ is a nondecreasing function in the interval $[0, L]$. Since $\eta(0)=\eta(L)=0$, we have $\eta\, d\eta/dv \equiv 0$ or $d\eta^2/dv \equiv 0$ in $[0, L]$.

Thus $\eta=\text{const}$, and $\eta\equiv 0$ in $[0, L]$ due to $\eta(0)=0$. Hence there are no conjugate points on the geodesic arc α.

The second conclusion of the theorem follows readily from Theorem 8.16.

8.19. Remark. Since on a connected surface S of nonpositive Gaussian curvature K no two geodesics can meet more than one point (Application 3.14b), any two points on S can be joined by a *unique* geodesic arc that is shortest among all its admissible arcs.

8.20. Theorem (O. Bonnet). *Let the Gaussian curvature K of a complete surface S satisfy the condition*

$$K \geqslant \frac{1}{k^2}, \tag{8.27}$$

where k is a positive constant. Then S is compact, and the diameter ρ (see Definition 1.4.4 *of Chapter* 1) *of S satisfies the inequality*

$$\rho \leqslant \pi k. \tag{8.28}$$

Proof. Since S is complete, by Theorem 8.10 any two points $\mathbf{p}, \mathbf{q} \in S$ can be joined by a geodesic arc α whose length L is equal to $d(\mathbf{p}, \mathbf{q})$; therefore the second variation $\delta^2 L$ of L is nonnegative. On the other hand, from (8.27) we have

$$\eta'^2 - \frac{1}{k^2}\eta^2 \geqslant \eta'^2 - K(0, v)\eta^2,$$

which together with (8.25) implies that

$$\int_0^L \left(\eta'^2 - \frac{1}{k^2}\eta^2 \right) dv \geqslant \delta^2 L \geqslant 0 \tag{8.29}$$

for any admissible function $\eta(v)$, and in particular for

$$\eta(v) = \varepsilon \sin \frac{\pi v}{L} \tag{8.30}$$

with small ε, which is admissible because $\eta(0)=\eta(L)=0$. Substituting (8.30) in (8.29) gives

$$0 \leqslant \int_0^L \left(\frac{\pi^2}{L^2} \cos^2 \frac{\pi v}{L} - \frac{1}{k^2} \sin^2 \frac{\pi v}{L} \right) dv = \frac{L}{2} \left(\frac{\pi^2}{L^2} - \frac{1}{k^2} \right). \tag{8.31}$$

From (8.31) it follows at once that $L \leqslant \pi k$, which means that $d(\mathbf{p}, \mathbf{q}) \leqslant \pi k$ for any two points $\mathbf{p}, \mathbf{q} \in S$. Thus S is bounded and its diameter $\rho \leqslant \pi k$.

Moreover, since S is complete, from property c of Theorem 8.8 it follows that S is compact.

8.21. Remarks. 1. The hypothesis $K \geqslant 1/k^2 > 0$ in Theorem 8.20 cannot be weakened to $K \geqslant 0$. In fact, the paraboloid

$$\{(x_1, x_2, x_3) \in E^3 \mid x_3 = x_1^2 + x_2^2\}$$

has Gaussian curvature $K > 0$ and is complete (see part 3 of Example 8.7), but it is not compact (see Theorem 1.5.5 of Chapter 1), since it is unbonded. Furthermore, the Gaussian curvature K of the paraboloid tends toward zero as the distance of the point $(x_1, x_2) \in E^2$ to the origin becomes arbitrarily large.

2. The estimate of the diameter $\rho \geqslant \pi k$ given in Theorem 8.20 is the best possible, as shown by the example of the unit sphere for which $K \equiv 1$ and $\rho = \pi$.

Exercises

1. Let S be a complete surface, and F a nonempty closed subset of S such that the complement $S - F$ is connected. Prove that $S - F$ is a noncomplete regular surface.

***2.** Show that the converse of Theorem 8.10 is false: give an example of a surface S such that any two points on S can be joined by a geodesic with the shortest length, but S is not complete.

3. Prove (8.26).

***4.** Let α be a geodesic on the outer equator of a torus with parametrization (1.18) of Chapter 3. Find s_1 for the first conjugate point $\alpha(s_1)$ of $\alpha(0) = \mathbf{p}$ on α.

***5.** Is the converse of Theorem 8.20 true? That is, if S is compact and has diameter $\rho \leqslant \pi k$, is $K \geqslant 1/k^2$?

APPENDIX 1

Proof of Existence Theorem 1.5.1, Chapter 2

The proof of Theorem 1.5.1 (Chapter 2) depends on the following existence and uniqueness theorem for systems of ordinary differential equations of the first order.

A.1. Theorem. *Consider a system of ordinary differential equations of the form*

$$\frac{dy_i}{dx}=f_i(x;y_1,\cdots,y_n), \qquad i=1,\cdots,n, \tag{A.1.1}$$

where the functions f_i are of class C^k in some neighborhood of $(x_0,0,\cdots,0)\in R\times E^n$, R being the real line. Then there exist neighborhoods U of the origin in E^n and I of x_0 in R such that for any $(y_{10},\cdots,y_{n0})\in U$ and all $x\in I$ there exists a unique C^{k+1} solution $y_i(x)$, $i=1,\cdots,n$, of (A.1.1) *satisfying the initial conditions*

$$y_i(x_0)=y_{i0}, \qquad i=1,\cdots,n. \tag{A.1.2}$$

Proof of Existence Theorem 1.5.1, *Chapter* 2. Consider Frenet formulas (1.3.11) of Chapter 2, which can be written as

$$\frac{dy_i}{ds}=\sum_{j=1}^{9}\mu_{ij}(s)y_j(s), \qquad i=1,\cdots,9, \tag{A.1.3}$$

where $(y_1,y_2,y_3)=\mathbf{e}_1,(y_4,y_5,y_6)=\mathbf{e}_2,(y_7,y_8,y_9)=\mathbf{e}_3$, and the functions $\mu_{ij}(s)$ are, within sign, either the curvature $\kappa(s)$, or the torsion $\tau(s)$, or zero. Then from Theorem A.1 it follows that this system (A.1.3) possesses a unique solution of class C^1, for which the initial values of the functions y_i are given. In other words, given a positively oriented frame $\mathbf{e}_1^0\mathbf{e}_2^0\mathbf{e}_3^0$ (Definitions 3.5.6 and 5.4.1 of Chapter 1) in E^3 and a value $s_0\in I\subset R$, there exists a unique family of trihedrons $\mathbf{e}_1(s)\mathbf{e}_2(s)\mathbf{e}_3(s)$, $s\in I$, with $\mathbf{e}_i(s_0)=\mathbf{e}_i^0$, $i=1,2,3$.

Now we first show that for each s the trihedron $\mathbf{e}_1(s)\mathbf{e}_2(s)\mathbf{e}_3(s)$ thus obtained is also a positively oriented frame. For this purpose we use Frenet formulas (1.3.11) of Chapter 2 to obtain the first derivatives, with respect to s, of the six functions of s

$$\mathbf{e}_1\cdot\mathbf{e}_2,\qquad \mathbf{e}_1\cdot\mathbf{e}_3,\qquad \mathbf{e}_2\cdot\mathbf{e}_3,\qquad \mathbf{e}_1\cdot\mathbf{e}_1,\qquad \mathbf{e}_2\cdot\mathbf{e}_2,\qquad \mathbf{e}_3\cdot\mathbf{e}_3,$$

namely,

$$\begin{aligned}
\frac{d(\mathbf{e}_1\cdot\mathbf{e}_2)}{ds} &= \tau(\mathbf{e}_1\cdot\mathbf{e}_3)-\kappa(\mathbf{e}_1\cdot\mathbf{e}_1)+\kappa(\mathbf{e}_2\cdot\mathbf{e}_2),\\
\frac{d(\mathbf{e}_1\cdot\mathbf{e}_3)}{ds} &= -\tau(\mathbf{e}_1\cdot\mathbf{e}_2)+\kappa(\mathbf{e}_2\cdot\mathbf{e}_3),\\
\frac{d(\mathbf{e}_2\cdot\mathbf{e}_3)}{ds} &= -\kappa(\mathbf{e}_1\cdot\mathbf{e}_3)-\tau(\mathbf{e}_2\cdot\mathbf{e}_2)+\tau(\mathbf{e}_3\cdot\mathbf{e}_3),\\
\frac{d(\mathbf{e}_1\cdot\mathbf{e}_1)}{ds} &= 2\kappa(\mathbf{e}_1\cdot\mathbf{e}_2),\\
\frac{d(\mathbf{e}_2\cdot\mathbf{e}_2)}{ds} &= -2\kappa(\mathbf{e}_1\cdot\mathbf{e}_2)+2\tau(\mathbf{e}_2\cdot\mathbf{e}_3),\\
\frac{d(\mathbf{e}_3\cdot\mathbf{e}_3)}{ds} &= -2\tau(\mathbf{e}_2\cdot\mathbf{e}_3).
\end{aligned}\tag{A.1.4}$$

It is easily seen that

$$\begin{aligned}
&\mathbf{e}_1\cdot\mathbf{e}_2\equiv 0, \qquad \mathbf{e}_1\cdot\mathbf{e}_3\equiv 0, \qquad \mathbf{e}_2\cdot\mathbf{e}_3\equiv 0,\\
&\mathbf{e}_1\cdot\mathbf{e}_1\equiv 1, \qquad \mathbf{e}_2\cdot\mathbf{e}_2\equiv 1, \qquad \mathbf{e}_3\cdot\mathbf{e}_3\equiv 1
\end{aligned}$$

is a solution of the system (A.1.4) with initial conditions $0,0,0,1,1,1$ for $s=s_0$. By the uniqueness part of Theorem A.1, the trihedron $\mathbf{e}_1(s)\mathbf{e}_2(s)\mathbf{e}_3(s)$ is a frame for every $s\in I$. Moreover, the frame $\mathbf{e}_1(s)\mathbf{e}_2(s)\mathbf{e}_3(s)$ for every $s\in I$ is positively oriented, since the determinant $|\mathbf{e}_1(S),\mathbf{e}_2(s),\mathbf{e}_3(s)|$ is a continuous function of s, and is equal to ± 1 and to $+1$ for $s=s_0$.

From the family $\mathbf{e}_1(s)\mathbf{e}_2(s)\mathbf{e}_3(s)$ we can obtain a curve $\mathbf{x}(s)$ by setting

$$\mathbf{x}(s)=\int_{s_0}^{s}\mathbf{e}_1(s)\,ds, \qquad s\in I.$$

It is clear that $\mathbf{x}'(s)=\mathbf{e}_1(s)$, where the prime denotes the derivative with respect to s, and that $\mathbf{x}''(s)=\kappa\mathbf{e}_2$, so that s and κ are, respectively, the arc length and the curvature of the curve $\mathbf{x}(s)$ at s. Moreover, since κ and $\mathbf{e}_2$ are differentiable,

$$\mathbf{x}'''(s)=\kappa'\mathbf{e}_2+\kappa\mathbf{e}_2'=-\kappa^2\mathbf{e}_1+\kappa'\mathbf{e}_2+\kappa\tau\mathbf{e}_3. \tag{A.1.5}$$

Since κ is of class C^1, τ is continuous, and the $\mathbf{e}_i$ are differentiable (hence continuous), $\mathbf{x}'''(s)$ is continuous, and the curve $\mathbf{x}(s)$ is of class C^3. Finally, from (A.1.5) it follows that the torsion of the curve $\mathbf{x}(s)$ is τ (cf. Exercise 3b, Section 1.3, Chapter 2). Hence $\mathbf{x}(s)$ is the required curve.

APPENDIX 2

Proof of the First Part of Theorem 7.3, Chapter 3

As the proof given in Appendix 1 depends on Theorem A.1, the proof given in this appendix depends on the following existence and uniqueness theorem for systems of partial differential equations of the first order.

A.2. Theorem. *Consider a system of partial differential equations of the form*

$$\frac{\partial y^k}{\partial u^\alpha} = f_\alpha^k(u^1,\cdots,u^m;y^1,\cdots,y^n), \qquad k=1,\cdots,n; \qquad \alpha=1,\cdots,m, \tag{A.2.1}$$

where the functions f_α^k are of class C^2 and satisfy the integrability conditions $\partial^2 y^k/\partial u^\alpha \partial u^\beta = \partial^2 y^k/\partial u^\beta \partial u^\alpha$ or

$$\frac{\partial f_\alpha^k}{\partial u^\beta} + \sum_{j=1}^{n} \frac{\partial f_\alpha^k}{\partial x^j}\frac{\partial y^j}{\partial u^\beta} = \frac{\partial f_\beta^k}{\partial u^\alpha} + \sum_{j=1}^{n} \frac{\partial f_\beta^k}{\partial x^j}\frac{\partial y^j}{\partial u^\alpha}, \qquad k=1,\cdots,n; \qquad \alpha,\beta=1,\cdots,m, \tag{A.2.2}$$

in some neighborhood of $(u_0^1,\cdots,u_0^m,0,\cdots,0)\in E^m\times E^n$. Then there exist neighborhoods U of the origin in E^n and V of $(u_0^1,\cdots,u_0^m)$ in E^m such that for any $(y_0^1,\cdots,y_0^n)\in U$ and all $(u^1,\cdots,u^m)\in V$ there exists a unique solution $y^i(u^1,\cdots,u^m)$, $i=1,\cdots,n$, of (A.2.1), satisfying

$$y^i(u_0^1,\cdots,u_0^m)=y_0^i, \qquad i=1,\cdots,n. \tag{A.2.3}$$

Proof of the First Part of Theorem 7.3, *Chapter* 3. Consider the system S of partial differential equations consisting of (7.5) and (7.1) of Chapter 3, where the coefficients Γ_{ik}^j are expressed by (7.20) of Chapter 3 in terms of C^2 functions E, F, G and C^1 functions L, M, N.

System S defines a system of partial differential equations

$$\begin{aligned}\frac{\partial y^1}{\partial u} &= f^1(u,v;y^1,\cdots,y^9),\\ &\cdots\cdots\cdots\\ \frac{\partial y^9}{\partial v} &= f^{15}(u,v;y^1,\cdots,y^9),\end{aligned} \tag{A.2.4}$$

where $(y^1,y^2,y^3)=\mathbf{x}_u,(y^4,y^5,y^6)=\mathbf{x}_v,(y^7,y^8,y^9)=\mathbf{e}_3$, and f^i, $i=1,\cdots,15$, are linear combinations of $y^1,\cdots,y^9$ with coefficients $0,\Gamma^j_{ik},L,M,N$, or any one of

$$\frac{FM-GL}{EG-F^2},\ \frac{FL-EM}{EG-F^2},\ \frac{FN-GM}{EG-F^2},\ \frac{FM-EN}{EG-F^2}. \tag{A.2.5}$$

Then from Theorem A.2 it follows that the system S possesses a unique C^2 solution $\mathbf{x}_u,\mathbf{x}_v,\mathbf{e}_3$ once the initial values for these solution vectors have been prescribed at a point (u_0,v_0). Now we choose the initial values to be

$$\mathbf{x}_u(u_0,v_0)=\mathbf{x}^0_u,\quad \mathbf{x}_v(u_0,v_0)=\mathbf{x}^0_v,\quad \mathbf{e}_3(u_0,v_0)=\mathbf{e}^0_3,$$

such that

$$\begin{aligned}\mathbf{x}_u\cdot\mathbf{x}_u(u_0,v_0)&=E(u_0,v_0),\\ \mathbf{x}_u\cdot\mathbf{x}_v(u_0,v_0)&=F(u_0,v_0),\\ \mathbf{x}_v\cdot\mathbf{x}_v(u_0,v_0)&=G(u_0,v_0);\end{aligned} \tag{A.2.6}$$

$$\mathbf{e}^0_3=\frac{\mathbf{x}_u\times\mathbf{x}_v}{\|\mathbf{x}_v\times\mathbf{x}_v\|}(u_0,v_0). \tag{A.2.7}$$

$\mathbf{e}^0_3$ can be chosen in this way, since $EG-F^2>0$ and (A.2.6) imply that $\mathbf{x}^0_u$ and $\mathbf{x}^0_v$ are linearly independent.

With the given solution $\mathbf{x}_u,\mathbf{x}_v,\mathbf{e}_3$, we form a new system of partial differential equations

$$\mathbf{x}_u=(y^1,y^2,y^3),\qquad \mathbf{x}_v=(y^4,y^5,y^6), \tag{A.2.8}$$

which clearly satisfies the integrability conditions since $\mathbf{x}_{uv}=\mathbf{x}_{vu}$. From Theorem A.2 again it follows that a unique solution $\mathbf{x}(u,v)$ of (A.2.8) results when its initial value $\mathbf{x}(u_0,v_0)$ is prescribed. It should be remarked that the solution $\mathbf{x}(u,v)$ is of class C^3. Now let $\mathbf{x}\colon \overline{V}\to E^3$ be a solution of (A.2.8), defined in a neighborhood $\overline{V}$ of (u_0,v_0), with $\mathbf{x}(u_0,v_0)=\mathbf{x}_0\in E^3$. We shall show that by contracting $\overline{V}$ and interchanging u and v, if necessary, $\mathbf{x}(V)$ is the required surface.

We first show that the solution $\mathbf{x}_u,\mathbf{x}_v,\mathbf{e}_3$ of (A.2.4) satisfies the following equations:

$$\begin{aligned}&\mathbf{x}_u\cdot\mathbf{x}_u=E,\qquad \mathbf{x}_u\cdot\mathbf{x}_v=F,\qquad \mathbf{x}_v\cdot\mathbf{x}_v=G,\\ &\mathbf{x}_u\cdot\mathbf{e}_3=\mathbf{x}_v\cdot\mathbf{e}_3=0,\qquad \mathbf{e}_3\cdot\mathbf{e}_3=1.\end{aligned} \tag{A.2.9}$$

To this end, by using (7.5) and (7.1) to express the partial derivatives of

$$\mathbf{x}_u\cdot\mathbf{x}_u,\ \mathbf{x}_u\cdot\mathbf{x}_v,\ \mathbf{x}_v\cdot\mathbf{x}_v,\qquad \mathbf{x}_u\cdot\mathbf{e}_3,\ \mathbf{x}_v\cdot\mathbf{e}_3,\ \mathbf{e}_3\cdot\mathbf{e}_3 \tag{A.2.10}$$

as linear combinations of the same six quantities, we obtain a system of 12 partial differential equations:

$$\begin{aligned}
\frac{\partial(\mathbf{x}_u\cdot\mathbf{x}_u)}{\partial u} &= F^1(u,v;\mathbf{x}_u\cdot\mathbf{x}_u,\cdots,\mathbf{e}_3\cdot\mathbf{e}_3),\\
\frac{\partial(\mathbf{x}_u\cdot\mathbf{x}_u)}{\partial v} &= F^2(u,v;\mathbf{x}_u\cdot\mathbf{x}_u,\cdots,\mathbf{e}_3\cdot\mathbf{e}_3),\\
&\cdots\cdots\cdots\cdots\\
\frac{\partial(\mathbf{e}_3\cdot\mathbf{e}_3)}{\partial v} &= F^{12}(u,v;\mathbf{x}_u\cdot\mathbf{x}_u,\cdots,\mathbf{e}_3\cdot\mathbf{e}_3),
\end{aligned} \tag{A.2.11}$$

where F^i, $i=1,\cdots,12$, are linear combinations of the six quantities of (A.2.10) with coefficients (except possibly for a factor 2) 0, Γ^j_{ik}, L, M, N, or any of the four quantities of (A.2.5). System A.2.11 obviously satisfies the integrability conditions, since $\mathbf{x}_u,\mathbf{x}_v,\mathbf{e}_3$ are of C^2. Hence by Theorem A.2 again, a solution (A.2.10) of system A.2.11 is uniquely determined if their initial values satisfy (A.2.6) and

$$\mathbf{x}_u\cdot\mathbf{e}_3(u_0,v_0)=\mathbf{x}_v\cdot\mathbf{e}_3(u_0,v_0)=0,\quad \mathbf{e}_3\cdot\mathbf{e}_3(u_0,v_0)=1. \tag{A.2.12}$$

Since A.2.11 is obtained from (A.2.4), using (7.5), (7.1), and (7.20), we can easily verify that (A.2.9) is a solution of (A.2.11) with initial conditions (A.2.6) and (A.2.12). Hence the uniqueness part of Theorem A.2 shows that the solution $\mathbf{x}_u,\mathbf{x}_v,\mathbf{e}_3$ of (A.2.4) satisfies (A.2.9).

From (A.2.9) it follows that

$$\|\mathbf{x}_u\times\mathbf{x}_v\|^2=\mathbf{x}_u^2\mathbf{x}_v^2-(\mathbf{x}_u\cdot\mathbf{x}_v)^2=EG-F^2>0.$$

Thus if $\mathbf{x}$: $\overline{V}\to E^3$ is given by

$$\mathbf{x}(u,v)=(x_1(u,v),x_2(u,v),x_3(u,v)),\qquad (u,v)\in\overline{V},$$

one of the three components of $\mathbf{x}_u\times\mathbf{x}_v$, say the Jacobian determinant $\partial(x_1,x_2)/\partial(u,v)$, is not zero at (u_0,v_0). Therefore we may invert the system formed by the first two component functions of $\mathbf{x}$, in a neighborhood $U\subset\overline{V}$ of (u_0,v_0), so that we obtain a mapping $F(x_1,x_2)=(v,v)$. By restricting $\mathbf{x}$ to U, the mapping $\mathbf{x}$: $U\to E^3$ is injective, and its inverse $\mathbf{x}^{-1}=F\circ\pi$ (where π is the projection of E^3 on the $\mathbf{x}_1\mathbf{x}_2$-plane) is of class C^3, since $\mathbf{x}$ is of class C^3. Therefore $\mathbf{x}$: $U\to E^3$ is a C^3 homeomorphism with $\mathbf{x}_u\wedge\mathbf{x}_v\neq 0$, hence is a regular surface.

From (A.2.9) it follows immediately that E, F, G are the coefficients of the first fundamental form of the surface $\mathbf{x}(U)$ and that $\mathbf{e}_3$ is a unit vector normal to the surface. Interchanging u and v, if necessary, we obtain

$$\mathbf{e}_3 = \frac{\mathbf{x}_u \times \mathbf{x}_v}{\|\mathbf{x}_u \times \mathbf{x}_v\|}. \tag{A.2.13}$$

Finally, from (A.2.13) and (7.5) we have

$$\mathbf{e}_3 \cdot \mathbf{x}_{uu} = L, \qquad \mathbf{e}_3 \cdot \mathbf{x}_{uv} = M, \qquad \mathbf{e}_3 \cdot \mathbf{x}_{vv} = N,$$

which show that L, M, N are the coefficients of the second fundamental form of the surface $\mathbf{x}(u, v)$, and the proof of the first part of Theorem 7.3, Chapter 3, is complete.

Bibliography

1. L. V. Ahlfors and L. Sario, *Riemann Surfaces*, Princeton University Press, Princeton, NJ, 1960.
2. T. M. Apostol, *Mathematical Analysis*, Addison-Wesley, Reading, MA, 1957.
3. A. Barber, "Note sur le problème de l'aiguille et le jeu du joint couvert," *J. Math. Pur. Appl.*, **5** (1860): 273–286.
4. L. Bieberbach, "Über eine Extremaleigenschaft des Kreises," *Jahrb. Dtsch. Math.-Verein.*, **24** (1915): 247–250.
5. W. Blaschke, *Vorlesungen über Differentialgeometrie*, Vol. I, 4th ed., Springer, Berlin, 1945.
6. K. Borsuk, "Sur la courbure totale des courbes fermées," *Ann. Soc. Polon. Math.*, **20** (1948): 251–265.
7. F. Brickell and C. C. Hsiung, "The total absolute curvature of closed curves in Riemannian manifolds," *J. Differ. Geom.*, **9** (1974): 177–193.
8. M. P. do Carmo, *Differential Geometry of Curves and Surfaces*, Prentice-Hall, Englewood Cliffs, NJ, 1976.
9. E. Cartan, *Les Systèmes Différentiels Extérieurs et leurs Applications Géometriques*, Actualités Scientifiques et Industrielles, Hermann, Paris, 1945.
10. S. S. Chern, "Some new characterizations of Euclidean spheres," *Duke Math. J.*, **12** (1945): 279–290.
11. ———, Lecture notes on differential geometry, University of Chicago, 1954 (unpublished).
12. ———, "An elementary proof of the existence of isothermal parameters on a surface," *Proc. Am. Math. Soc.*, **6** (1955): 771–782.
13. ———, "A proof of the uniqueness of Minkowski's problem for convex surfaces," *Am. J. Math.*, **79** (1957): 949–950.
14. ———, "Curves and surfaces in Euclidean space," *Studies in Global Geometry and Analysis*, *Studies in Mathematics*, Vol. 4, Mathematical Association of America, 1967, pp. 16–56.
15. E. B. Christoffel, "Über die Bestimmung der Gestalt einer krummen Oberfläche durch lokale Messungen auf Derselben," *J. Reine Angew. Math.*, **64** (1865): 193–209.
16. R. Courant, *Differential and Integral Calculus*, Vol. I, Nordemann, New York, 1938.
17. F. H. C. Crick, "Linking numbers and nucleosomes," *Proc. Natl. Acad. Sci. U.S.A.*, **73** (1976): 2639–2643.
18. R. H. Crowell and R. H. Fox, *Introduction to Knot Theory*, Ginn, Boston, 1963.
19. I. Fary, "Sur la courbure totale d'une courbe gauche faisant un noeud," *Bull. Soc. Math. Fr.*, **77** (1949): 128–138.

20. W. Fenchel, "Über Krümmung und Windung geschlossener Raumkurven," *Math. Ann.*, **101** (1929): 238–252.

21. ———, "On differential geometry of global space curves," *Bull Am. Math. Soc.*, **57** (1951): 44–54.

22. F. B. Fuller, "The writhing number of a space curve," *Proc. Natl. Acad. Sci. U.S.A.*, **68** (1971): 815–819.

23. H. Geppert, "Sopra una caratterzione della spera," *Ann. Mat. Pur. Appl.* **20** (1941): 59–66.

24. G. Herglotz, "Über die Starrheit der Eiflächen," *Abh. Math. Sem. Hansischen Univ. (Hamburg)*, **15** (1943): 127–129.

25. H. Hopf, "Über die Drehung der Tangenten und Sehnen ebener Kurven" *Compositio Math.*, **2** (1935): 50–62.

26. ———, Selected topics in differential geometry in the large, New York University, 1955 (unpublished).

27. ———, Lectures on differential geometry in the large, Stanford University, 1955 (unpublished).

28. H. Hopf and W. Rinow, "Über den Begriff der Vollständigen differentialgeometrischen Fläche," *Comment. Math. Helv.*, **3** (1931): 209–225.

29. H. Hopf and K. Voss, "Ein Satz aus der Flächentheorie im Grossen," *Arch. Math.*, **3** (1952): 187–192.

30. C. C. Hsiung, "A theorem on surfaces with a closed boundary," *Math. Z.*, **64** (1956): 41–46.

31. ———, "Some global theorems on hypersurfaces," *Can. J. Math.*, **9** (1957): 5–14.

32. ———, "A uniqueness theorem for Minkowski's problem for convex surfaces with boundary," *Ill. J. Math.*, **2** (1958): 71–75.

33. ———, "Isoperimetric inequalities for two-dimensional Riemannian manifolds with boundary," *Ann. Math.*, **73** (1961): 213–220.

34. A. Huber, "On the isoperimetric inequality on surfaces of variable Gaussian curvature," *Ann. Math.*, **60** (1954): 237–247.

35. A. Hurwitz, "Sur quelques applications géométriques des séries de Fourier," *Ann. École Norm.*, **19** (1902): 357–408.

36. S. B. Jackson, "Vertices of plane curves," *Bull. Am. Math. Soc.*, **50** (1944): 564–578.

37. W. Klingenberg, *A Course in Differential Geometry*, translated by D. Hoffman, Springer, New York–Heidelberg–Berlin, 1978.

38. R. S. Millman and G. D. Parker, *Elements of Differential Geometry*, Prentice-Hall, Englewood Cliffs, NJ, 1977.

39. J. W. Milnor, "On the total curvature of knots," *Ann. Math.*, **52** (1950): 248–257.

40. H. Minkowski, "Volumen und Oberfläche," *Math. Ann.*, **57** (1903): 447–495.

41. S. Mukhopadhyaya, "New methods in the geometry of a plane arc," *Bull. Calcutta Math. Soc.*, **1** (1909): 31–37.

42. A. H. Newman, *Elements of the Topology of Plane Sets of Points*, 2nd ed., Cambridge University Press, Cambridge, 1951.

43. L. Nirenberg, "The Weyl and Minkowski problems in differential geometry in the large," *Comment. Pure Appl. Math.*, **6** (1953): 337–394.

44. B. O'Neill, *Elementary Differential Geometry*, Academic Press, New York, 1966.

45. A. V. Pogorelov, *Differential Geometry*, Noordhoof, Groningen, 1966.

46. G. de Rham, "Sur la réductibilité d'un espace de Riemann," *Comment. Math. Helv.*, **26** (1952): 328–344.

47. A. Rosenthal and O. Szász, "Eine Extremaleigenschaft der Kurven konstanter Breite," *Jahrb. Dtsch. Math.-Verein.*, **25** (1916): 278–282.

48. H. Rutishauser and H. Samelson, "Sur la rayon d'une sphère dont la surface contient une courbe fermée," *C. R. Acad. Sci. Paris*, **227** (1948): 755–757.

49. H. Samelson, "Orientability of hypersurfaces," *Proc. Am. Math. Soc.*, **22** (1969): 301–302.

50. E. Schmidt, "Über das isoperimetrische Problem in Raum von n Dimensionen," *Math. Z.*, **44** (1939): 689–788.

51. I. J. Schoenberg, "An isoperimetric inequality for closed curves convex in even-dimensional Euclidean spaces," *Acta Math.*, **91** (1954): 143–164.

52. B. Segre. "Sui circoli geodetici di una superficie a curvatura totale costante, che contengono nell'interno una linea assegnata," *Boll. Unione Mat. Ital.*, **13** (1934): 279–283.

53. ———, "Sulla torsione integrale delle curve chiuse sghembe," *Atti Accad. Naz. Lincei Rend. Cl. Sci. Fis. Mat. Natur.*, **3** (1947): 422–426.

54. M. Spivak, *A Comprehensive Introduction to Differential Geometry*, Vols. I–V, Publish or Perish, Boston, 1970.

55. J. J. Stoker, *Differential Geometry*, Wiley-Interscience, New York, 1969.

56. D. J. Struik, *Lectures on Classical Differential Geometry*, 2nd ed., Addison-Wesley, Reading, MA, 1960.

57. J. A. Thorpe, *Elementary Topics in Differential Geometry*, Springer, New York–Heidelberg–Berlin, 1979.

58. L. Vietoris, "Ein einfacher Beweis des Vierscheitelsatzes der ebenen Kurven," *Arch. Math.*, **3** (1952): 304–306.

59. K. Voss, "Einige differentialgeometrische Kongruenzsätze für geschlossene Flächen und Hyperflächen," *Math. Ann.*, **131** (1956): 180–218.

60. H. Weyl, "Über die Bestimmung einer geschlossenen konvexen Fläche durch ihr Linienelement," *Vierteljahrschr. Naturforsch. Ges., Zürich*, **61** (1916): 40–72.

61. T. J. Willmore, *An Introduction to Differential Geometry*, Oxford University Press, Oxford, 1959.

62. Y. C. Wong, "A global formulation of the condition for a curve to lie in a sphere," *Monatsh. Math.*, **67** (1963): 363–365.

Answers and Hints to Exercises

CHAPTER 1

Section 1.1

2. (a) Empty set. (b) The whole plane.

Section 1.2

1. The union of the closed disks $\{\mathbf{p}\in E^2 \mid d(\mathbf{p},\mathbf{0})\leqslant 1-1/n, \mathbf{0}=(0,0)\}, 2\leqslant n<\infty$, is the open disk $\{\mathbf{p}\in E^2 \mid d(\mathbf{p},\mathbf{0})=1\}$.

2. Hint: (a) Take the union of all open subsets of S. (b) Take the intersection of all closed sets containing S.

3. Hint: Treat C as our universe and take the complement of U in C. The set $C-V$ is closed relative to C for any set C, closed or not.

7. Example: Take $U=[0,1), V=[1,2], T=E^2$, and assume that

$$f(x)=\begin{cases}(x,0) & \text{for} \quad x\in U,\\ (x,3) & \text{for} \quad x\in V.\end{cases}$$

Section 1.3

1. Open, bounded, connected, and the boundary is the sphere $\{\mathbf{p}\in E^n \mid d(\mathbf{p},\mathbf{p}_0)=\rho\}$.

2. Open, unbounded, disconnected, and the boundary consists of two intersecting lines $x-y=0$ and $x+y=0$.

4. No. Draw some pictures.

Section 1.5

1. Hint: Assume the contrary and use Exercise 4 of Section 1.2 to derive a contradiction.

3. ε = absolute minimum of f by Theorem 1.5.9.

4. Absolute minimum $48\sqrt[3]{4}$.

5. Since for $0<x<\pi/2$, $f(x)>0$, $f(0)=\infty$, and $f(\pi/2)=\infty$, there must be an absolute minimum; it is 125 at the point $x=\arcsin\frac{3}{5}$.

6. Absolute maximum: $\frac{3}{2}$; absolute minimum: -3.

Section 2.1

1. (b) Hint: For $h>0$, $0<\theta<1$ implies that $1+\theta h<1+h$ and therefore that $1/(1+h)<1/(1+\theta h)<1$. For $-1<h<0$, $0<\theta<1$ implies that $1>1+\theta h>1+h>0$ and therefore that $1/(1+h)>1/(1+\theta h)>1$.

2. Hint: Use Theorem 2.1.1.

Section 2.2

1.
$$\begin{aligned} f(x,y)=17&+30(x-1)-13(y+2)\\ &+18(x-1)^2-25(x-1)(y+2)+3(y+2)^2\\ &+4(x-1)^3-12(x-1)^2(y+2)+6(x-1)(y+2)^2+R(x\,y). \end{aligned}$$

Section 2.3

1. Relative maximum at $(-4,3)$, relative minimum at $(0,3)$, saddle points at $(0,3)$ and $(-4,-3)$.

3. $(0,1,\frac{1}{2})$, $(0,-1,-\frac{1}{2})$.

Section 2.4

1. (a) The maximum is 3 at $(\sqrt{2}/2,\sqrt{2}/2)$ and $(-\sqrt{2}/2,-\sqrt{2}/2)$; the minimum is -1 at $(\sqrt{2}/2,-\sqrt{2}/2)$ and $(-\sqrt{2}/2,\sqrt{2}/2)$.

(b) The maximum is $106\frac{1}{4}$ at $(\frac{3}{2},4)$ and $(-\frac{3}{2},-4)$; the minimum is -50 at $(2,-3)$ and $(-2,3)$.

Section 3.3

6. Hint: Use Exercise 4.

7. Hint: Use the unit vector $\mathbf{e}=\mathbf{v}\times\mathbf{w}/\|\mathbf{v}\times\mathbf{w}\|$.

Section 3.4

2. (a) Independent. (b) Dependent. (c) Dependent.

5. $\{(1,1,3), (1,-1,0), (2,1,3)\}$.

6. Dimension is 2; $\{(1,2,0), (2,1,-1)\}$ is a basis.

Section 3.5

1. (a) $\mathbf{u}_1(\mathbf{p})+5\mathbf{u}_2(\mathbf{p})-4\mathbf{u}_3(\mathbf{p})$.
(b) $\mathbf{v}_p=((1,-1,2),\ (0,-3,3))$, $-2\mathbf{w}_p=((1,-1,2),\ (-1,-7,6))$, $\mathbf{v}_p+\mathbf{w}_p=((1,-1,2),\ (1,0,1))$.

2. (b) $\dfrac{x_1-x_2}{x_1^2+1}\mathbf{v}_1+\dfrac{-x_1+x_2+(x_1^2+1)x_3}{2(x_1^2+1)}\mathbf{v}_2+\dfrac{x_1^3+x_2}{x_1^2+1}\mathbf{v}_3$.

3. $\mathbf{v}=\dfrac{7\sqrt{6}}{6}\mathbf{e}_1-\dfrac{7\sqrt{2}}{2}\mathbf{e}_2+\dfrac{4\sqrt{3}}{3}\mathbf{e}_3$.

4. Hint: Use (3.3.8).

Section 3.6

1. -11.

2. (a) $x_3^2+x_1x_2$. (b) $x_2^3x_3^2-3x_1^2x_3+x_1x_2^4$. (c) $-x_1^2+3x_2x_3$.

3. Hint: Evaluate $\mathbf{v}=\Sigma_{i=1}^3 v_i\mathbf{u}_i$ on x_j.

Section 4

1. (a) $(0,0)$. (b) $(2,-1), (-2,1)$. (c) $(0,0), (\frac{1}{2},\frac{1}{2}), (\frac{1}{2},-\frac{1}{2})$.

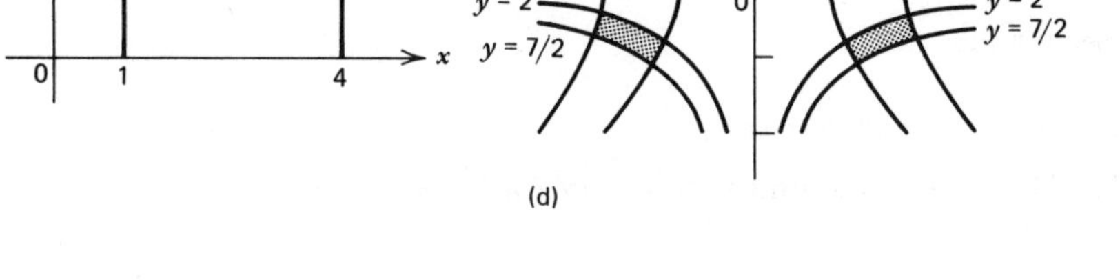

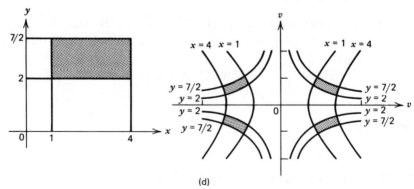

(d)

2. Let $F(u,v)=(x,y)$.
(a) The region bounded by the two parabolas $y^2=4(x+1)$ and $y^2=16(x+4)$.
(b) The unit disk $x^2+y^2\leqslant 1$.
(c) The half-plane $x\geqslant 0$.
Hint: Begin by finding the images of the boundary curves of S.

5. (a) $(3,-2,0)$. (b) $(3,-\pi,-2)$.

6. (a) No. (b) Yes.

9. Let $F(u,v)=(x,y)$. (a) $F^{-1}(x,y)=(ye^{-x},x)$; F is a diffeomorphism. (b) $F^{-1}(x,y)=(x+\sqrt[3]{y},\sqrt[3]{y})$; F is not a diffeomorphism, since F^{-1} is not differentiable when $v=0$. (c) $F^{-1}(x,y)=((3x+y-1)/7,(-x+2y+5)/7)$; F is a diffeomorphism.

Section 5.1

2. Hint: Use $\sum_{i=1}^{3}a_{ij}A_{ik}=\delta_{jk}$ to prove (5.1.16).

3. (a) $\begin{bmatrix}\frac{1}{2} & -\frac{\sqrt{3}}{2}\\ \frac{\sqrt{3}}{2} & \frac{1}{2}\end{bmatrix}$. (b) Not an orthogonal matrix.

(c) $\begin{bmatrix}\frac{3}{5} & \frac{4}{5}\\ \frac{4}{5} & -\frac{3}{5}\end{bmatrix}$.

4. $\begin{bmatrix}\frac{5}{13} & 0 & \frac{12}{13}\\ -\frac{48}{65} & \frac{3}{5} & \frac{4}{13}\\ -\frac{36}{65} & -\frac{4}{5} & \frac{3}{13}\end{bmatrix}$, $\begin{bmatrix}\frac{1}{3} & \frac{2}{3} & \frac{2}{3}\\ \frac{2}{3} & -\frac{2}{3} & \frac{1}{3}\\ \frac{2}{3} & \frac{1}{3} & -\frac{2}{3}\end{bmatrix}$.

5. $-\psi,-\theta,-\phi$.

Section 5.3

2.

	translation part	orthogonal part
$F\bar{F}$	$G(\mathbf{d})+2\mathbf{c}$	$G\bar{G}$
$\bar{F}F$	$\bar{G}(\mathbf{c})+2\mathbf{d}$	$\bar{G}G$

3. (a) $\begin{bmatrix} 2-\sqrt{2} \\ 1 \\ -1+4\sqrt{2} \end{bmatrix}$. (b) $\begin{bmatrix} \frac{7\sqrt{2}}{2} \\ -5 \\ \frac{5\sqrt{2}}{2} \end{bmatrix}$. (c) $\begin{bmatrix} -\frac{\sqrt{2}}{2} \\ 1 \\ \frac{9\sqrt{2}}{2} \end{bmatrix}$

4. In (b) and (d) F is not an isometry.

	translation part	orthogonal part
(a)	identity	$\begin{bmatrix} -1 & 0 & 0 \\ 0 & -1 & 0 \\ 0 & 0 & -1 \end{bmatrix}$
(c)	by $(-3,-2,1)$	$\begin{bmatrix} 0 & 0 & 1 \\ 0 & 1 & 0 \\ 1 & 0 & 0 \end{bmatrix}$

8. (b)

$$H(\mathbf{p})=\mathbf{p}+\left(\frac{1}{2},-1,1\right), \qquad G=\begin{bmatrix} \frac{1}{\sqrt{2}} & \frac{1}{\sqrt{2}} & 0 \\ 0 & 0 & 1 \\ \frac{1}{\sqrt{2}} & -\frac{1}{\sqrt{2}} & 0 \end{bmatrix}.$$

9.

$$H(\mathbf{p})=\mathbf{p}+(3,-2,1), \qquad G=\begin{bmatrix} \frac{1}{\sqrt{2}} & 0 & \frac{1}{\sqrt{2}} \\ -\frac{2}{3} & \frac{1}{3} & \frac{2}{3} \\ \frac{1}{3\sqrt{2}} & \frac{4}{3\sqrt{2}} & -\frac{1}{3\sqrt{2}} \end{bmatrix}.$$

Section 5.4

4. Hint: At first use Theorem 5.1.4 to show that F has a characteristic root $+1$, so there is a point $\mathbf{p}\neq 0$ such that $F(\mathbf{p})=\mathbf{p}$.

5. Hint: For the first part of this exercise use the definition of an orthogonal transformation and the identity (3.3.8) of Lagrange.

Section 6.1

1. (a) 8. (b) −19. (c) −21.
3. (a) $y^2z - xz^2$. (b) $y^2z + (x+y)z^2$. (c) $yz = xz^2/y + y^2z/x$.
4. (a) $5f^4\,df$. (b) $\frac{1}{2}f^{-1/2}\,df$.
5. (a) $df = z^2\,dx - 2yz\,dy + (2xz - y^2)\,dz$, -7.
   (b) $df = yze^{xz}\,dx + e^{xz}\,dy + xye^{xz}\,dz$, $27e^2$.
   (c) $df = (y\cos xy - z\sin xz)\,dx + x\cos xy\,dy - x\sin xz\,dz$, $17\cos 5 - 5\sin 2$.
6. (a) $dx_2 - dx_3$. (b) Not a 1-form. (c) $x_2\,dx_1 + x_1\,dx_3$. (d) $2x_1\,dx_1 - 2x_2\,dx_2 + 2x_3\,dx_3$. (e) 0. (f) Not a 1-form.

Section 6.2

1. (a) $-y^2z\,dx\wedge dy - x\,dx\wedge dz + y\,dy\wedge dz$. (b) $-z\,dx\wedge dy - y\,dx\wedge dz$.
2. (a) $2\,df\wedge dg$. (b) 0. (c) $(1-f)\,df\wedge dg$.

Section 7

2. (a) $y = C_1x^{-1} + C_2$. (b) $y = C_1\sin(4x - C_2)$.
   (c) $y = -\dfrac{1}{4}x^2 + C_1x + C_2$. (d) $y = C_1e^x + C_2e^{-x} + \dfrac{1}{2}\sin x$.

CHAPTER 2

Section 1.1

1. (a) $x_1 = 3t - 5$, $x_2 = -2t + 5$.
   (b) $x_1 = \ln t$, $x_2 = t$.
   (c) $x_1 = \sin t$, $x_2 = \cos 2t$.
   (d) $x_1 = \dfrac{3t}{1+t^3}$, $x_2 = \dfrac{3t^2}{1+t^3}$.
2. $\mathbf{x}(t) = (\sin t, \cos t)$, $0 \leqslant t \leqslant 2\pi$.
4. $\mathbf{x}(t) = (e^t + 1, -t^2 + 3, t^3/3)$.
5. $\mathbf{x}(t) = c_1t + c_2$, a straight line.
7. $u \to (2,0,0) + u(0,2,1)$, $u \to (\sqrt{2}, \sqrt{2}, \frac{1}{4}\pi) + u(-\sqrt{2}, \sqrt{2}, 1)$.

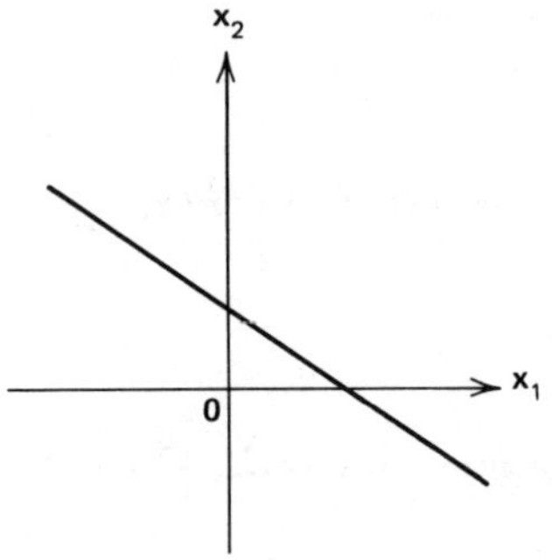

(a)

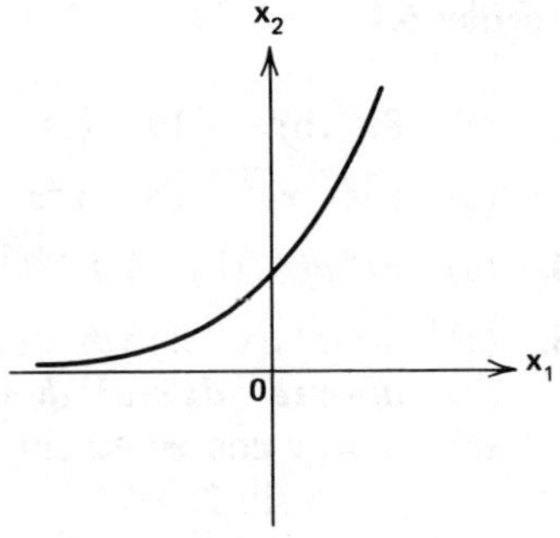

(b)

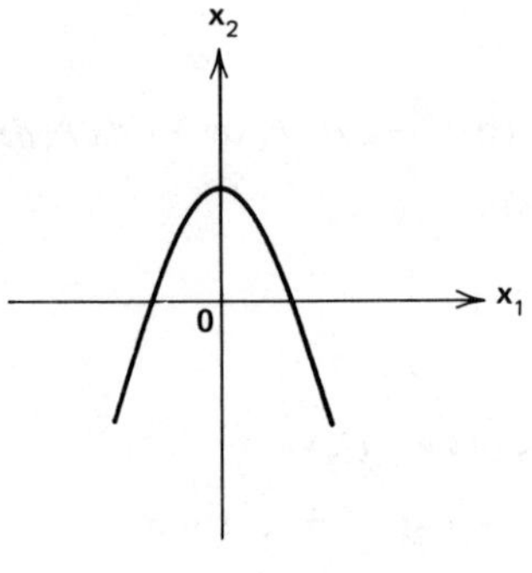

(c)

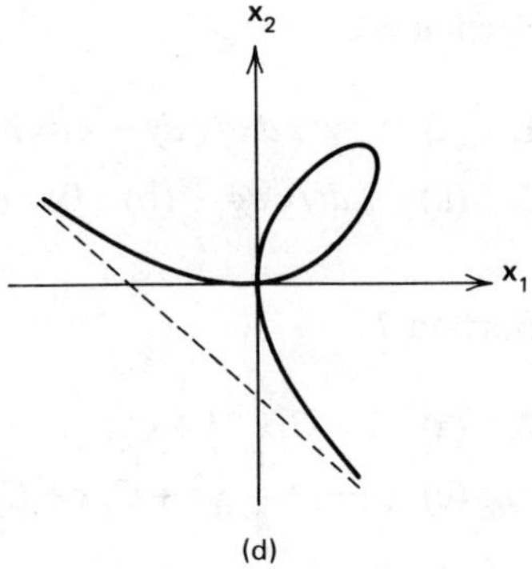

(d)

8. $\mathbf{y}(\tau)=(2(1-\tau^2),\ 2\tau\sqrt{1-\tau^2}\ ,2\tau),\ 0\leqslant\tau\leqslant 1.$

Section 1.2

1. (a) $\mathbf{x}(s)=\left(a\cos\frac{s}{c},\ a\sin\frac{s}{c},\ \frac{bs}{c}\right),\ c=\sqrt{a^2+b^2}\ .$

(b) $\left(-\frac{a}{c}\sin\frac{s}{c},\ \frac{a}{c}\cos\frac{s}{c},\ \frac{b}{c}\right).$ (c) $2\pi c.$

3. $8a.$

4. (a) $-e^t\mathbf{u}_1-e^{-t}\mathbf{u}_2-\sqrt{2}\ t\mathbf{u}_3.$

(b) $-2e^{-t}\mathbf{u}_2+\sqrt{2}\ \mathbf{u}_3.$

(c) $\pm\left(\frac{1}{1+e^{2t}}\mathbf{u}_1-\frac{e^{2t}}{1+e^{2t}}\mathbf{u}_2-\frac{\sqrt{2}\ e^t}{1+e^{2t}}\mathbf{u}_3\right).$

5.

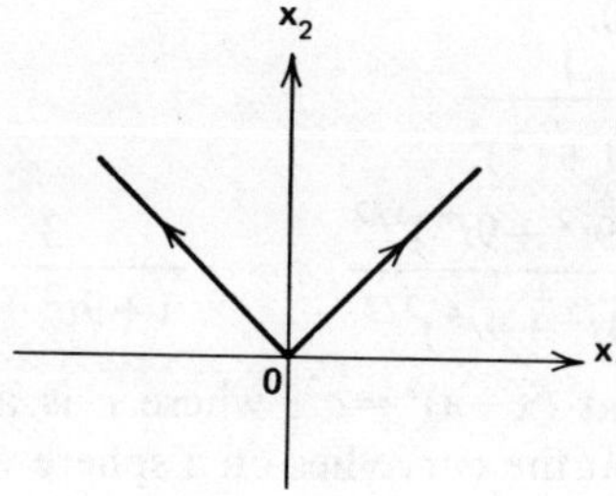

7. (a)

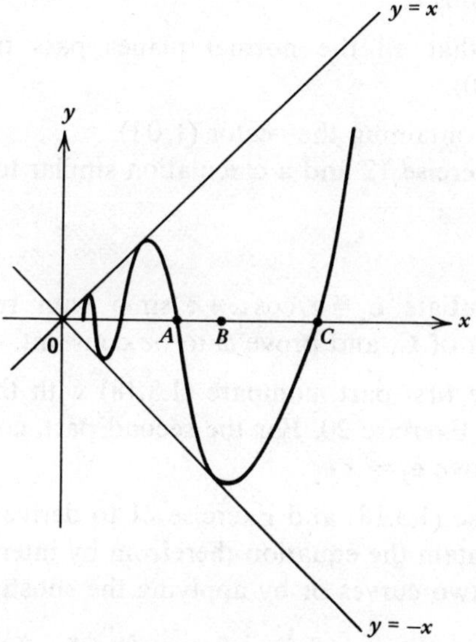

where $A = 1/(n+1)$, $B = 1/\left(n + \dfrac{1}{2}\right)$, $C = 1/n$.

8. Hint: Use mean value theorem 2.1.1, Chapter 1, and Exercise 2 of Section 2.1, Chapter 1.

Section 1.3

1. (a) $\kappa = a/c^2$, $\tau = b/c^2$.

(b) $\left(-\cos\dfrac{s}{c}, -\sin\dfrac{s}{c}, 0\right)$, $\left(\dfrac{b}{c}\sin\dfrac{s}{c}, -\dfrac{b}{c}\cos\dfrac{s}{c}, \dfrac{a}{c}\right)$.

(c) $b\sin\dfrac{s}{c}\,x_1 - b\cos\dfrac{s}{c}\,x_2 + ax_3 - ab\dfrac{s}{c} = 0,$

$a\sin\dfrac{s}{c}\,x_1 - a\cos\dfrac{s}{c}\,x_2 - bx_3 + b^2\dfrac{s}{c} = 0,$

$\cos\dfrac{s}{c}\,x_1 + \sin\dfrac{s}{c}\,x_2 - a = 0.$

2. (a) $\kappa=\pm 1,\ \tau=0.$

(b) $\kappa=-\tau=\dfrac{1}{3(1+t^2)^2}.$

(c) $\kappa=\pm\dfrac{2(1+9t^2+9t^4)^{1/2}}{(1+4t^2+9t^4)^{3/2}},\ \tau=-\dfrac{3}{1+9t^2+9t^4}.$

9. Hint: Prove that $(\mathbf{x}-\mathbf{a})^2=c^2$, where c is a positive constant; this shows that the curve lies on a sphere with center at the point $\mathbf{a}$ and radius c.

10. Hint: Prove that all the normal planes pass through the point $(-1,0,0)$.

13. (b) The line containing the vector $(1,0\,1)$.
Hint: Use Exercise 12 and a calculation similar to that for Exercise 2c.

16. $f(t)=c_1t+c_2$.

20. Hint: Differentiate $\bar{\mathbf{e}}_1=\mathbf{e}_1\cos\omega+\mathbf{e}_3\sin\omega$ with respect to the arc length s of C, and prove ω to be constant.

21. Hint: For the first part, compare (1.3.18) with the equation in the hint of Exercise 20. For the second part, construct $\bar{\mathbf{x}}=\mathbf{x}+a\mathbf{e}_2$ and prove $\bar{\mathbf{e}}_2=\pm\mathbf{e}_2$.

22. Hint: First use (1.3.18) and Exercise 21 to derive $d\bar{s}/ds=a\tau/\sin\omega$; then obtain the equation therefrom by interchanging the roles of the two curves or by applying the substitution

$$\begin{pmatrix} s & a & \omega & \kappa & \tau \\ s & -\varepsilon a & -\varepsilon\omega & \bar{\kappa} & \bar{\tau} \end{pmatrix},$$

where $\varepsilon=\pm 1$.

24. Hint: For "if" part, construct $\bar{\mathbf{x}}=\mathbf{x}+a\mathbf{e}_2$ and prove that both $\bar{\mathbf{e}}_1$ and $\bar{\mathbf{e}}_2$ are linear combination of $\mathbf{e}_1$ and $\mathbf{e}_3$.

26. (a) Hint: Let C be given by $\mathbf{x}(s)$ with arc length s and Frenet frame $\mathbf{e}_1(s)\mathbf{e}_2(s)\mathbf{e}_3(s)$, and write a parallel curve of C as $\bar{\mathbf{x}}=\mathbf{x}+a\mathbf{e}_1+b\mathbf{e}_2+c\mathbf{e}_3$.

(b) All circles in the plane $x_3=c$ for every constant c concentric with the circle of intersection of the plane and the cylinder $x_1^2+x_2^2=r^2$.

27. Hint: Use (1.3.3), (1.3.7), and the equations similar to (1.3.3) for the frame $\mathbf{x}\mathbf{e}_1\mathbf{e}_2^*\mathbf{e}_3^*$ to compute $d\mathbf{e}_2/ds$ in two ways; then obtain $q_{23}^*ds=q_{23}ds-d\theta$ by comparing the coefficients of the corresponding terms of the two expressions.

Section 1.4

2. Hint: Use Exercise 2c of Section 1.3. Let P_i be given by (t_i, t_i^2, t_i^3), $i=1,2,3$. For b, prove that the three vectors QP_i, $i=1,2,3$, are related by

$$QP_1=\left(\frac{t_1-t_3}{t_2-t_3}\right)^3 QP_2-\left(\frac{t_1-t_2}{t_2-t_3}\right)^3 QP_3.$$

Section 1.5

1. (a) $\mathbf{x}(s)=\left(\frac{1}{a}(1-\cos as), -\frac{1}{a}\sin as, 0\right)$.

   (b) $\mathbf{x}(s)=(x_1(s), x_2(s), x_3(s))$, where

   $$x_1(s)=\frac{1}{2}(s+m)\left(\sin\ln\frac{s+m}{m}-\cos\ln\frac{s+m}{m}\right)+\frac{m}{2},$$

   $$x_2(s)=-\frac{1}{2}(s+m)\left(\sin\ln\frac{s+m}{m}+\cos\ln\frac{s+m}{m}\right)+\frac{m}{2},$$

   $$x_3(s)=0.$$

   (c) $\mathbf{x}(s)=(\sqrt{1+s^2}-1, -\ln(s+\sqrt{1+s^2}), 0)$.

Section 2

2. $f(t)=A+B\cos t+C\sin t$; central conics.

3. (a) $A=0, B=0, C=2, D=0$; $I=1$.
   (b) $A=0, B=-1, C=-2, D=0, E=1, F=2, G=1$; $I=0$.
   (c) $A=1, B=2, C=3, D=0$; $I=3$.
   (d) $A=-2, B=-1, C=-2, D=-2, E=0$; $I=-4$.
   (e) $A=1, B=-1, C=0, D=1$; $I=1$.
   (f) $A=-1, B=-2, C=0, D=-2, E=-1$; $I=0$.

5. (a) Hint: Use the Jordan curve theorem (Theorem 2.1.3) and an argument in Proof 1 of the four-vertex theorem (Theorem 2.6).

8. No. Use the isoperimetric inequality.

9. Hint: First observe that the area bounded by H is greater than or equal to the area bounded by C and that the length of H is smaller than or equal to the length of C. Then apply Section 2.6.5 to H.

11. (a) Hint: Differentiate $(\mathbf{y}-\mathbf{x})\cdot\mathbf{e}_2$ with respect to s.
    (b) Hint: Differentiate (*) and use Theorem 1.3.3.

12. Hint: First use Exercise 23d of Section 1.3 to show that the distance between two opposite points Q, R is

$$\begin{aligned} QR &= PQ + PR \\ &= \operatorname{arc} PB + DB + DB + DB + \operatorname{arc} AB - \operatorname{arc} AC + \operatorname{arc} CP \\ &= \operatorname{arc} BC + \operatorname{arc} AB - \operatorname{arc} AC + 2DB. \end{aligned}$$

13. Hint: Let $D(\theta) = p(\theta) + p(\theta + \pi)$. Use (2.42) to show that there is at least one root of $D(\theta) - D(\theta + \frac{1}{2}\pi) = 0$ in the interval $0 \leqslant \theta < \pi$.

14. $\kappa(\theta) = ab(a^2 \sin^2 \theta + b^2 \cos^2 \theta)^{-3/2}$.

CHAPTER 3

Section 1

2. $\mathbf{x}(u, v) = (\cos v, \sin v, u)$.

3. (a) The vertex (0,0,0). (b) All points on the circle $x_1^2 + x_2^2 = 1, x_3 = 0$. (c) All points on the $\mathbf{x}_3$-axis.

4. (b) $c \neq -4$.

6. (a) Yes. (b) No. $\mathbf{x}$ is not one-to-one. (c) No. $\mathbf{x}$ is not one-to-one. (d) Yes.

8. (a) $\mathbf{x}(u, v) = (\cosh u, u \sin v, u \cos v)$.
(b) $\mathbf{x}(u, v) = ((3 + 2\cos u)\cos v, 2\sin u, (3 + 2\cos u)\sin v)$.
(c) $\mathbf{x}(u, v) = (u, v, u^2 + v^2)$.

10. $0 < u < 2\pi$, $0 < v < 2\pi$; $0 < u, v < \pi/2$; $\pi/2 < u, v \leqslant 2\pi$; $0 \leqslant u, v < \frac{3}{2}\pi, \frac{3}{2}\pi < u, v \leqslant 2\pi$.

11. (b) Hint: Use the parametrization [from the parametrization (1.8) in Example 1.4]

$$\mathbf{x}(u, v) = (\sin\theta \cos\phi, \sin\theta \sin\phi, 1 + \cos\theta)$$

for S^2 minus the south pole to show that

$$\pi_*(\mathbf{x}_u) = \mathbf{y}_u, \quad \pi_*(\mathbf{x}_v) = \mathbf{y}_v,$$

where $\pi(\mathbf{x}) = \mathbf{y}$, are orthogonal and nonzero, hence linearly independent.

12. (a) Hint: Let f_0: $[a, b] \to E^2$ be a closed curve, and let f_1: $[a, b] \to \mathbf{p} \in E^2$ be any constant closed curve at $\mathbf{p}$. Then we have the homotopy

$$F(t, \tau) = (1 - \tau) f_0(t) + \tau f_1(t).$$

(b) Hint: Let $\mathbf{p}=(x_0, y_0)$. Then the procedure in part a fails, since for t satisfying $\tan t = -y_0/x_0$ we can find τ such that

$$(1-\tau)(\cos t, \sin t)+\tau(x_0, y_0)=(0,0).$$

Section 2

7. $2b_0u_0(\cos v_0)x_1+2au_0(\sin v_0)x_2-abx_3-abu_0^2=0.$

8. Hint: Obtain the equation $f_{x_1}dx_1+f_{x_2}dx_2+fx_3\,dx_3=0$ and interpret it geometrically.

9. $f_{x_1}(\bar{x}_1,\bar{x}_2,\bar{x}_3)(x_1-\bar{x}_1)+f_{x_2}(\bar{x}_1,\bar{x}_2,\bar{x}_3)(x_2-\bar{x}_2)-(x_3-\bar{x}_3)=0.$

11. $2x_1+x_2-3x_3=0.$

13. Hint: Let π be the plane containing the given line r and passing through any point $\mathbf{p}$ of the given surface S, and let ξ be the plane through $\mathbf{p}$ and orthogonal to r. Use Exercise 11 of Section 1.3, Chapter 2, to show that the curve of intersection of ξ and S is a circle, so that every point $\mathbf{p}\in S$ has a neighborhood contained in some surface of revolution with r as axis.

15. Hint: Use Theorem 1.6 and the chain rule.

18. (a) $E=-a, F=0, G=-a\sin^2\theta;\ L=-a, M=0, N=-a\sin^2\theta.$

(b) $E=1+f'^2, F=0, G=u^2;$

$$L,M,N=\frac{uf'',0,u^2f'}{|u|\sqrt{1+f'^2}},$$

where the primes denote the derivative with respect to u.

(c) $E=\dfrac{b^2}{b^2-(u-a)^2},\ F=0,\ G=u^2;$

$$L=\frac{-b}{b^2-(u-a)^2},\ M=0,\ N=\frac{u(a-u)}{b}.$$

(d) $E=1+f'^2,\ F=af',\ G=u^2+a^2;$

$$L,M,N=\frac{uf'',-a,u^2f'}{\sqrt{u^2(1+f'^2)+a^2}}.$$

19. (a) $x_1^2+x_2^2=1.\ x_3=0.$

(b) Vector $(\pm1/\sqrt{3}, \pm1/\sqrt{3}, \pm1/\sqrt{3})$.

(c) $x_1^2+x_2^2+x_3^2=1$.
(d) Plane $x_3=\pm 1/\sqrt{2}$.

20. The image curve of the meridian $v=\text{const}$ is the great circle on the plane $x_2-x_1\tan v=0$, and that of the parallel $u=\text{const}$ is the circle of intersection of the unit sphere and the plane $x_3=\sin u$. Points $\mathbf{x}(0,v)$ and $\mathbf{x}(\pi,v+\pi)$ have the same image point $(\cos v, \sin v, 0)$.

21. $\frac{1}{4}$.

Section 3

1. Hint: Let $\mathbf{e}_3$ and $\bar{\mathbf{e}}_3$ be the unit normal vectors of S and $\bar{S}$ at $\mathbf{p}$, respectively. Compute $\|\kappa_N\bar{\mathbf{e}}_3-\bar{\kappa}_N\mathbf{e}_3\|$ in two different ways, for one of which use (2.27), Exercise 4 of Section 3.3 of Chapter 1, and the fact that $\mathbf{e}_3\times\bar{\mathbf{e}}_3$ is along the tangent direction of C at $\mathbf{p}$.

2. Hint: Use (3.11).

3. (x_1,x_2,x_3), where $x_2=0$, $x_1^2=\dfrac{a^2(a^2-b^2)}{a^2-c^2}$, $x_3^2=\dfrac{c^2(b^2-c^2)}{a^2-c^2}$.

4. $(\pm 2, \pm 2, \pm 2)$.

Section 4

1. $\dfrac{f'}{|u|\sqrt{1+f'^2}}$ and $\dfrac{uf''}{|u|(1+f'^2)^{3/2}}$ for the surface (1.12).

10. $K=-169/4$, $H=0$.

18. Hint: Use (4.22).

20. Hint: Apply Exercise 18 to two special surfaces.

21. It is true only on a plane.

26. Hint: Use Exercise 4 of Section 2 and Exercise 25.

Section 5

2. Hint: Use (2.32).

3. (a) No effect on meridians, but a reflection with respect to the $\mathbf{x}_1\mathbf{x}_2$-plane for parallels.
(b) For meridians: clockwise rotations through an angle $\pi/2$, first about the $\mathbf{x}_1$-axis and then about the $\mathbf{x}_3$-axis. For parallels: a clockwise rotation about the $\mathbf{x}_1$-axis through an angle $\pi/2$.

(c) For meridians: a rotation in the $\mathbf{x}_1\mathbf{x}_2$-plane through an angle $\pi/4$. For parallels: a reflection with respect to the $\mathbf{x}_1\mathbf{x}_2$-plane.

4. (a) The translation taking the point $(0,2,-3)$ to the point $(1,0,0)$.
(b) A rotation about the $\mathbf{x}_3$-axis through an angle $\pi/4$.
(c) An isometry.

5. (b) No.

8. Hint: Use (1.8) and $(\cos\phi, \sin\phi, \theta)$ for the parametrizations of M and C, respectively.

9. $g(v) = \text{const}$, and the surface of revolution is a plane.

12. (a) Hint: Use $a\mathbf{x}^2 + b\mathbf{x} + c = 0$, $b^2 - 4ac > 0$, for the vector equation of a sphere.
(b) Hint: Let $E\,du^2 + 2F\,du\,dv + G\,dv^2$ and $E^*\,du^2 + 2F^*\,du\,dv + G^*\,dv^2$ be, respectively, the first fundamental forms of a surface S and its image surface $S^* = f(S)$. Then $E^* = E\mathbf{x}^{-4}$, $F^* = F\mathbf{x}^{-4}$, $G^* = G\mathbf{x}^{-4}$.
(c) Hint: Let $L\,du^2 + 2M\,du\,dv + N\,dv^2$ and $L^*\,du^2 + 2M^*\,du\,dv + N^*\,dv^2$ be the second fundamental forms of S and S^*, respectively. Use Exercise 4 of Section 3.3, Chapter 1, and

$$L = \mathbf{e}_3 \cdot \mathbf{x}_{uu}, \qquad M = \mathbf{e}_3 \cdot \mathbf{x}_{uv}, \qquad N = \mathbf{e}_3 \cdot \mathbf{x}_{vv},$$

where $\mathbf{e}_3$ is the unit normal vector of S, to show that

$$L^* = -\frac{1}{\mathbf{x}^4}\left[\mathbf{x}^2 L + 2E(\mathbf{x}\cdot\mathbf{e}_3)\right],$$

$$M^* = -\frac{1}{\mathbf{x}^4}\left[\mathbf{x}^2 M + 2F(\mathbf{x}\cdot\mathbf{e}_3)\right],$$

$$N^* = -\frac{1}{\mathbf{x}^4}\left[\mathbf{x}^2 N + 2G(\mathbf{x}\cdot\mathbf{e}_3)\right].$$

(d) Hint: This also follows readily from parts (b) and (c) and Exercise 2 of Section 3.

Section 7

3. Hint: Use (7.19), (7.10), and (7.16).

6. Hint: Use Exercise 3 for (7.25), and (7.25) and (7.18) for (7.26).

Section 8

2. The surface is developable.

4. (a) Hint: In the parametrization (8.1) for the hyperboloid, $\mathbf{y}(s) = (a\cos s, b\sin s, 0)$ is a parametrization of the ellipse $x_1^2/a^2 + x_2^2/b^2 = 1$, $x_3 = 0$, and $\mathbf{z}(s)$ is either $\mathbf{y}'(x) + c\mathbf{e}_3$ or $-\mathbf{y}'(s) + c\mathbf{e}_3$, where the prime denotes the derivative with respect to s, and $\mathbf{e}_3$ is the unit vector in the positive direction of the $\mathbf{x}_3$-axis.

(b) Hint: In (8.1), take $\mathbf{y}(s) = (s, 0, 0)$ and $\mathbf{z}(s) = (0, 1, ks)$, which is a vector along the line of intersection of the saddle surface and the plane $x_1 = s$ at the point $(s, 0, 0)$. By interchanging x_1 and x_2 we obtain another family of rulings.

(c) Hint: The plane $x_3 = 0$ intersects the paraboloid at two lines L_1: $x_1/a + x_2/b = 0$, $x_3 = 0$, and L_2: $x_1/a - x_2/b = 0$, $x_3 = 0$. Take L_1 to be the curve $\mathbf{y}(s)$ in (8.1) so that $\mathbf{y}(s) = (s, -(b/a)s, 0)$. Through L_1 we can determine a unique plane such that the other line of intersection L_2^* of this plane and the surface passes through the point $(s, -(b/a)s, 0)$. Take $\mathbf{z}(s)$ to be the vector $(1, b/a, 4s/a^2)$ along L_2^*. Similarly, by using L_2 we can obtain another family of rulings.

5. (a) Hint: Use Exercise 9 of Section 4.

6. $f(x_1) = \dfrac{1}{c}\sec^2(cx_1 + c_1) + c_3,$

$F(x_2) = -\dfrac{1}{c}\sec^2(-cx_2 + c_2) + c_4,$

where all c's are constant.

7. Hint: Suppose a minimal surface S is not a plane. Then by Exercise 7 of Section 4, Lemma 4.11, and Definition 1.3.7 of Chapter 2, all the asymptotic curves orthogonal to each asymptotic tangent, which lies entirely on S by Lemma 4.10, are Bertrand curves and are therefore circular helices by Corollary 1.3.12 of Chapter 2. Since the torsion of a circular helix is constant, we can easily see that the whole surface S is part of a helicoid.

Section 9

1. Hint: Use (9.2) and $\mathbf{v}(0) = \mathbf{x}_1(\theta_0, 0)$, $y^1(0) = 1$, $y^2(0) = 0$. Then (9.7) becomes

$$\frac{dy^1}{d\phi} - y^2 \sin\theta_0 \cos\theta_0 = 0, \qquad \frac{dy^2}{d\phi} + y^1 \cot\theta_0 = 0,$$

and by solving these differential equations with the initial conditions we can obtain the required result.

Section 10

1. Hint: Use (2.27) and (10.7).
3. (a) A line of curvature. (b) A line of curvature and a geodesic. (c) A line of curvature, an asymptotic curve, and a geodesic.
4. (b) Yes. (c) Yes.
5. (c) No.
6. (b) Hint: (1) Show that the differential equation of the geodesics on a sphere S with parametrization (1.8) is

$$(*)\ \phi'' = -2\phi' \cot\theta - \phi'^3 \sin\theta\cos\theta,$$

where the primes denote the derivative with respect to the arc length along the geodesics.
(2) Show that (*) is the equation of all great circles on S by differentiating the equation

$$A\sin\theta\cos\phi + B\sin\theta\sin\phi + C\cos\theta = 0$$

of these circles twice with respect to θ and eliminating the arbitrary constants A, B, C.

8. $(u-c_2)^2 = 4c_1^2(v-c_1^2)$, where c_1 and c_2 are constant.
9. The generators of the cylinder, the circles of intersection of the cylinder and the planes orthogonal to the generators of the cylinder, and the helices on the cylinder.
11. Hint: Use Theorem 1.3.4 of Chapter 1, Lemma 4.11, and (7.1) to find τ_1 and τ_2.

CHAPTER 4

Section 1

4. Hint: Use Exercise 3.

Section 2

2. Hint: Choose $\kappa_1 > \kappa_2$. Since κ_1 is a monotonically nonincreasing function of κ_2, κ_2 has a minimum at the point **p** where κ_1 has a maximum. If there exist nonumbilical points on S, then **p** is not an umbilical point. For otherwise, $\kappa_1 = \kappa_2$ at **p**, and at a nonumbilical point **p*** the principal curvatures κ_1^*, κ_2^* satisfy $\kappa_1^* \leqslant \kappa_1, \kappa_2^* \geqslant \kappa_2$, so that $\kappa_1^* \leqslant \kappa_2^*$, which contradicts the assumption $\kappa_1^* > \kappa_2^*$. Hence $\kappa_1 \neq \kappa_2$ at **p**, and S is a sphere by virtue of

the argument at **p** in the proof of Theorem 2.4. For the second part of the exercise, first notice that (2.25) can be written as

$$(a\kappa_1+b)(a\kappa_2+b)=b^2-ac>0,$$

and then differentiate the equation above.

Section 3

2. Hint: Use the Gauss-Bonnet formula.

4. 0.

Hint: Let S be the torus. Then use

$$\int\int_S K\,dA=\lim_{\varepsilon\to 0}\int_\varepsilon^{2\pi-\varepsilon}\int_\varepsilon^{2\pi-\varepsilon}K\sqrt{EG-F^2}\,du\,dv,$$

Exercise 18c of Section 2, Chapter 3, and Exercise 6 of Section 4, Chapter 3.

5. Hint: Show that the mapping $\bar{x}_1=x_1$, $\bar{x}_2=x_2^2$, $\bar{x}_3=x_3^3$ gives a homeomorphism of the sphere $\bar{x}_1^2+\bar{x}_2^2+\bar{x}_3^2=1$ onto the surface $x_1^2+x_2^4+x_3^6=1$.

7. (a) 1.

Hint: Restrict **v** to the unit circle $\mathbf{y}(t)=(\cos t,\sin t)$, $t\in[0,2\pi]$. The angle that $\mathbf{v}(t)$ makes with the $\mathbf{x}_1$-axis is t. Thus $2\pi I=2\pi$; hence $I=1$.

(b) 1.

(c) -1.

(d) -2.

Hint: Restricting **v** to the circle $\mathbf{y}(t)=(\cos t,\sin t)$, $t\in[0,2\pi]$, we obtain

$$\mathbf{v}=(\cos^2 t-\sin^2 t,\ -2\cos t\sin t)$$

$$=(\cos 2t,\ -\sin 2t).$$

Thus $I=-2$.

Section 4

3. Hint: Use (4.7)* and (4.17) to compute de_1^* in two ways; then obtain $\omega_{12}^*=\omega_{12}+d\Delta$ by comparing the coefficients of the corresponding terms of the two expressions.

Section 5

3. Hint: Cf. Exercise 4 of Section 3.

4. Hint: First, use the argument at the beginning of the proof of Theorem 2.4 to show that $K=c^2$, where c is a positive constant. Then divide the proof into two cases according as $H>0$ or $H<0$, and use (5.21), (5.16), and (5.19) to show that for $H>0$,

$$-cA \leqslant \iint_S pK\,dA \leqslant -cA,$$

so that

$$\iint_S pK\,dA = -cA = -\iint_S H\,dA,$$

which implies that $H^2-K=0$. Similarly, for $H<0$ we have

$$cA \leqslant \iint_S pK\,dA \leqslant cA.$$

Section 8

2. Hint: S: $u^2+v^2<1$ with the Euclidean metric.

4. $s_1=\pi\sqrt{r(R+r)}$.
   Hint: Cf. Exercise 6 of Section 4, Chapter 3.

5. No.
   Hint: Take S to be the torus with parametrization (1.18), which has diameter $\rho=\pi(R+r)$.

Index

Abelian group, 24
Accumulation point, 5
Admissible arc, 301
Affine group, 53
Affine transformation, 52
  isometric, 56
Angle:
  between curves on surface, 199
  between vectors, 25
  exterior, 252
  interior, 252
    of geodesic triangle, 264
Antipodal, point, 303
  mapping, 246
Arc length, 83, 84, 88
  of curve on surface, 176
  reparametrization of curve by, 84
Area, element of, 271
Associated Bertrand curves, 93
Asymptotic curve, 192-193
  of developable surface, 208
Asymptotic direction, 184
Axiom of completeness, 9
Axis:
  of helix, 79
  of revolution, 158

Basis, natural, 28
  orthonormal, 28
  of a space, 28
Beltrami-Enneper theorem, 197
Bertrand curve, 92-96, 97, 98
Binormal, 91
  vector, 92
Bolzano, intermediate value theorem of, 14
Bonnet's theorems, 212, 272, 305, 309
Boundary of set, 3
Boundary point of set, 3
Bounded set, 9
Buffon's needle problem, 128, 130

Cartesian product, 3
Catenary, 168, 220
Catenoid, 168, 220
  as minimal surface of revolution, 220
Cauchy, convergence condition, 10
  sequence, 10
Cauchy-Crofton formula, 128
Cauchy-Riemann equations, 202
Cauchy's formula, 115
Center of curvature, 95, 103, 104, 185
Chain rule, 20
Christoffel's problem, 286
  uniqueness theorem for, 289
Christoffel symbols, 210
Circular helix, 95, 104 (Ex. 4)
Closed form, 71
Closed plane curve, 85, 86
  diameter of 113, 123
  exterior of, 110
  interior of, 110
Closed set, 4
Closure of set, 5
Cluster point, 5
Cohn-Vossen's theorem, 275
Compact space, 11
Complement of set, 5, 38
Completely integrable system, 212
Complete surface, 249, 294, 298, 299, 305
Component:
  of set, 8
  of vector, 23
Cone, 161
  local isometry of, to plane, 198
  as ruled surface, 215
Confocal quadrics, 204, 206

Conformal mapping, 198
  Liouville's theorem for, 206
  local, 199
    of planes, 202
    of Poincaré half-plane to Euclidean plane, 251
    of spheres to planes, 200, 201
    of surfaces, 200
Conjugate directions, 194
  net, 194, 197
Conjugate point, 302-305
Connected space, 7
Conoid, 202
Contact of order k with curve, 101
Continuous mapping, 6
Convergence of sequence, 10
Convex hull, 137, 289
Convex set, 137
Convex surface, 275
  of constant Gaussian curvature, 280
  of constant mean curvature, 285
  in direction, 284
Coordinate, neighborhood, 151
Coordinate functions:
  of curve on surface, 172
  of vector field on E^3, 32
Coordinate system, 151
  isothermal, 200, 222
  spherical, 204
Covariant differential, 225
Covering, 11
  open, 11
  finite, 11
  sub-, 11
Critical point, 19, 180
Crofton's theorem, 143
Cubic parabola, 104
Curvature:
  center of, *see* Center of curvature
  of curve, 90, 91, 96
  Gaussian, 189, 196, 211, 270
    intrinsic property of, 211, 272
    of torus, 194
  geodesic, 299
  lines of, 192, 204, 206
    differential equation of, 189
  mean, 189, 196, 270
  normal, 177, 183, 184
  principal, 184, 189
  radius of, 91, 116
Curve:
  asymptotic, *see* Asymptotic curve
  Bertrand, *see* Bertrand curve
  of class C^k, 85
  closed, 85, 86
  of constant width, 113, 114, 138
  continuous, 85
  convex, 113, 134, 135, 137
  curvature of, 90, 91, 96
  Mannheim, 98
  oriented, 89
  periodic, 86
  piecewise (sectionally) regular (smooth), 82, 86, 133
  plane, 92, 93, 136
  rectifiable, 83
  regular, 82
  reparametrization of, 80-81, 83
    by arc length, 84
  simple, 85
  smooth, 82, 86
  spherical, 104
  on surface, 171-172
  torsion of, 91, 92, 96
  vertex of, 123, 138, 139
Cusp, 79
Cycloid, 87
Cylinder, 4, 157, 244
  first fundamental form of, 179
  geodesics of, 239
  local isometry of, to plane, 198
  parametrization of, 158
  as ruled surface, 215
Cylindrical helix, 97, 107

Darboux frame, 176
Decomposition of space, 38
Deformation of curve, 147
Derivative mapping, 41
Determinant of three vectors, 25
Developable surface, 208
Diameter:
  of closed plane curve, 113, 123
  of set, 9
Diffeomorphism, 43
  area-preserving, 202
  orientation-preserving or exterior-reversing, 245

Differential:
  form, *see* Form, differential
  of function, 64
  mapping, 41
Dimension of space, 28
Direction:
  asymptotic, 184
  principal, 184, 189
Directional derivative, 33, 173
Direct sum of spaces, 38
Disconnected space, 7
Distance:
  in E^3, 2, 24
  on surface, 293-294
Divergence of sequence, 10
Dual basis, 40, 65
Dual space, 39
Dupin:
  indicatrix, 184
  theorem, 204

Element:
  of arc, 176
  of area, 218, 271
Ellipsoid, 157, 186, 244
Empty set, 4
Enneper's minimal surface, 223
Envelope:
  of family of curves, 112
  of family of tangent planes, 227
Equator, 154
Equivalent knots, 86
Erdmann's theorem, 76
Euclidean coordinate functions:
  of form, 65
  of mapping, 35
  of vector field, 30, 85
Euclidean group of rigid motions, 56
Euclidean space, 1
Euler angles, 50
Euler characteristic, 255, 256
Euler equation, 76
Euler's formula, 184
Evolute, 98
Exact form, 71
Existence and uniqueness theorem:
  for system of ordinary differential equations of first order, 307
  for system of partial differential equations of first order, 309
Existence theorem for curves, 105, 307
Exterior, of closed plane curve, 110
Exterior angle, 252
  derivative, 69
Exterior differential form, 64, 267
Exterior multiplication, 67
Exterior point, 3
Exterior product, 69
Extremal curve of integral, 76
Extremum, 18
  absolute, 20
  relative (local), 18, 20
Extrinsic property, 227

Fary-Milnor theorem, 145
Fenchel's theorem, 140
Field:
  frame, 31
  Frenet frame, 90
  vector, *see* Vector field
Finite covering, 11
First fundamental form, 175, 176
  coefficients of, 177
First variation:
  of integral, 75
  of length, 300
Folium of Descartes, 82
Form:
  closed, 71
  on E^3, 64
  exact, 71
  exterior differential, 64, 267
  fundamental, 175-178, 269
  structural equations for, 75, 269
  on surface, 173, 174
Four-vertex theorem, 123
Frame, 31
  Darboux, 176
  moving, 88
  positively or negatively oriented, 60, 85
  right-or left-handed oriented, 60, 85
Frenet formulas, 90, 109

Frenet frame field, 90
Function:
of class C^k, 170
continuous, 6
harmonic, 222
height, 145
Laplacian of, 222
periodic, 19
regular value of, 155, 244
Fundamental forms, 175-178, 269
relation among, 208
Fundamental theorem:
for curves, 105
for surfaces, 212, 272, 309

Gauss:
equations, 210, 211
mapping, 176
trihedron, 176
Gauss-Bonnet formula:
global, 257
local, 256
Gauss-Bonnet theorem, 256
applications of, 261-264
Gaussian curvature, *see* Curvature, Gaussian
Generalized uniqueness theorem for curves, 105
General linear group, 47
Genus of surface, 256
Geodesic circles, 236
Geodesic curvature, 229-231, 239
Geodesic disk, 293
Geodesic parallels, 232
Geodesic polar coordinates, 234
area of geodesic circle in, 237
first fundamental form in, 233
Gaussian curvature in, 236
perimeter of geodesic circle in, 236
Geodesics, 229
characterizations of, 231
of circular cylinder, 239
closed, 261, 262
differential equations of, 231
existence and uniqueness of, 231
field of, 232
of plane, 239
of Poincaré half-plane, 251
as shortest arcs:
in large, 298, 299
in small, 233, 304
of sphere, 239
Geodesic torsion, 237-238
of asymptotic curves, 240
Geodesic triangle, 264
excess of, 264
Gradient:
on E^3, 72
on surface, 181
Graph, 13
of C^3 function, 155, 160
area of, 218
Gaussian curvature of, 196
mean curvature of, 196
principal curvature of, 196
tangent plane of, 180
Greatest lower bound, 9
Green's formula, 258, 268
Group, 47
Abelian, 24
affine, 53
general linear, 47
of isometrics, 56
orthogonal, 48

Hausdorff space, 11
Height function, 145
Heine-Borel theorem, 11, 132
Helicoid, 181
as ruled minimal surface, 223
Helix, 79, 87, 96
axis of, 79
circular, 95, 104
cylindrical, 97, 107
pitch of, 79
Hilbert's theorem, 249
Homeomorphic spaces, 6
Homeomorphism, 6
Homotopy of curves, 169
Hopf-Rinow's theorem, 295
Hopf-Voss symmetry theorem, 284
Hopf-Voss translation theorem, 280
Hyperbolic paraboloid, 169, 223
Hyperboloid:
of one sheet, 223
of two sheets, 157, 244

Identity, Lagrange, 26

Index:
rotation, 111, 131, 134, 135, 136
of vector field, 264-267
Indicatrix:
binormal, 97
principal normal, 262, 263
tangent, 97, 140
Inequality, isoperimetric, 118
Infimum, 9
Inner product, 24, 31
Integrability condition, 212, 269
Interior:
of closed plane curve, 110
of set, 3
Interior point, 3
Intermediate value theorem of Bolzano, 14
Intrinsic property, 226
Inverse function theorem, 44
Inversion, 203
Involute, 98, 138
Isometric affine transformation, 56
Isometry:
of E^3, 54
local, 198, 246
orientation-preserving, 61, 105, 127
orientation-reversing, 61
of surfaces, 197
Isoperimetric inequality, 118
Isothermal coordinate system, 200, 222

Jacobian matrix, 42
Jacobi equation, 302
Jacobi theorem, 263
Joachimstahl's theorem, 196
Jordan curve theorem, 110

Kernel, 36
Knot, 86
cloverleaf, 86
figure-eight, 86
four, 86
Listing's, 86
overhand, 86
polygonal, 87
tame, 87
trivial, 86
wild, 87
Kronecker delta, 31

Lagrange identity, 26
Lagrange multiplier, 21
Least upper bound, 9
Left-handed oriented frame field, 60, 85
Length:
of curve, *see* Arc length
of vector, 24
Levi-Civita parallelism, 225
property of, 227
Liebmann's theorem, 247
Limit point, 5
Linear combination, 27
Linear dependence, 28
Linear independence, 27
Linear space, 24
Linear transformation, 36, 47
kernel of, 36
nonsingular, 47
nullity of, 38
rank of, 38
Liouville's theorem, 206
Local canonical form of curve, 100
Lower bound, 9

Maclaurin's formula, 18, 100
Mainardi-Codazzi equations, 212
Mannheim curve, 98
Mapping:
antipodal, 246
area-preserving, 202
of class C^k, 35
conformal, 198
continuous, 6
derivative, 41
differential, 41
Gauss, 176
position, 111
principal normal, 262
regular, 43
tangential, 111
Matrix:
Jacobian, 42
orthogonal, 48
Maximum:
absolute, 13, 20
relative (local), 18, 20

Mean curvature, 189, 196, 270
  vector, 222
Mean value theorem:
  of differential calculus, 15
  of integral calculus:
    first, 15
    generalized first, 16
    second, 16
Measure of set of lines, 126
Mercator projection, 201
  stereographic, 169, 200
Meridian:
  of sphere, 154
  of surface of revolution, 158
Meusnier's theorem, 177
Minding's theorem, 246
Minimal surface, 217
  Enneper's, 223
  with isothermal parameters, 223
  of revolution, 220
  ruled, 223
  Scherk's, 223
  as solution to variational problem, 217
Minimum:
  absolute, 13, 20
  relative (local), 18, 20
Minkowski's formulas, 277, 279
Minkowski's problem, 285
  uniqueness theorem for, 286
Möbius band, 163
  nonorientability of, 244
  parametrization of, 164
Monge parametrization, 155
Motion:
  improper, 61
  proper, 61
  rigid, 54
Multiplication, exterior, 67
Multiplication wedge, 67

Natural basis, 28
Natural coordinate functions, 2
Natural frame field on E^3, 30
Negatively oriented frame, 60, 85
Neighborhood, 2
  open spherical, 2
Normal coordinates, 234
  first fundamental form in, 236
Normal curvature, 177, 183, 184
Normal line to curve, 91
Normal plane to curve, 91
Normal principal, 91
Normal section, 177, 185
Normal vector field, 243, 245
Normal vector to surface, 174, 180
Norm of vector, 24
Nullity, 38

Open covering, 11
Open set, 4
Opposite points, 137
Orientable surface, 242-243
Orientation:
  of frame, 59
  of surface, 241-245
Orientation-preserving (-reversing) isometry, 61, 105, 127
Oriented curve, 89. *See also* Positively oriented
Orthogonal group, 48
Orthogonal matrix, 48
Orthogonal system of surfaces, 203
Orthogonal trajectory, 99
Orthogonal transformation, 48
Orthonormal basis, 28
Osculant of curve, 101
Osculating circle, 103
Osculating plane, 91, 101-102, 104
Osculating sphere, 103
Overdetermined system, 225

Paraboloid, 294, 306
Parallel curves, 99, 116
  Steiner's formulas for, 116
Parallelism, Levi-Civita, 225
Parallels:
  of colatitude, 154
  geodesic, 232
Parallel translate, 226
  existence and uniqueness of, 226
  geometric interpretation of, 227
Parallel vector field along a curve, 225
  differential equations for, 226
  path-independence of, 228
Parameter of curve, 78
Parameters:
  change of, for surfaces, 165

isothermal, 200
Parametric representation of plane curve, 115
Parametrization of surface, 151
isothermal, 200, 222
Parseval's formulas, 122
Period, 19, 86, 88
Periodic function, 19
Piecewise regular (smooth) curve, 82
Plane:
first fundamental form of, 178
normal, 91
osculating, *see* Osculating plane
rectifying, 91
tangent, 173, 180
Plateau's problem, 217
Poincaré half-plane, 226, 250-251
Poincaré theorems, 70, 265
Point:
accumulation, 5
antipodal, 303
of application, 29
conjugate, 302-305
critical, 19, 180
elliptic, 182, 185, 186
hyperbolic, 182, 185, 186
limit, 5
parabolic, 182, 185, 186
planar, 183, 186
saddle, 19
umbilical, 184, 186, 190
Polar tangential coordinates, 115
Position:
mapping, 111
vector, 24
Positively oriented:
boundary of simple region, 254
frame, 60, 85
simple closed curve, 110
tangent, 110
Principal:
curvature, 184, 189
direction, 184, 189
normal, 91
Product:
inner, 24, 31
scalar, 24
vector, 25
Projection, Mercator, 201
stereographic, 169, 200
Pseudosphere, 195

Quadratic form, 21

Radius:
of curvature, 91
of torsion, 91
Rank of linear transformation, 38
Rectifiable curve, 83
Rectifying plane, 91
Reflection, 49, 61
Region:
regular, 253
simple, 254, 260
Regular curve, 82
Regular mapping, 43
Regular value of function, 155, 244
Relative topology, 3
Reparametrization of curve, 80-81, 83
by arc length, 84
orientation-preserving (-reversing), 84
Riemann symbols, 213, 214
Right-handed oriented frame, 60, 85
Rigid affine transformation, 56
Rigidity of sphere, 249
Rigid motion, 54
Rodriques, equation of, 192
Rotation, of E^2, 49, 61
Rotation index, 111, 131, 134, 135, 136
Ruled surface, 214
directrix of, 214
ruling of, 214

Saddle point, 19
Saddle surface, 223
Scalar product, 24
Schur's theorem for curves, 147
Screw surface, 203
Second fundamental form, 175, 177, 269
coefficients of, 178
Second variation of length, 302
Section, normal, 177, 185

Sectionally regular (smooth) curve, 82
Sequence:
  Cauchy, 10
  convergent, 10
  divergent, 10
Set:
  bounded, 9
  closed, 4
  empty, 4
  open, 4
Sign of isometry, 60
Simple curve, 85
Simple region, 254, 260
Simple surface, 155
Singular point of curve, 82
Space:
  compact, 11
  connected, 7
  disconnected, 7
  Hausdorff, 11
  linear, 24
Spaces, homeomorphic, 6
Sphere:
  first fundamental form of, 179
  meridian of, 154
  parametrizations of, 152-155, 169
Stationary integral, 75
Steiner's formulas for parallel curves, 116
Stereographic projection, 169, 200
Straight line, 78, 92
  length-minimizing property of, 87
Structural equations for forms, 75, 269
Subcovering, 11
Subspace, 3
Supremum, 9
Surface:
  complete, *see* Complete surface
  of constant Gaussian curvature, 246-252
    Hilbert's theorem for, 249
    Liebmann's theorem for, 247
    Minding's theorem for, 246
  convex, *see* Convex surface
  developable, 208
  equation of, 183
  genus of, 256
  minimal, *see* Minimal surface
  Monge parametrization of, 155
  parametrization of, 151
    by lines of curvature, 192
  parametrized, 167
  of revolution, 158, 244
    axis of, 158
    of constant Gaussian curvature, 251
    meridian of, 158
    minimal, 220
    parallels of, 158
    parametrization of, 159
  ruled, *see* Ruled surface
  saddle, 223
  screw, 203
  simple, 155
  star-shaped, 279
    of constant Gaussian curvature, 280
    of constant mean curvature, 280
  Weingarten, 252
    special, 252
Symbols:
  Christoffel, 210
  Riemann, 213, 214

Tangent:
  indicatrix, 97, 140
  line of curve, 82
  plane, 173, 180
  space, 30
  surface, 216
    local isometry of, to plane, 216
  vector:
    to curve, 80
    to E^3, 29
    to surface, 173
Tangential mapping, 111, 140
Tangents, theorem on turning, 131
Taylor's formulas, 17
Third fundamental form, 175, 176
Topological product, 4
Topological space, 3
Topology, 3
  relative, 3
Torsion:
  of curve, 91, 92, 96
  geodesic, 237-238
    of asymptotic curves, 240
  radius of, 91
  total, 146
Torus, 4, 161

Gaussian curvature of, 194
parametrization of, 162, 181
Total curvature of curve, 139
Fary-Milnor theorem on, 145
Fenchel's theorem on, 140
Total twist number of curve, 100
Tractrix, 195
Transformation:
affine, 52
isometric, 56
linear, *see* Linear transformation
orthogonal, 48
Translation, 52, 61
Trefoil, 86
Triangle, geodesic, 264
Triangle inequality, 2
Triangulation, 254-255
Trihedron:
Gauss, 176
right-handed rectangular, 1
Triply orthogonal system, 203-205
Dupin's theorem for, 204

Umbilical point, 184, 186, 190
Unimodular affine group, 54
Uniqueness theorem for curves, 105
Unit vector, 24
Upper bound, 9

Variation:
first:
of area, 219
of length, 300
second, of length, 302
of vector field, 257
Variations:
calculus of, 75
normal, of surfaces, 218
Vector, 23
curl of, 73
divergence of, 73
mean curvature, 222
product, 25
space, 24
tangent, to curve, 80
unit, 24
velocity, 80
Vector field, normal:
on curve, 85
on E^3, 30
on surface, 264
index of, 264-267
normal, 243, 245
singular point of, 264
tangent, 173
Vertex, of plane curve, *see* Curve, vertex of
Vertices of piecewise regular curve, 82

Wedge multiplication, 67
Weingarten formulas, 207
Weingarten surface, special, 252
Weyl's problem, 285
uniqueness theorem for, 275, 286
Width of closed curve:
constant, 113, 114, 138
in direction, 113
Winding number, 111, 136
Wirtinger's lemma, 121